GEELONG'S CHANGING LANDSCAPE

Ecology, Development and Conservation

EDITORS: DAVID S. JONES AND PHILLIP B. ROÖS

During the passage of compiling this book, much-loved member of the Geelong community Tandop *David Tournier passed away.*

This book is dedicated in loving memory of Tandop *David John Tournier who entered the Dreaming on 31 December 2016.*

GEELONG'S CHANGING LANDSCAPE

Ecology, Development and Conservation

EDITORS: DAVID S. JONES AND PHILLIP B. ROÖS

CSIRO PUBLISHING

A catalogue record for this book is available from the National Library of Australia.

ISBN: 9780643103603 (pbk.)
ISBN: 9780643103610 (epdf)
ISBN: 9780643103627 (epub)

Published by:

CSIRO Publishing
36 Gardiner Road, Clayton VIC 3168
Private Bag 10, Clayton South VIC 3169
Australia

Telephone: [+613] 9545 8555
Email: csiropublishing@csiro.au
Website: www.publishing.csiro.au

Front cover: (top) Aerial view of Geelong (photo: Donna Squire); (bottom) perspective of Geelong to the south from the 1988 report *Geelong: City by the Bay – The Bay Link*

Back cover: (left to right) Aerial view of Barwon Heads (photo: Donna Squire); small stand of Buloke (*Allocasuarina luehmannii*) just north of Little River (photo: Geoff Carr); Warralily Park (photo: David Rowe)

Set in 10.5/12 Minion and Stone Sans
Edited by Adrienne de Kretser, Righting Writing
Cover design by Andrew Weatherill
Typeset by Thomson Digital
Index by Bruce Gillespie
Printed by Ingram Lightning Source

CSIRO acknowledges the Traditional Owners of the lands and waters that we live and work on across Australia and pays its respect to Elders past and present. CSIRO recognises that Aboriginal and Torres Strait Islander peoples have made and will continue to make extraordinary contributions to all aspects of Australian life including culture, economy and science.

Feb26_RP_ILS

Foreword

Like the authors, I am a relative newcomer to Geelong, moving to this beautiful city from Perth in Western Australia when I became Deakin University's Vice-Chancellor in 2010. When I arrived, I found a city rich in natural endowments, creative talent and infrastructure. But I also found a region in despair with long-established companies such as Ford and Alcoa in a downward spiral, and local downstream industries collapsing as their supply chain broke down. Businesses everywhere faced a confluence of disruptive forces with accelerating technological innovation, a rapidly changing global demography and the challenge of predicting future workforce skills.

It was a grim time. Some suggested Deakin University should move its business up the Geelong road to Melbourne. But challenges are always accompanied by opportunities and universities through the ages have always served the communities in which they found themselves. Why would Deakin be different? Our *LIVE the Future* strategy, Deakin's grand ambition to grow a university town in Geelong in the tradition of a sustainable ecosystem of cradle-to-grave education, became our challenge. Our dream for a clever and creative city, building on the growing knowledge economy, to survive and thrive down the generations became a primary focus.

Fundamental to the *LIVE the Future* strategy was a commitment to the communities we serve. Deakin developed the Geelong Future Economy Precinct, joining the dots between community, industry and academia to underpin innovation in areas as diverse as health, carbon fibre, nanotechnology, energy and data analytics. Our manufacturing innovation hub ManuFutures, developed in partnership with the Victorian government, offers its tenants access to Deakin's research expertise as well as business support and corporate facilities. Innovation for Deakin means high-value, high-productivity jobs and we anticipate that by 2020 the precinct will have created well over 3000 new jobs.

This is important because it's estimated that, by 2025, two-thirds of all jobs will require post-secondary education backed by the application of skills. Our region's greatest challenge is to persuade the people of Geelong to see TAFE and university study as desirable and achievable goals. The essential ingredient for a strong future for Geelong is a highly skilled, educated and innovative population and workforce. With rapidly changing workforce needs, Geelong must come to grips with Industry 4.0, the fourth industrial revolution, which brings together advanced manufacturing techniques and transformative technologies to connect the physical and digital worlds. The knowledge economy relies on the production, distribution and careful use of information and ideas rather than physical resources – IP and ideas are today's top consumables, not minerals and money.

Being recognised as a UNESCO 'City of Design' has been an important milestone for Geelong. It affirms the region's rich legacy of innovation, from its *Wadawurrung* Aboriginal heritage to wool production and automotive manufacturing through to the more recent innovations in advanced manufacturing and virtual reality. It will do much to shape Geelong's future, connecting it with the great cities of design across the world. The underpinning design thinking enables us to take shiny new ideas through to innovative products, delivering total solutions to the challenges of a rapidly changing economy.

I am delighted the authors have undertaken this endeavour – to draw together the rich tapestry of the Geelong story – raising awareness, asking questions and opening up discourse and debate on this astoundingly beautiful, stylish, clever and creative city that we all call home.

Professor Jane den Hollander AO
Vice-Chancellor, Deakin University (2010–2019)

Contents

Acknowledgement of Country

We wish to respectfully acknowledge the Elders, families and forebears of the *Wadawurrung* peoples of the Geelong–Bellarine–Ballarat region, of the *Kulin* Nation surrounding the Port Phillip Bay region, past, present and future, who are the Traditional Owners and custodians of these lands, waters and skies for many centuries.

Acknowledgements

We wish to acknowledge the many people and organisations who offered support and encouragement in realising this project.

Thanks are due to the organisations of City of Greater Geelong Council, Geelong Art Gallery, Geelong Heritage Centre, Geelong Regional Library, Barwon Water, Deakin University Library, Alfred Deakin Prime Ministerial Library, State Library of Victoria, National Library of Australia, *Wadawurrung* (Wathaurung Aboriginal Corporation), Wathaurong Aboriginal Co-operative, Golden Plains Council, the Surf Coast Council and staff at CSIRO Publishing.

Special thanks are also due to the following individuals: Dr Mick Aberton, Uncle Reg Abrahams, Sue Anderson, Professor Lee Astheimer, Damein Bell, Shivani Bhatnagar, Dr Susan Bird, N'Arweet Carolyn Briggs, Monique Cahill, Dr Peter Coutts, Paul Davis, Jen Dearnaley, Corrina Eccles, Professor Hisham Elkadi, Shelly Fanning, Dr Beth Gott, Dr Fiona Gray, Karen Hall, Frank Hanson, Tim Hellsten, Dr John Henshall, Melinda Kennedy, Kevin Krastins, Lakshmanan Madhu, Aunty Lyn McInnes, Jeremy Minter, Dr Joanne Mitchell, Veema Mooniapah, Isobel Paton, Trevor Pescott, Gavin Pocock, Gareth Powell, Dr Gary Presland, Dr Jim Rhoads, Dr David Rhodes, Susan Ryan, Jeanette Spittle, Michael Stitt, Manita Stokes, Heather Threadgold, Kirstin Tyler, Dr Geoff Wescott, Leanne Willis, Emily Wong and Liz Wood.

List of contributors

Editors

David S. Jones
Dr David S. Jones is Foundation Professor of Planning and Landscape Architecture at Deakin University. A student of McHarg, Seddon and Sinatra, he taught at the universities of Melbourne, RMIT, Adelaide, Pennsylvania and Tasmania before arriving at Deakin. At the University of Adelaide, he led the renovation and growth of the School of Architecture, Landscape Architecture & Urban Design, and drew *Kaurna* engagement and respect into the School. He is the author of the *Adelaide Park Lands and Squares Cultural Landscape Assessment Study* (2007) that provided the evidence for the successful National Heritage nomination of the 'City of Adelaide Plan and layout' to the Commonwealth's National Heritage Register – the first designed landscape to be included. His recent research and teaching implicate Indigenous knowledge systems, a continued engagement with several Aboriginal communities (*Boon Wurrung, Gunditjmara, Kaurna, Quandamooka, Wadawurrung*) in Country topics, and the position of Indigenous knowledge systems in Australia's tertiary-level professionally accredited built environment courses (Jones *et al.* (2018) *Indigenous Knowledge in the Built Environment: A Guide for Tertiary Educators*).

Phillip B. Roös
Dr Phillip B. Roös is the Director of the Live+Smart Research Laboratory, and the Associate Head of School – Industry Engagement at the School of Architecture and Built Environment, Deakin University. His teaching and supervision focus on Architecture, Environmental Design and Planning, Ecological Urbanism, and Biophilic Design. His work spans architecture, urban design and planning, landscape architecture, environmental design, teaching and research, as well as writings and art. He is known internationally as a leader in Environmental Design and has been working as a design professional and architect for 30 years on an extensive range of large-scale projects in Europe, Africa and Australasia. His work is influenced by whole systems thinking and his application of environmental design is closely related to the ordering of the large-scale aspects of the environment by means of architecture, engineering, landscape architecture, urban design and ecological planning. His teaching and research interests are centred on the human–nature relationship and the identification of optimised design processes based on a regenerative pattern language theory. This approach incorporates the principles of biophilia and regenerative design as well as an adaptive pattern language that re-establishes our wholeness with nature, and considers the vulnerabilities of a changing landscape.

Authors

Beau B. Beza

Dr Beau Beza is Associate Head of School, Teaching and Learning at the School of Architecture and Built Environment, Deakin University. He holds undergraduate and postgraduate degrees in landscape architecture, urban planning and a PhD (Architecture and Design) from the University of California (Davis), the University of Melbourne and the Royal Melbourne Institute of Technology, respectively. He has years of professional and academic experience working on a variety of projects in Australia, Bosnia-Herzegovina, the United States, Norway, Mexico, Colombia, Nepal and the Hashemite Kingdom of Jordan. His research revolves around developing an understanding in how people derive meaning from the natural and built environment. Much of his perception and planning work is based in (rural) urbanising communities. Beau has published a range of journal and conference papers along with co-editing and contributing four chapters to the Routledge publication, *Sustainability Citizenship in Cities: Theory and Practice.* This book explains how sustainability citizenship can manifest in urban built environments as both responsibilities and rights. He also co-edited and contributed three chapters to *The Public City: Essays in Honour of Paul Mees* (Melbourne University Press). This publication provides a blueprint for the improvement of civic and institutional purpose in the creation of the public city.

Geoff Carr

Dr Geoff Carr is an ecologist, taxonomist and horticulturalist and founder, in 1981, of the biological consultancy Ecology Australia, based in Melbourne. He grew up in Geelong and from 1965 to 1974 worked in the nursery at Geelong Botanic Garden. Employment in the Botany Department, La Trobe University, as a member of the technical staff followed for seven years, during which time he completed a Bachelor of Science degree (botany, zoology and art history) at the university, graduating in 1979. Since then he has been a full-time consultant. Geoff has conducted numerous formal and informal studies on vegetation of south-eastern Australia, many in the Geelong region, and has published widely. Particular interests and expertise include environmental weed invasions and their management; conservation management of vegetation and rare or threatened plant species; and taxonomic research, principally on orchids (*Caladenia*), *Dianella* (in Australia, Malaysia and Indonesia), *Correa*, *Begonia* (in Indonesia) and the naturalised exotic Australian flora. For 65 years he has been an avid gardener.

Peter Dahlhaus

Associate Professor Peter Dahlhaus is Principal Research Fellow at the Centre for eResearch and Digital Innovation at Federation University Australia. His career spans 40 years in engineering geology, environmental geology and hydrogeology in both the private and public sectors, investigating the geology, geomorphology, soils, groundwater and geohazards in south-west Victoria. He has been influential in applying his scientific knowledge to direct policy on urban development and catchment management. Peter is well known as a science communicator by his students and community groups, and as an independent adviser to catchment management authorities, water authorities and municipalities in the region. During the 1980s, Peter completed his Masters thesis on the topic of engineering geological mapping for urban development, which introduced him to the seminal works by Robert F. Leggett (*Cities and Geology*), Ian L. McHarg (*Design with*

Nature) and George Seddon (*Sense of Place*). Understanding geomorphological processes, designing with nature, and respecting landscapes were insights that Peter extended into his PhD thesis on managing salinity risk in the Corangamite region, through applying a holistic systems science approach. With colleagues, Peter's current research focuses on spatial data interoperability and visualisation to ensure that environmental data, information and knowledge are globally and publicly available. Peter's passion is for establishing data democracies to ensure free, timely and equitable access to the information required for designing resilient landscape futures.

Yolanda Esteban

Dr Yolanda Esteban holds a PhD and a combined degree of Architecture and Building with Honours from Deakin University. Over the past decade, she has developed an extensive teaching expertise across design disciplines in architecture, urban design and planning, at undergraduate and postgraduate levels at the School of Architecture and Built Environment at Deakin University where she is a core member of the Planning, Urban Design and Landscape (PUDLA) and the Architecture course teams. She is currently course director of the Masters of Planning (Professional) and the Landscape Architecture coursework degrees. Yolanda has industry experience in the areas of urban design and strategic planning and has worked on several key visioning projects across Victoria in both regional and metropolitan contexts. She also served as an urban design adviser between 2004 and 2007. Her research has focused on the changing profile of urban places, with specific expertise on the development of built form scenarios based on demographic modelling, and the nexus between vision-making and policy implementation.

Murray Herron

Dr Murray Herron possesses an in-depth knowledge of strategic land use planning and urban design, gained as a local government strategic planner in Australia and through multiple postgraduate qualifications. In his PhD entitled 'Climate Change Resilience and Development Growth of Two Australian Cities' (2016), his research offered internationally relevant applied findings that have been presented at and are applicable for national and international land use planning, modelling and visual scenario-making and assessment applications. He has presented at multiple national and international conferences on climate change and urban and regional sustainability, as well as written 15 Australia and international peer-reviewed conference or journal papers.

Stephanie Ho

Stephanie Ho has a Bachelor of Design (Architecture), a Diploma of Project Management (Holmesglen TAFE) and is currently completing a Master of Architecture at Deakin University. She has experience in the design and management of medium-density residential developments and currently works for a commercial builder, integrating her design background and project management experience. Stephanie has a passion for travel and a specific interest in the interiors of hotels and hospitality fit-outs around the world.

Louise Johnson

Dr Louise Johnson is Professor of Australian Studies and Geography at Deakin University. A human geographer, she has researched the gendered nature of suburban houses, changing manufacturing workplaces and the dynamics of Australian regional economies. Her PhD was on the Geelong textile industry and she has maintained an ongoing research interest in this region. Major publications include *Suburban Dreaming* (DUP 1994), *Placebound: Australian Feminist Geographies* (OUP 2000), *Cultural Capitals: Revaluing the Arts. Remaking Urban Spaces* (Ashgate 2009) and *Planning in Indigenous Australia: From Imperial Foundations to Postcolonial Futures* (Taylor and Francis 2018, with Sue Jackson and Libby Porter) along with over 60 academic articles and book chapters. She is currently researching social polarisation and inclusion in Geelong, affordable housing and post-colonial planning. Having developed and taught Australian Studies, Cultural Studies and Women's Studies, she currently teaches Human and Urban Geography as well as Australian Studies.

Gavin Keeney

Gavin Keeney is an independent scholar currently engaged in research regarding transmedia and author rights. Recent publications include *Knowledge, Spirit, Law: Book 1, Radical Scholarship* (2015) and *Knowledge, Spirit, Law: Book 2, The Anti-capitalist Sublime* (2017). His research PhD, 'Visual Agency in Art and Architecture' (2014) resulted in two monographs, *Dossier Chris Marker: The Suffering Image* (2012) and *Not-I/Thou: The Other Subject of Art and Architecture* (2014), plus curation of two multi-media exhibitions, 'Shadow-lands I' (2012) and 'Shadow-lands II' (2014). He conducted a series of seminars on the subject of academic publications in Ljubljana, Slovenia, March–May 2015, under the auspices of the US Fulbright Specialist Program. He was Teaching Fellow at CEPT University, Ahmedabad in 2016–17, Visiting Research Fellow at Birkbeck Institute for the Humanities, London in May 2017, and Research Resident at the Giorgio Cini Foundation, Venice in May–July 2017.

Kate Kerkin

Dr Kate Kerkin has worked in the area of urban planning and social change for over 25 years. She lectured in community development, urban planning and urban design at university level during the 1980s and 1990s before establishing a social planning team in a large Australian-based private sector company in the early 2000s. Since 2009, Kate has run her own social planning company, K2 Planning Pty Ltd. Kate has a passionate commitment to planning for places that support local communities. Her work focuses on the provision of community facilities that are the right size, in the right location and offer the right mix of opportunities for local residents. She has worked across many of the urban growth areas of Melbourne, including Whittlesea, Hume and the City of Casey, as well as across regional Victoria and in the Armstrong Creek growth area in Geelong. Kate's projects focus on planning early years facilities, such as kindergartens and maternal and child health services, youth facilities, spaces that allow specialist services to support new residents and other potentially isolated members of the community as well as older residents and residents with limited mobility. Kate has developed plans for intergenerational community hubs, integrated early learning centres, multi-purpose community facilities and libraries, many of which are now built and actively used by their community. Kate is a strong advocate of the benefits of regional living and moved with her family to Geelong in 2003.

Kirsten Kilpatrick

Kirsten Kilpatrick is an experienced town planner and recently established NovoPlanning, a town planning and strategic advice consultancy, based in Geelong. In 2017, Kirsten was awarded the Planning Institute of Australia's Victorian Planner of the Year. Kirsten holds a BA (Urban Planning), a Grad. Dip. (Urban Planning) and an MBA. Her experience includes local government, state government and private sector, and over the past 20 years she has provided town planning services for a range of key infrastructure and development projects throughout Victoria. She is a strong advocate for regional Victoria. Kirsten is a Board Member of the Committee for Geelong. In 2016, Kirsten participated in the Committee for Geelong's global study tour of cities and now is the Chair – Winning from Second Sub-Committee. She is the former Chair of UDIA Geelong Region Chapter. She is also a Trust Member of the Geelong Performing Arts Centre. Kirsten is passionate about Geelong and excited by its future.

Stephen Murphy
Stephen Murphy has a BSc and graduate diplomas in Geology and Environmental Management, and has been a nurseryman and a designer of natural landscapes for over 30 years. Through his work as revegetation adviser to farmers he developed and published the highly regarded revegetation design manual *Recreating the Country: A Blueprint for the Design of Sustainable Landscapes* (2009). He is an active member of Landcare and founder of three 'Friends' groups that work with Parks Victoria and councils to protect and restore bushland reserves. He continues to write about natural history, indigenous landscape restoration and sustainable biorich design in his popular monthly blog. These are posted on his website www.recreatingthecountry.com.au.

Matt Novacevski
Matt Novacevski is a writer, planner, researcher, designer and place-maker who has worked across local government, academia and the private sector. He is fuelled by good coffee, fresh air and curiosity about what makes great places tick. Having grown up in Geelong during a period of significant change, spent many hours on the terraces of Kardinia Park watching his beloved Geelong Cats football team and worked much of his life in regional areas, Matt retains strong links to history and landscape. This has developed into a strong focus on an ecological place-centred practice that seeks to bring out the layered stories of places across scales. Matt's Masters thesis on place identity in Victoria's peri-urban growth towns won a commendation from the Planning Institute of Australia in 2016. Aside from attaining his Masters in Planning (Professional) at Deakin University, Matt has worked as a Research Fellow and sessional lecturer at Deakin, while being a practising town planner in regional local government specialising in small town settlement planning and urban design. Matt has begun work on a PhD at The University of Melbourne, focusing on theoretical approaches to the evaluation of place-making projects across Australasia, through a scholarship awarded under an Australian Government Research Program Training Initiative. He is a Member of the Planning Institute of Australia and Placeleaders Asia-Pacific.

Owen O'Carroll
Owen O'Carroll is an Irish-trained architect experienced in mixed-use urban projects in Asia, Europe and the US. He is founding partner of a 70-strong urban and architectural design practice in London, part-time lecturer and visiting critic at London Southbank University Department of Architecture, and has devised and led specialist planning and design seminars for several London boroughs and the Greater London Authority. He is a former elected member and Vice President of the RIBA 2011–13 and co-author of specialist institute guidance on design contests and procurement. He is a founding director of ProjectCompass CIC, which provides professional support and linkage to architects practising

in the UK and EU. He is currently undertaking a PhD, focusing on Urban Design Methodology and Practice, at CEPT University, Ahmedabad, and also teaches in the undergraduate architecture program at Heriot-Watt University in Dubai.

Uncle Bryon Powell

Uncle Bryon Powell is a *Wadawurrung* Elder and has served as Chair (2004–2017) and a director of the *Wadawurrung* (Wathaurung Aboriginal Corporation), and has been a member of the Corporation since its inception in 1998. He has been involved in native title and cultural heritage issues for many years, working as a Projects Officer with the Corporation in Ballarat. He is a passionate advocate for Aboriginal cultural heritage recognition and equality, and an educator of the wider community in Aboriginal issues. He is also a musician and an ex-roadie, has an interest in history, is a world traveller, is a cultural observer, a pilot, and a proud father and a grandfather. As a grandson of *Wadawurrung* women Mabel Violet Robinson (1899–1984), his family can trace its descent from the Traditional Owners of the *Wadawurrung* Country including its lands and waters around the Ballarat region.

John Rollo

Dr John Rollo is the discipline leader for urban design within the School of Architecture and Built Environment at Deakin University. With a focus on place-making and the perception and visualisation of the built environment, he has been providing public sector based urban design research for over 15 years. He completed his Bachelor of Architecture with Honours at Deakin University, and holds qualifications at both Master's and Doctorate levels from the Graduate School of Architecture and Urban Planning at the University of California, Los Angeles, and the Martin Centre for Land Use and Built Form Studies at Cambridge University. John initiated the Urbanheart Surgery urban design studio and has worked with his long-time colleague Yolanda Esteban, to help facilitate discourse on contemporary urban issues between governments, communities, the design professions and higher education. With over 1500 students having participated in the studio since its inauguration in 1999, Urbanheart Surgery has facilitated over 25 planning and design forums. Six of the studios have focused exclusively on the City of Greater Geelong, including both its inner-city area and the neighbourhood activity centres of Belmont and Waurn Ponds. Eight have collaborated with other centres in regional Victoria, in particular the City of Warrnambool, Surf Coast Shire, the City of Greater Bendigo and the City of Ballarat, and 13 have collaborated with the Melbourne metro areas of the City of Port Phillip, City of Melbourne, City of Wyndham and City of Maribyrnong.

David Rowe

Dr David Rowe is Director of Authentic Heritage Services Pty Ltd, a heritage consultancy practice based in Geelong that specialises in built heritage and architectural history. Over the past 25 years, David has prepared a large range of local government heritage studies, thematic histories, conservation management plans and other heritage assessments on thousands of places in Geelong and throughout Victoria. He has authored several history and heritage publications, and has presented several talks and lectures on Geelong's post-European contact history and heritage. Since the early 2000s, David has been the consultant Heritage Adviser to the City of Greater Geelong. He is also Heritage Adviser to the Golden Plains Shire and Surf Coast Shire. In these roles David provides expert heritage and design advice on the conservation and redevelopment of heritage places, working collaboratively to ensure viable and heritage-sensitive outcomes for architectural regeneration. David is a presenter of heritage training programs for PLANET offered by the Planning Institute of Australia (Victoria), and gives regular university lectures on heritage principles, processes and management.

Chris Smyth

Chris Smyth has a private consultancy specialising in marine and coastal research. Chris' interest in marine and coastal management and planning was sparked at university when he directed a research project on Victorian coastal management, and continued during his time as a geography teacher in Victorian regional coastal communities where he was also closely involved in local environment groups. Chris spent time at the Gould League of Victoria as an environmental education consultant and its editor and marketing coordinator in the early 1990s, followed by his first stint at the Australian Conservation Foundation as the editor of *Habitat* magazine and ACF diaries, and later as marketing manager and general manager. He left the ACF in 1999 and spent a year with the Merri Creek Management Committee as its Education Manager before joining the Victorian National Parks Association in 2000 to coordinate its campaigns for Victoria's marine national parks network and an integrated national park at Point Nepean. He returned to the ACF as its healthy oceans campaigner in 2003, focusing on the establishment of the national marine reserves network, ecosystem-based regional marine planning, national oceans law reform and sustainable seafood assessment, until 2012 when he established his consultancy. From 2015 to 2017, Chris again worked with the Victorian National Parks Association as its marine and coastal coordinator, acted for eight months as its executive director, edited several issues of its *Park Watch* magazine and helped coordinate the redevelopment of its website. Chris now lives in Queenscliff where he takes an active interest in the future of the Geelong region's marine and coastal environments.

Donna Squire

Donna Squire is a photographer with over 20 years experience in the field. Working in the education industry for the last two decades has given her a broad view of life, opening up endless opportunities: 'With the photography I do, every day is different, always a new adventure, meeting new and interesting people'. Specialising in supporting researchers has allowed her to capture pictures of seaweed, chemiluminescence reactions, aerial surveys ... the list is long. Photography is not only her job, it is her passion – a passion that has given her an excuse to travel the world. When not out and about, she shares her life in Geelong with her husband and son and a menagerie of animals. On the weekend she likes to take photos.

***Tandop* David Tournier**

Tandop David Tournier (1955–2016) was a *Yorta Yorta* man from his father's Country (around the Deniliquin area in New South Wales and the Victoria region), and *Ngarrindjeri* from his mother's Country (the Coorong area in South Australia), with familial connections through his great-great-great-grandfather within *Wadawurrung* Country at Bacchus Marsh. For many years he was the Cultural Heritage Education Language Officer for the Wathaurong Aboriginal Co-operative in North Geelong. When he became a Koorie educator he was looking at *Wiradjuri* language then came across the *Wemba Wemba* language before starting to incorporate the *Wathaurong/Wadawurrung* language into schools. 'It's brilliant stuff. We all know that language deals with Country and this is what I'm doing, trying to explain to people, if you want to know your Country, you want to learn your language. Language is land, land is Country, Country is you'. Tandop David's skills and responsibilities at the Co-op included performing 'Welcome to Country' ceremonies, delivering cultural sessions to schools, story-telling at kindergartens, sessions with individuals 'wanting to know' and conducting cultural education programs for various organisations, such as Barwon Health and the Department of Health. He also worked with the *Wathaurong/Wadawurrung* language to develop resources and strategies to implant the language back into the *Wathaurong/Wadawurrung* community.

Mark Trengove

Mark Trengove established the Geelong Indigenous Nursery in 1988. Since then his main interest has been the care of the indigenous vegetation of the region, through seed collection and propagation, vegetation management, and the documentation of remnants by compiling species lists and through photography. Since 2002 he has maintained a full-time ecological consultancy centred in the Barwon region. Mark has prepared vegetation management plans for many significant areas of native vegetation and has contributed to publications such as *Indigenous Trees and Shrubs of Western Port Phillip Region* and the City of Greater Geelong Council's *Biodiversity Management Plan* (2013). Mark spends part of every year on the south coast of Kangaroo Island, where he has developed a passion for coastal Mallee vegetation.

Dennis Williamson

Dr Dennis Williamson is a landscape planner/landscape ecologist and geographer with a range of specialist skills in landscape planning, landscape ecology, natural resource management, visual resource assessment and tourism planning. He has a PhD in Architecture (Landscape Ecology), a Masters in Landscape Architecture and a BA in Geography. His PhD research thesis focused on the topic of Australian wheatbelt farmers' and natural resource managers' responses to key ecological principles applied to habitat fragmentation, climate change, agricultural productivity and species loss. Dennis is a director of Melbourne-based Scenic Spectrums Pty Ltd and its Geoscene International division (www.scenicspectrums.com.au). He has held positions in multiple land use planning, landscape management, scenic and recreation planning roles with the US Forest Service, the Pinchot Institute for Conservation and the former departments of the Forests Commission Victoria and Department of Conservation, Forests and Lands in Australia. He was a founding member of the Five Valleys River Park Association in western Montana (now the Five Valleys Land Trust, http://www.fvlt.org/), which in 2008 was among the first land trusts in the US to be accredited by the Land Trust Accreditation Commission. It now protects 140 miles of stream frontage and 64 000 acres of land, wetlands and wildlife habitat. Dennis has conducted nomination assessments for several World Heritage Areas that are now listed in Australia and P.R. China. He is currently the Scientific Adviser for the World Heritage Advisory Committee at Purnululu National Park in Western Australia. Dennis has directed and been a member of various multi-disciplinary environmental planning projects, including coastal management and action plans, catchment management plans, and river, lake and wetland management plans in Australia. He has served as project manager on many significant land use development, planning, landscape design and environmental protection projects with interdisciplinary planning and design teams.

Ross Wissing

Ross Wissing has spent most of his professional career in the design, management and involvement of government and the community in improving the health of rivers and wetlands. He began his professional journey working with Barwon Water in the creation of wetlands and revegetation of the Barwon River in Geelong, then established the Corangamite Waterwatch Program in the Barwon catchment. Waterwatch is a community water monitoring program that assists local communities to learn about and actively improve the waterways where they live. Ross then managed the South Australian and National Waterwatch programs before returning to Victoria where he was the program manager for the first Port Phillip and Western Port Water Quality Improvement Plan, also known as the Better Bays and Waterways Plan. Ross has worked for the Wildfowl and Wetland Trust in the UK, the organisation that built the London Wetland Centre and the Hong Kong Wetland Park. Today Ross lives near the Barwon River in Geelong and is engaged in public open space planning for the Surf Coast Shire, which includes a significant part of the Barwon catchment. Ross is completing a PhD at Deakin University that investigates the sustainability of different eras of urban form in Geelong and how local residents can sustainably meet their basic needs in their backyards.

Josh Zeunert

Dr Joshua Zeunert is a Senior Lecturer at the University of New South Wales and a Registered Landscape Architect in Australia (AILA) and the US (ASLA). He has worked at five universities internationally, as well as in industry with leading landscape architecture and urban design practices for multiple award-winning projects. His research interests focus on the nexus of academia and professional practice in environmental design, planning, strategy and food systems. He is the author of *Landscape Architecture and Environmental Sustainability* (Bloomsbury, 2017), lead editor of *The Routledge Handbook of Landscape and Food* (2018), and numerous journal and professional articles and book chapters. His PhD explores multi-dimensional sustainability and public urban agriculture in landscape architecture.

1

Geelong: *Djilang* – a tapestry of histories, voices and ecologies

David S. Jones and Phillip B. Roös

Introducing the tapestry

> This is the description of the country: *'Beautiful land, and all good sheep country, rather sandy, but the sand black and rich, covered with kangaroo grass about ten inches high, and as green as a field of wheat, beautiful plains, excellent land, &c.'*
>
> *On the Saturday he has more to say in praise. 'Good hay could be made, and in any quantity. I never saw anything equal to the land in my life. I was never so astonished in my life.' On the Friday they had walked but twelve miles, but on the Saturday twenty. They had anchored about St. Leonard's, wandered over the Bellerine hills, and beheld the magnificent Geelong plains. Not astonishing that he said: 'From what I have seen I am quite delighted with Port Phillip'.*
>
> – Batman in Bonwick (1867, p. 14).

In traversing the uplands of the Bellarine Peninsula and thence skirting Corio Bay in 1837, adventurer John Batman recorded the above observations of a verdant Elysian landscape fit for pastoral colonisation. This was the landscape of the *Bengalat balug* and *Wada wurrung balug* clans of the *Wadawurrung* peoples whom had long resided here (Clark 1990; CoGG 2014), and a place that in the future would witness western innovation and industry creating the Geelong we know today.

The following year surveyors Garrard and Shaw (Plate 1.1) prepared and published a map of the town and suburbs of Geelong comprising the lands in the parishes of Ghering-hap, Moorpanyal, Barrabool, Duneed, Moolap, Bellerine and Pawit, together with the positions of the bar and the proposed improvements, that sets forth the first colonisation development vision for this landscape. Senbergs has offered a frightening visual projection of what Geelong could have been like if it had become the capital of Victoria (Plate 1.2).

This book seeks to translate the historical evolution of Geelong, with a particular emphasis upon its natural and cultural values, their tensions with accommodating growth and 'advancement', and the ideas and visions that created contemporary Geelong and are shaping its trajectory today. The book narration involves a survey of these tapestry threads supported by extensive referencing to enable readers to journey more deeply. Its chapters are written by a team of authors who are passionate about their topics, how they fit into the tapestry, and understand the need to respect these topics as Geelong ventures into the future. In moving forward, irrespective of institutional approved strategic plans, of scientific appraisals of ecological attributes and the sensitivity of places and ecologies, and of the cultural values of people and the meanings they ascribe to places, we need to appreciate the past of a place, appraise the human and ecological values and attributes of a place, and consider the decisions that informed place transformation. We need to take heed of these while at the same time setting forth a vision scaffolded carefully from all these variables. This approach is necessary so that we do not compromise the human ecological values and characteristics of place that would otherwise result in its despoliation.

This book offers insights and a translation of these human and ecological values and characteristics as they pertain to the wider Geelong landscape, as well as offering several warnings to be heeded.

What is evident is that the Geelong–Bellarine Peninsula–Surf Coast region is collectively experiencing exponential urbanisation. With urbanisation comes tensions as to physical and social infrastructure and provision, stresses upon terrestrial and aquatic ecosystems and their wildlife residents, and haphazard governance decision-making influenced by political expediency rather than long-term visions and equity. What suffers is the landscape, the place, Country, which more often is compromised in the name of development and sustainability despite the best intentions of vision-makers, land use planners and their plans. Thus, the very essence of what makes the qualities of Geelong can be compromised by these processes and deliberations, wilting the very reasons why humans consciously seek to shift and reside there, enjoy its lifestyle and believe in its future.

North American academic Ian McHarg (1920–2001) once stated, 'Man is a blind, witless, low brow, anthropocentric clod who inflicts lesions upon the Earth' (McHarg 1965). The legacy of our human actions in shaping and transforming planet Earth, let alone this corner of the Australian continent, have all been subject to human interventions, 'management', and 'care' irrespective whether pre-colonisation or post-colonisation. A key facet in our collective vision should be to care, to renourish, and to look to the health of our landscapes, our Country, our peoples.

Wadawurrung Elder Uncle Bryon Powell has observed, 'From the headwaters down to where it joins with the Barwon River, the Moorabool River tells us if Country is well. Look at the sky, look for *Bunjil*, where the eagles fly we know the river is healthy'. As Geelong residents and visitors, we have an individual and collective responsibility to read and listen to this Country, to sensitively enable development at the same time as healing and caring for the place (Powell 2018).

As North American author Wallace Stegner (1909–93) has written: 'We are the most dangerous species of life on the planet, and every other species, even the Earth itself, has cause of fear our power to exterminate. But we are also the only species which, when it chooses to do so, will go to great effort to save what it might destroy' (Stegner in Fradkin 2009, p. 184).

The responsibility is in our hands. We reside within the legacy of the past – the vision of the future and the wildlife of our landscapes are what we leave to our children.

Narrating the tapestry

As co-authors of this publication, we both came to the Geelong region with fresh eyes and a desire to understand its landscape mosaic or tapestry. In many ways, this learning was a struggle. Not because of what was physically evident, planted, constructed, painted, dying, moving or plastered on signboards, but because there was a dearth of textual and oral resources that coalesced in explaining 'why it is so'.

The lack of comprehensive or accessible literature, reports, scientific assessments or longevity of thought and insightful memories and oral coffee sessions was both frustrating and daunting.

This is not to say that individual local municipal area histories were not comprehensive (Brownhill 1990; Hundt 1971; Seaton 1978, 1983; Wynd 1971, 1981, 1988, 1992), nor that there was a lack of natural science overviews of the region (Dahlhaus *et al.* 2007; Pescott 1983, 1988, 1995, 2017; Wood 1878–1959), or that there were no isolated and exceedingly helpful specialised publications about Geelong-based topics (CoGG 2014) or that there were not several insightful architectural (Buchan Group 1980; Page 1990), landscape architectural (Lethlean 2013), *Wadawurrung* culture (CoGG 2014; Ferrier 1992; Pascoe 1997, 2003, 2007), archaeology (Rhoads 1986; Richards and Jordan 1999), cultural heritage (McLean 2005; Rowe 2002, 2016; Rowe and Huddle 2000; Rowe and Jacobs 2007, 2010, 2014–2016, 2016; Mary Sheehan and Associates 2003; Willingham 1986), natural resource management (CoGG 2018a) and urban planning-related (Almeida Correia and Denham 2016; CoGG 1994; G21 GRA 2018) references, but that they were more often topic-specific, lacked a holistic perspective, were increasingly dated, or were little known and of limited copy issue despite their locations on various websites.

This became the kernel of the rationale of this book – the need to synthesise much of this little accessible knowledge into one text to aid residents, visitors and students alike in better 'reading' and understanding this burgeoning place.

The scope of this discussion treats Geelong as encompassing the volcanic plains up to the You Yangs and Werribee River, across the middle and lower catchments of the Barwon and Moorabool rivers, from the eastern portion of the Otway Ranges from Aireys Inlet east, and the larger lands and waters of the Bellarine Peninsula. By colloquial terminology, it is 'Geelong and the Surf Coast'.

In the *Wadawurrung* lens, is it the clan territories of the *Bengalat balug, Wada Wurrung balug, Neerer balug, Yaawangil balug, Gerarlture balug* and the *Mon Mart balug* peoples (CoGG 2014), and to us it is 'home'.

Surveying the tapestry

Geelong, or *Djilang* in *Wadawurrung* language, is nestled on the south-western edge of Port Phillip Bay, inside Corio Bay. Uniquely in Australia, it is a city with a north-facing marine outlook.

Its *Wadawurrung* legacy lies in intensive Living Stations of occupancy encampments that offered reliable and rich food harvests, and thus the need for infrequent movements and greater permanency of residency. Some 40 000–60 000 years ago, the *Wadawurrung* settled across this region. They established a particular civilisation that had little reliance on physical manifestations and buildings, instead formulating a sustainable landscape management relationship built upon careful resource use and respectful human engagements through commonality.

From ~1800 European explorers, military, whalers and sealers, and adventurers set foot on this landscape with aspirations, sheep, cattle, guns, axes and a desire to 'develop', blinkered with personal economic gain. Names like Murray, Flinders, Hume, Hovell, Collins, Batman and Buckley litter contemporary regional and Port Phillip Bay histories and nomenclature as the 'discoverers' who 'found this landscape'. Key places often mentioned in these histories are Geelong, Sorrento, Freshwater Creek, Indented Head, Queenscliff, Hovells Creek, Breamlea, Corio Bay, Point Henry and the You Yangs. It was in March 1836 that squatters formally arrived in Geelong with flocks of sheep, thus commencing 'settlement' and land clearance there. In October 1838, Assistant Surveyor W.H. Smythe surveyed the town plan of Geelong, a north–south oriented grid rectangle stretching from the bay to the river, using the mathematical survey template previously applied in the east–west survey of Melbourne. From the mid 1830s personalities like Fisher, Manifold, Strachan, Russell, Thomson, Smythe, Fyans and La Trobe forged Geelong's initial development, enabling a governance system, the construction of wharfs, warehouses, the breakwater, bridges, steamship services, wool export businesses and lime extraction.

Invention and initiative were propitious at this time. Newspapers were established, wool treatment systems were pioneered, the world's first ether vapour compression cycle ice-maker and refrigerator were invented, the first shipping channel was dredged in Port Phillip Bay, the first rabbits were introduced into Australia via Geelong, and the smell of gold at Ballarat opened Geelong up to the international shipping community.

Over the course of the 1850s to 1900 Geelong established itself as a city, a key wool export port and woollen mill venue, a regional railway node, a viticulture venue and an educational venue (with the advent of the Gordon Memorial Technical College, commemorating General Charles Gordon who died at Khartoum in January 1885). The threads of industrialisation and industrial processing were evident in this period even before the construction of major installations from 1900 to the 1970s that established its reputation as a blue-collar industrial centre. Sewerage, lights, trams and harbour dredging all foreshadowed the 1920s accelerated growth of woollen mills, fertiliser plants, the Ford Motor Co.'s first vehicle plant at Norlane, salt extraction fields at Moolap and Avalon, and whisky distilling. In the years preceding World War II, International Harvester opened its doors in Norlane, Corio Quay was constructed to host major grain exporting, and Shell Australia commenced oil refining. In 1962, Alcoa constructed its recently closed alumina smelter at Point Henry. To accommodate industrial expansion, employees were needed. Thus new suburbs grew up in North Shore, East Geelong, Belmont, Norlane and the Highton localities close to these industrial complexes or near railway stations. At the same time the city invested in the Eastern Beach foreshore, the botanic garden, along river corridors, in golf courses, in parking meters and in Johnstone Park as part of civic infrastructure.

From the 1970s, national and global economic changes affected Geelong severely. Most woollen mills closed, large tracts of huge storage warehouses adjacent to Corio Bay became vacant, demonstrating the major shifts in technological cost-efficiencies that later enveloped the Ford Co., Alcoa, International Harvester and the railway marshalling and repair yards in the 1990s and 2000s. Innovations like the creation of Apex International (Page 1990) and the maturation of the Buchan Group (Buchan Group 1990) also occurred during this period.

In conscious response, the federal and state governments signalled and enabled strategic economic workforce-building initiatives in the region in the 1970s to 1990s. While the federal Department of Urban & Regional Development seeded regional transformative investments, the Victorian government established the Geelong Regional Planning Authority and its successor the Geelong Regional Commission (McLean 2005). Collectively these were major visionary investments, of which the Commission's legacy is perhaps the least understood today. The investments encouraged Target to establish its headquarters in Geelong, the construction of the Geelong Performing Arts Centre, the creation of Deakin University (Hay *et al.* 2002), the creation of a National Wool Museum, and the relocation of the Commonwealth Animal Health Laboratory to Geelong. The legacy of the Geelong Regional Commission included forward plans for the Armstrong Creek–Warralily suburbs (CoGG 1994), a future for Avalon Airport as a flight node, the reservation of the Geelong Ring Road, reconstruction of former warehouses to accommodate the Transport Accident Commission, Deakin and Westfield, the transformation of the Geelong Waterfront, and a strategic plan for taking Geelong towards 2020. The latter is now the mantle of the G21 Geelong Region Alliance (G21 GRA 2018) in partnership with the City of Greater Geelong Council and adjacent councils.

To many residents and outside observers, Geelong has been hit hard economically. Its blue-collar workforce has faced the onslaught of changes in national and international tariffs, trade agreements and industry policy shifts. But what is also evident is that an innovative transformation is afoot in Geelong, shifting its employment base from traditional industrial to innovative industrial (digital carbon fibre, liquids), to education resourcefulness, to health futures and equality, to design quality. All of these changes are resulting in suburban and inner-city gentrification, a new sense of city liveability, the emergence of a vibrant coffee and student culture, and a place where is far cheaper to raise a family than in the burgeoning metropolis 'an hour up the road'. Projects like the Green Spine (CoGG 2018b), #Vacant Geelong (Lozanovska *et al.* 2017), GPAC expansion (GPAC 2018), Barwon Water (Edmonds 2004) office consolidation, new hotels and apartment complexes, Armstrong Creek–Warralily, Johnstone Park reinvention and water futures (Jones and Whitehead 2005; Lovell Chen 2016; O'Connor 2016), major expansion of Deakin University (Hay *et al.* 2002) at both campuses, and new suburbs on the Bellarine Peninsula behind Torquay, at Lovely Banks and Batesford, and new venues at Avalon Airport, all signal confidence. They also significantly evidence a major shift in the region's employment backbone and culture, and an urgent need to protect, conserve, care for, identify and zone lands, waters and places to ensure their integrity, qualities and characteristics are maintained for future generations. We need to ensure that the very nature of the place that is hosting these visions is not compromised.

The recent designation of Geelong as a UNESCO Creative City/Réseau des villes créatives 'City of Design' – the only such designation of a city in Australia – is a further signal of a new industry direction unfolding in Geelong (UNESCO 2016).

Pieces in the tapestry

The strategy in this book is two-fold. First, to provide foundation resources and reading in each chapter, as well as providing clues as to where to obtain more information. Second, to instigate a discussion about where is Geelong going, and the need for us to make wise, culturally and ecologically sensitive decisions and judgements about this direction so as not to destroy that which we hold so valuable as representing the essence of this place, this landscape, this Country.

Section 1 (Chapters 2–5) sets the context of this discussion. In Chapter 2, Peter Dahlhaus surveys the geological formation of the regional landscape. In Chapter 3, Geoff Carr takes readers into the ecological history and plant profile of the landscape. Uncle Bryon Powell and *Tandop* David Tournier provide a significant *Wadawurrung* voice and insight into past, present and future through their eyes in Chapter 4. In Chapter 5, David Rowe surveys the post-colonisation transformation of Geelong.

Section 2 (Chapters 6–9) focuses on translating the ecological fabric of this tapestry. Mark Trengove deeply examines the botanical communities of the region's landscape in

Chapter 6. Stephen Murphy provides an ecological review of the Bellarine Peninsula in Chapter 7. Ross Wissing and Chris Smyth discuss the riverine and coastal segments of the landscape respectively in Chapters 8 and 9. Appendix Plates 1 and 2 graphically depict the vegetation units for the region, which can aid in orienting the reader to the region's vegetative characteristics.

Section 3 (Chapters 10–18) considers human agency of change and vision. In Chapter 10, Dennis Williamson considers the landscape through natural resource management policies and models. In Chapter 11, Phillip B. Roös has coalesced the key at-risk ecological and human variables confronting this landscape. In Chapter 12, Louise Johnson takes readers through the geography of suburban transformation of the Geelong landscape, before Beau Beza, Josh Zeunert and Murray Herron in Chapter 13 discuss and analyse emergent suburban models and opportunities for this landscape. In Chapter 14, Kate Kerkin surveys the social fabric of the place and the social values, considerations and patterns that are evident. Matt Novacevski, in Chapter 15, discusses the tangible–intangible effects of the current mainstream industrial transformation of the landscape, and in Chapter 16 Gavin Keeney and Owen O'Carroll provide an insightful alternate view of this transformation. In Chapter 17, John Rollo and Yolanda Esteban review and critique the visioning that has historically occurred and is still occurring in Geelong's central business district. Kirsten Kilpatrick, in Chapter 18, reflects upon the key planning challenges facing Geelong in the next 10–20 years.

The final chapter, Chapter 19, by David S. Jones and Phillip B. Roös, synthesises the book, summarising key patterns and future challenges for curating this landscape from its demise.

As stated, the second intention of this book is to raise awareness, to raise questions, and to raise discourse and debate. If we have opened up new ways of thinking, if we have stirred discussions, if we have caused some angst and some new ways of thinking about topics, irrespective how large or small, then we have achieved the purpose of this book and it will not recede to collect dust on bookshelves.

References

Almeida Correia J, Denham T (2016) *Winning from Second: What Geelong can Learn from International Second Cities*. Committee for Geelong, Geelong. http://citiesprogramme.org/wp-content/uploads/2016/12/winning-from-second-web.pdf (accessed 1 June 2018).

Bonwick J (1867) *John Batman: The Founder of Victoria*. Samuel Mullen, Melbourne. Reproduced in http://gutenberg.net.au/ebooks13/1303761h.html (accessed 1 March 2018).

Brownhill WR (1990) *The History of Geelong and Corio Bay. Geelong Advertiser*, Geelong.

Buchan Group (1980) *Ninety Years of Architecture in Victoria: Buchan Laird and Buchan, a History 1890–1980*. Buchan Group, Melbourne.

Clark ID (1990) *Aboriginal Languages and Clans: An Historical Atlas of Western and Central Victoria, 1800–1900*. Department of Geography and Environmental Science, Monash University, Melbourne.

CoGG (1994) *Mount Duneed Armstrong Creek Urban Development Study*. Report prepared by Henshall Hansen Associates for City of Greater Geelong, Geelong.

CoGG (2014) *Karreenga Aboriginal Action Plan 2014–2017*. City of Greater Geelong, Geelong. https://www.geelongaustralia.com.au/common/Public/Documents/8d1ac695c3a580a-Karreenga%20Aboriginal%20Action%20Plan%202014-2017.pdf (accessed 1 June 2018).

CoGG (2018a) *Indigenous Plants of the Geelong Region*. City of Greater Geelong, Geelong. https://www.geelongaustralia.com.au/indigenousplants/article/item/8ce589e1bce0fe8.aspx (accessed 1 March 2018).

CoGG (2018b) *Malop Street Green Spine*. City of Greater Geelong, Geelong. https://www.geelongaustralia.com.au/common/public/documents/8d4be2e7fa85b3e-RCG_GreenSpine_JUNE17_web.pdf (accessed 1 March 2018).

Dahlhaus PG, Billows CA, Carey SP, Gwyther J, Nathan EJ (2007) *Lake Connewarre Values Project: Literature Review*. Corangamite Catchment Management Authority, Colac, Victoria.

Edmonds L (2004) *Living by Water. A History of Barwon Water and its Predecessors*. Barwon Water, Geelong.

Ferrier S (*c*1992) *Wathaurong Medicines*. Wathaurong Aboriginal Co-operative, Geelong.

Fradkin PL (2009) *Wallace Stegner and the American West*. University of California Press, Berkeley.

G21 GRA (2018) *G21 Geelong Region Alliance*. http://www.g21.com.au/ (accessed 1 March 2018).

GPAC (2018) *Ryrie Street Redevelopment*. Geelong Performing Arts Centre, Geelong. https://gpac.org.au/redevelopment?stage=Stage (accessed 1 March 2018).

Hay R, Lowe D, Gibb DM, Anderson B (2002) *Breaking the Mould: Deakin University, the First Twenty-five Years*. Deakin University, Geelong.

Hundt R (1971) *Early History of the South Barwon Shire*. Gordon Institute of Technology, Geelong.

Jones & Whitehead (2005) *Johnstone Park Conservation Management Plan*. Jones & Whitehead Pty Ltd, Melbourne.

Lethlean P (2013) *02 Braided Pathways: A Practice Sustained by Difference – It's Hard getting Messy when you're Compositional*. PhD thesis, RMIT University, Melbourne. https://practice-research.com/portfolio-item/its-hard-getting-messy-when-youre-compositional-braided-pathways-a-practice-sustained-by-difference-by-perry-lethlean/ (accessed 1 March 2018).

Lovell Chen (2016) *Heritage Impact Statement: Johnstone Park, 24–28 Gheringhap Street, Geelong*. Lovell Chen Architects & Heritage Consultants, Melbourne.

Lozanovska M, Beynon D, Bishop C, Wilson AS (2017) *#Vacant Geelong – Iconic Industry: Exploring the Industrial Built Fabric of Geelong*. Deakin University, Geelong.

McHarg ID (1965) *Design with Nature*. Natural History Press, Garden City, New Jersey.

McLean GA (2005) The history of the Geelong Regional Commission. Unpublished PhD thesis, Deakin University, Geelong.

O'Connor A (2016) *Preliminary Aboriginal Cultural Heritage Study: Johnstone Park, Geelong, Victoria*. Ecology & Heritage Partners, Geelong.

Page MF (1990) *An Architectural Apex*. Buchan Laird International, Melbourne.

Pascoe B (1997) *Wathaurong: Too Bloody Strong – Stories and Life Journeys of People from Wathaurong*. Pascoe Publishing, Apollo Bay, Victoria.

Pascoe B (2003) *Wathaurong: The People who said No*. Wathaurong Aboriginal Co-operative, Geelong.

Pascoe B (2007) *Convincing Ground: Learning to Fall in Love with your Country*. Aboriginal Studies Press, Canberra.

Pascoe B (2016) *Dark Emu: Black Seeds: Agriculture or Accident?* Magabala Books, Broome, WA.

Pescott T (1983) *Birds of Geelong*. Neptune Press, Melbourne.

Pescott T (1988) *From Buckleys to the Break: A History and Natural History of the Barwon River through Geelong*. Geelong Field Naturalists Club, Geelong.

Pescott T (1995) *The You Yangs Range*. Yaugher Print, Geelong.

Pescott T (2017) *Birds and Botanists: A Field Naturalist's History of Geelong*. Trevor Pescott, Geelong.

Powell B (2018) *Discover the Living Moorabool*. Corangamite Catchment Management Authority, Colac, Victoria. http://www.ccmaknowledgebase.vic.gov.au/moorabool/ (accessed 1 March 2018).

Rhoads JW (1986) *Bellarine Peninsula: Archaeological Site Assessment and Management Study*. Department of Anthropology, University of Sydney, Sydney.

Richards T, Jordan J (1999) *Aboriginal Archaeological Investigations in the Barwon Drainage Basin*. Aboriginal Affairs Victoria, Melbourne.

Rowe D (2002) *The Pleasure Grounds of the Barwon Coast: A History*. Barwon Coast Committee of Management, Barwon Heads, Victoria.

Rowe D (2016) *Lara Heritage Review: Thematic Environmental History*, vol. 2. City of Greater Geelong, Geelong.

Rowe D, Huddle L (2000) *Greater Geelong Outer Areas Heritage Study Stage 2*. City of Greater Geelong, Geelong.

Rowe D, Jacobs W (2007) *Belmont Heritage Review*. City of Greater Geelong, Geelong.

Rowe D, Jacobs W (2010) *Ashby Heritage Review*, vol. 1. City of Greater Geelong, Geelong.

Rowe D, Jacobs W (2016) *Lara Heritage Review*. City of Greater Geelong, Geelong.

Rowe D, Jacobs W (2014–2016) *West Moorabool Shire Heritage Study Stage 2A*. Moorabool Shire Council, Ballan, Victoria.

Seaton G (1978) *The Ashby Story: A History of Geelong West*. Geelong West City Council, Geelong West.

Seaton G (1983) *Model Borough: An Introduction to Old Newtown and Chilwell*. City of Newtown, Geelong.

Mary Sheehan and Associates (2003) *Colac Otway Shire Heritage Study Thematic History*, vol.1. Shire of Colac Otway, Colac, Victoria.

UNESCO (2016) *Creative Cities Network: Réseau des villes creatives*. UNESCO, Paris. www.en.unesco.org/creative-cities/ (accessed 1 March 2018).

Willingham A (1986) *Geelong Region Historic Buildings and Objects Study*. Geelong Regional Commission, Geelong.

Wood PJ (1878–1959) *Geelong's Birdlife in Retrospect: A Selection of Geelong Advertiser Articles by P.J.W 1945–1958*. Yaugher Print, Geelong.

Wynd I (1971) *Geelong: The Pivot – A Short History of Geelong and District*. Cypress Books, Melbourne.

Wynd I (1981) *So Fine a Country: A History of the Shire of Corio*. Shire of Corio, Geelong.

Wynd I (1988) *Balla-wein: A History of the Shire of Bellarine*. Shire of Bellarine, Drysdale, Victoria.

Wynd I (1992) *Barrabool: Land of the Magpie*. Barrabool Shire, Torquay, Victoria.

PART 1

ENVIRONMENTAL HISTORY OF THE GEELONG REGION

2

The lay of the land: the geological evolution of the landscape

Peter Dahlhaus

> *The development of cities and new settlements is in so many cases too rushed to pay due regard to natural environment and particularly to geological conditions. It is very often forgotten that the geological history of land is so frequently an important guide to the planning of its future use.*
>
> – Quido Zaruba, Foreword, *Cities and Geology* (Legget 1973, p. ix).

The landscapes of the Geelong region have been fashioned by geological and geomorphological processes over many hundreds of millions of years, resulting in the topography of today. They comprise both the oldest and youngest rocks in Victoria, spanning a period of more than 600 million years. Yet, despite its unchanging normality when viewed through human eyes, Geelong's terrain is constantly changing through erosion, deposition and tectonic forces. Many geologists regard some of the Geelong landscapes as among the most active landscapes in Australia (Gibson and Brown 2003).

Geomorphologists understand that there is a cause for every shape in the landscape. Landforms such as river valleys, mountains or lakes are not shaped by accident, but reflect some underlying reason why they occur where they do and why they are shaped as they are. Such underlying reasons may be geological boundaries between hard and soft rocks, weaknesses in the rocks created by geological faults and joints, or even the activity of humans using bulldozers or shovels. When we inhabit a landscape, we can avoid many of the hazards – such as landslides, earthquakes, floods and erosion – by recognising the land-forming processes associated with constantly evolving landscapes.

The landscape physiography is also a determining factor in ecology. Plant and animal communities evolve in response to the available water, sun or shade, soil profile and nutrients at a particular place in the landscape. Humans too have chosen to settle in places because of the landforms, soils, rocks, minerals, springs, lakes, rivers, estuaries and coastal embayments. In fact, all of these components played a role in attracting humans to the Geelong region, as reflected in place names like *Dijlang*, the *Wadawurrung* name for the local landforms of Geelong (Pascoe and Krishna-Pillay 1968–2007), and the place names later used by European settlers, such as Clifton Springs, Limeburners Bay, Waurn Ponds, Lovely Banks and Little River. Hence, the geological evolution could be seen as the basic reason for its initial habitation by *Wadawurrung* people and eventual evolution to a city.

In the beginning

Geology, being the science of the Earth's history, is calibrated against the geological time-scale (Table 2.1), listed as Epochs, which are subdivisions of Periods, which are in turn subdivisions of Eras (Gradstein *et al.* 2012). Three basic concepts underpin the Geelong region's geological evolution:

1. the stratigraphy (the succession of rock units from oldest to youngest)
2. the tectonics and structural geology (the movement of continents and modification of stratigraphy by subsequent tectonic forces to produce features such as faults and folds)
3. the palaeoclimatic changes (past climates changes responsible for changes in the rates of erosion and deposition and sea-level rise and fall).

Forming the ancient basement rock units

The oldest rocks in the Geelong district (and Victoria) are represented by small erosion-resistant outcrops of metagabbro, a generally fine-grained dark green metamorphic rock known as the Ceres Gabbro (Morand 1995), which occur in the Barrabool Hills and near Dog Rocks (Plate 2.1). In the past the rock was quarried by the *Wadawurrung* people for stone axes which were highly prized and traded throughout south-east Australia (HCV 2017). It was briefly quarried for building stone in the late 19th century, since the rock is very hard and holds a good polish, but ultimately proved too expensive to quarry and dress (Mines Department 1949).

Although the rock unit has not been precisely dated, it is believed to be Neoproterozoic in age, based on its structural setting (Cayley *et al.* 2002). The outcrops of Ceres Gabbro are thought to represent an ancient block of the Earth's continental crust that underlies central Victoria, Bass Strait and parts of Tasmania. This block of crustal rocks that comprise the regional basement rocks is named the Selwyn Block, in honour of Alfred Selwyn, the founder of the Geological Survey of Victoria (Vandenberg *et al.* 2000).

Table 2.1. The portion of the geological timescale represented in Geelong landscapes

Era	Period	Epoch	Numerical age of epoch (millions of years ago)
Cainozoic	Quaternary	Holocene	0–0.01
		Pleistocene	0.01–3
	Neogene	Pliocene	3–5
		Miocene	5–23
	Palaeogene	Oligocene	23–34
		Eocene	34–56
		Palaeocene	56–66
Mesozoic	Cretaceous	Early Cretaceous	101–145
Palaeozoic	Devonian	Late Devonian	359–419
	Ordovician		444–485
Neoproterozoic			541–1000

At the time of the Selwyn Block formation, this corner of Australia looked very different from today, as it was a crustal fragment coalescing with other fragments of the Earth's crust to form the supercontinent of Gondwana. Throughout most of the Palaeozoic Era eastern Australia formed part of the coastline of Gondwana, including the late Palaeozoic when Gondwana had collided with another continental crust to form the supercontinent termed Pangea. In the mid Mesozoic Era, Pangea separated again into the supercontinents of Gondwana and Laurasia. Throughout the late Mesozoic and early Cainozoic eras, Gondwana slowly fragmented into smaller continents, including the continent of Australia.

The next oldest rocks represented in the Geelong district are those of the Ordovician Period when the region lay within ocean basins off the edge of Gondwana. These rocks were formed by the deposition of sediment in a deep marine trough that extended north–south across central Victoria (Fergusson and Vandenberg 2003). The sediment source was the ancient highlands of South Australia and western Victoria, known as the Delamarian Highlands. Much of the sediment was deposited by turbidity currents caused by subsea landslides, and is termed turbidite deposits. Turbidites are recognised by the cyclic sedimentation of coarse-grained to fine-grained sediments as the gravel, sand, silt and clay settle through the water column, as well as scours and ripples preserved in the strata. The rocks are dated through the abundant fossils of graptolites – organisms that colonised the Earth's seas at the time and whose various species represent distinct time zones. From these fossils, geologists deduce that Victoria was closer to the equator, experiencing a tropical to subtropical climate (Webb 1991).

A subsequent series of continental collisions deformed and compressed the turbidite sediments (several kilometres thick) to form sedimentary rocks, firmly attaching the rocks to the continental crust. These tectonic events, or orogenies, folded and faulted the strata into high mountain ranges that were eroded over the following 400 million years to expose the rocks that we see in the Brisbane Ranges and in the Moorabool and Leigh valleys today. Typically, they comprise interbedded sequences of quartz-rich grey sandstone and mudstone with minor conglomerate and black shale. In the near surface they are weathered into pale, bleached mudstones, siltstones and sandstones that are usually orange, yellow, white and red colours.

During the orogenic mountain-building events, the sedimentary rocks were infused with hot mineralised groundwater that percolated into the cracks and openings of the deformed rocks to form quartz veins, some of which contained gold. The gold was exploited during the mid 19th century by gold miners around Steiglitz and Morrisons when Geelong became a major gateway to the gold fields of Ballarat.

Further tectonic activity during the Late Devonian injected the Ordovician sedimentary rocks with magma that seeped upwards from blocks of crust that were being subducted into the Earth's mantle (Vandenberg *et al.* 2000). The magma cooled deep below the surface to form the granite rock now exposed in the You Yangs and at Dog Rocks. The granite has been dated as 365 ± 2 million years old and it is thought that 2–5 km of overlying rock has been eroded away to expose the granite at today's surface. Much of the erosion of these basement rocks probably occurred during the Permian Period (250–300 million years ago) when southern Australia resided in the Antarctic Circle and was subjected to glacial erosion. Being a harder crystalline igneous rock, the granite is more erosion-resistant than the surrounding rocks, and rises to 319 m elevation at Flinders Peak in the You Yangs. The granite has been used for local building construction.

Foundation for the present landscapes

Towards the end of the Mesozoic Era, the basement rocks had been eroded to a more planar surface. Much of Gondwana had fragmented and, by the Early Cretaceous, the separation of Australia and Antarctica, the last remaining continents to split, had started (Duddy 2003). The separation formed a rift valley around 600 km long and 100 km wide along the southern margin of Australia. Volcanic eruptions commenced with the rifting and lasted for ~20 million years, providing a source for the sediment deposited in the rift valley by wide braided river systems. Around 3 km of sediment was subsequently lithified into layers of sandstone and mudstone rocks, with some thin seams of black coal. Located within the Antarctic Circle at the time, fossils confirm that the sediment was deposited in valleys inhabited by diverse plants and animals, including small dinosaurs.

As the rifting continued and Australia drifted further northwards, the associated tectonic forces caused parts of the Early Cretaceous rocks to gently fold, fault and uplift. The uplifted blocks now form the Otway Ranges, Barrabool Hills and part of the Bellarine Peninsula. The rocks, known as the Otway Group, are relatively poor in quartz but rich in feldspar and fragments of volcanic rocks, making the dark grey-green sandstone and mudstone prone to weathering to brownish-yellow clays. It was widely used as a building stone in the 19th century – many grand buildings in Geelong and Melbourne were constructed with 'Barrabool Sandstone' (King and Weston 1997).

By the mid Palaeogene Period of the Cainozoic Era, Australia had completely separated from Antarctica and continued its drift northwards. Initially, the rate was less than 1 cm per year, but it increased to around 5–6 cm per year ~40 million years ago (Webb 1991). The geological separation from Gondwana changed the course of the region's landscape evolution, ultimately shaping the geomorphic features of today.

Shaping the landscape

As the Australian continent separated from Antarctica, the ancient Mesozoic surface in the north of the region was gently uplifted to create the beginnings of the Great Dividing Range. The uplift renewed erosion of the land and created broad valleys of scree, outwash and braided river deposits between meridional ridges of Palaeozoic bedrock. The general form of this topography has persisted from the early Palaeogene Period, establishing the proto-valleys of the Moorabool and Leigh rivers (Holdgate *et al.* 2006). The remnants of the Palaeogene fluvial sediments form a quartz–pebble conglomerate, often cemented with iron or silica. They are observed in the northern part of the region (Plate 2.1) as isolated deposits overlying Palaeozoic rocks and broad tablelands fringing the overlying basalt.

Contemporaneously, in the southern part of the region known as the Torquay Basin, mostly non-marine sediments were accumulating to form the mudstones, sandstones and brown coal beds of the Eastern View Group rocks (Holdgate and Gallagher 2003). The brown coal units were exploited from the mid 20th to early 21st centuries for electricity generation at Anglesea, powering the aluminium smelter in Geelong.

As the rate of sea-floor spreading increased, a wide seaway formed between Australia and Antarctica, contributing to a major change in global climate and large turnovers in marine and terrestrial biota. The formation of large ice sheets on the Antarctic continent and changes in oceanic currents created a massive drop in global temperatures in the early Oligocene (Williams *et al.* 1993). In the north of the Geelong region, the onset of cool, drier conditions reduced stream flows, resulting in smaller streams following the broad valleys of the ancient Moorabool and Leigh rivers. Dendritic drainage patterns developed and the streams gradually incised the deposits of coarser gravels deposited in the earlier, higher-energy environments, to expose the highly weathered bedrock. By the late Oligocene these valleys, incised into weathered Palaeozoic rocks, had been well developed (Taylor *et al.* 1996).

The end of the Palaeogene Period is marked by a major influx of the sea, or marine transgression, which significantly changed the sediment deposition regimes throughout the entire region. As the sea levels rose and the coast crept into the river valleys, the erosive capacity of the streams reduced and alluvial sediment began to fill the valleys of the dissected northern highlands. In the south, the deposition of terrestrial sediment gave way to marine sediment, mostly biological carbonates initially deposited in shallow seas, but by the early Neogene Period, deep water conditions had formed (Holdgate and Gallagher 2003). The subsequent lithification of these sediments formed the marls and limestones now visible in the coastal cliffs at Jan Juc and Torquay.

By the mid to late Miocene Epoch the depositional environment had changed to intermediate depths as the sea began to retreat and limestone was deposited in the southern part of the region. The Fyansford Formation limestone has been extensively quarried since the mid 19th century for cement manufacture, building stone and agricultural lime. In the quarries near Batesford, abundant fossils of shells, bryozoans, echinoids, foraminifera and shark's teeth in the Batesford Limestone indicate that it was formed as a fringing reef around Dog Rocks, which was an island in the ancient sea (Gourley and Gallagher 2004).

Meanwhile, sporadic volcanic eruptions continued. The Older Volcanics are basalts and fragmental rocks (pyroclastics) that were erupted across the region in either the late Palaeogene or the Neogene. The basalt landscapes of Mount Bellarine represent the largest outcrop in the region, with smaller outcrops at Maude. The stone is quarried for crushed rock supplies (Olshina and Jiricek 1999).

Forming the structural elements

A change in the direction of regional tectonic stress occurred during the Miocene as the Australian Plate collided with the Pacific Plate (Edwards *et al.* 1996). The previous north–south tensional forces were replaced by compressional forces along a north-west–south-east axis, reactivating some old geological faults and creating new ones. These faults shape the current-day landscape elements such as the direction of the coastlines and bays, which are oriented more or less perpendicular to the axis of compression (Plate 2.2).

The Otway Ranges began uplifting between two reverse faults: the westerly dipping Torquay Fault (which shapes the present-day coast from Barwon Heads to Cape Otway) and the easterly dipping Bambra Fault to the west (Sandiford 2003). The uplift renewed erosion, creating deeply incised and steep river valleys. The forces in the Earth's crust also raised the Brisbane Ranges west of the Rowsley Fault, the Barrabool Hills along the Barrabool Fault, and raised the Bellarine Block, while the Moolap Block has subsided.

Uplift of the Bellarine Block occurs along three structures, the Curlewis Monocline in the north, the Bellarine Fault in the east and an inferred fault in the west. The effect is that the Bellarine Block is tilted to the south, with the greatest uplift along the Curlewis Monocline. Hence, areas around the Bellarine coastline include steep cliffs as well as coastal terraces.

Laying a veneer of sand and volcanic rocks

During the Pliocene a marine transgression commenced in the south-east, and shallow seas eventually covered most the region (Wallace *et al.* 2005). In the valleys of the northern highlands, sands and gravels were locally deposited in response to the changes in stream base level. As the sea retreated to the south-west towards the end of the Pliocene, a widespread thin veneer of coastal sand was left behind. Known as the Moorabool Viaduct Formation, it is a thin unit (generally no more than a few metres thick) of sand and sandstone cemented with iron oxide which is widespread across south-western Victoria. Isolated deposits of Neogene sands, often cemented with iron (ferricrete) or silica (silcrete) also occur in the north of the region.

Volcanic eruptions commenced around the same time as the marine transgression (Price *et al.* 2003). Initially sporadic, the volcanism increased to its greatest activity by around 2 million years ago. The volcanic activity is associated with up to 150 m of uplift in the northern part of the region, which preserved the shoreline deposits of the Pliocene seas near Elaine from being covered by the volcanic lavas (Carey and Hughes 2002). Elsewhere the erupted lavas flowed into the valleys, filling all of the drainage systems of the pre-volcanic landscapes, displacing the water flows to the edges of the lava flows. Disruption to the drainage systems forced the drainage divide of the Moorabool and Leigh rivers further north to its present position and formed large lakes on the plains to the west.

The Newer Volcanics basalts are the most widespread geological unit in the Geelong region (Fig. 2.1). The final eruption points form prominent scoria cones, or composite scoria/lava cones, such as The Anakies (Rosengren 1994). In the south, the Mount Duneed basalt flow encroaches upon the Lake Connewarre Complex and is a critical component in the development of the Lower Barwon wetlands and estuary (Rosengren 1973). North of

Fig. 2.1. Folded Ordovician sandstone near Meredith. Photo: Peter Dahlhaus.

Corio Bay, the widespread cover of basalt has buried the ancient Neogene landscapes, including the ancient drainage systems of the Moorabool and Barwon rivers (Evans 2006). On the western side of Corio Bay and across the Moolap Sunklands the former channel of the Barwon River was filled with lava from the volcanic eruptions, directing the course of the present-day Barwon River.

Current processes shaping the landscapes

The youngest geological units are sediments that have been deposited in the last 3 million years, including stream alluvium, hillslope colluvium, swamp and lagoon deposits, coastal and inland dunes, and human-derived fill. These relatively thin deposits of the Quaternary Period include sands, silts, clays and limestone, as well as anthropogenic rubble and rubbish.

Much of the sedimentation in the Quaternary was influenced by sea-level changes in response to global climate changes. Approximately 125 000 years ago the sea levels were around 7.5 m higher than present, and the Bellarine Block was an island in a shallow sea (Marsden 1988). As the sea levels dropped in response to the onset of the last Ice Age, a high coastal barrier dune developed along the coast of southern Australia. The remnants of this barrier now form the headland at the mouth of the Barwon River, known as The Bluff (Fig. 2.2). It comprises limestone of the Bridgewater Formation, deposited as wind-blown lime sand that formed coastal dunes before the last glaciation, and has since been cemented.

Sea levels continued to recede, dropping to at least 150 m below present levels by ~20 000 years ago (Holdgate *et al.* 2001). During this time the climate was arid and cold, with evidence of frequent high wind events. The Geelong region was situated well inland, with the shoreline west and south of Tasmania. It was during this time that the first humans arrived in the region, possibly up to 40 000 years ago, and migrated across the land to Tasmania (Mulvaney and Kamminga 1999). As the Ice Age ended, sea levels rose at varied rates, peaking at 2.5 m per century (or 25 mm per year). By 6000 years ago the sea levels were approximately 2 m higher than present and covered a considerable portion of the Moolap Sunkland. The sea levels fell to their present level around 3000 years ago. A sand blockage at the entrance of Port Phillip Bay is believed responsible for Port Phillip Bay (and Corio Bay) nearly completely drying between 2800 and 1000 years ago, as recorded in Aboriginal stories (Holdgate *et al.* 2011). In the past century, the sea levels have risen around 20 cm (BoM and CSIRO 2019).

Hence, the majority of the sediments and landforms of the coastal and low-lying areas around Corio Bay and the Bellarine Peninsula have very recent origins. Swan Bay is a shallow open lagoon fringed by salt marshes and seagrass beds of high environmental value. The eastern side has been greatly modified since European settlement, including the growth of the spit in the north-eastern section known as Edwards Point (Bird 1993). Salt Lake

Fig. 2.2. The Bluff, Barwon Heads – an ancient large dune formed during the last Ice Age. Photo: Peter Dahlhaus.

occupies an ancient embayment between St Leonards and Indented Heads which is now isolated as salt marsh terrain bordered by a barrier of shelly sand. Lake Victoria and Point Lonsdale occupy a low-lying coastal plain area between the Buckley Park Foreshore Reserve and the uplifted block of the Bellarine High. This area of distinctive terrain includes thick shell beds which are remnants of the abundant marine life in the warm inlets and lagoons of the Holocene maximum sea levels, 6000–3000 years before present (Gill 1948). These shell beds have been extensively mined since the first European settlement of the area, greatly modifying the landscape through drainage, landscaping and engineering works.

Similarly, across the Lower Barwon wetlands and estuary, the recent geological evolution is recorded in the landforms, such as the broad barrier spit at Ocean Grove that has been shaped by the waves refracting around The Bluff. The spit has trapped sediment carried by the Barwon River flows and the tides to form broad wetlands colonised by mangrove, salt marsh and seagrass plants (Rosengren 2009). A delta composed of silt islands has formed in Lake Connewarre where the tidal channel enters, and has grown in area over the past 150 years. Lake Connewarre is estimated to be half as deep as it was in the mid 1800s.

Erosion too continues, especially in areas along the coast such as Clifton Springs and Point Lonsdale (Bird 1993). Uplift continues through the earthquakes recorded in the past century or more, indicating that the north-east trending faults (perpendicular to the current axis of compression) are still active. A preliminary assessment of the seismic risk for the City of Greater Geelong (Miner and Dahlhaus 2000) indicated that at least eight faults in the area can be considered active. These include the Rowsley Fault, the Lovely Banks Monocline and the inferred fault from Point Henry to Lake Connewarre (Plate 2.1). The Bellarine Fault has recorded earthquakes to Richter magnitude 3.5.

Looking to the future

Geology and geomorphology have long been recognised as critical in the development of civilisation, providing access to water, the raw materials for tools and buildings, and the strategic landforms for defence and trade. They are the underlying reasons why cities and towns are founded where they are, and why they are abandoned when the geological resources are exhausted. But it was not until the environmental revolution of the 1960s that these subjects were recognised as critical in landscape and urban planning. The seminal publication *Design with Nature* (McHarg 1969) recognised that landscapes were dynamic and therefore that designing with nature (as opposed to designing against nature) avoided many of the pitfalls that had beset cities over the centuries of civilisation.

The *Wadawurrung* inhabitants of the Geelong region recognised that their landscapes were dynamic, changing with the seasons and millennia (McNiven 1998; Holdgate *et al.* 2011) and adapted their lifestyle to account for that. European settlers also recognised the dynamic processes, but responded with engineering, such as constructing the breakwater in the Barwon River only a few years after settlement, to control tidal inflows. In hindsight, few (if any) of the engineered changes heeded the landscape dynamics. Many human-made structures have since been destroyed by nature, have irreversibly altered natural processes or have created a costly legacy of environmental degradation and maintenance.

Looking to the future, the Geelong region faces many challenges from:

- exponential population growth and urbanisation
- a changing climate and the consequent acceleration of sea-level rise and erosion
- accelerated demand for geological resources such as water and construction materials.

In the face of such challenges, it is critical that future developments are designed with nature, taking into account the dynamic processes of landscape evolution, if they are to succeed.

References

Bird ECF (1993) *The Coast of Victoria: The Shaping of Scenery.* Melbourne University Press, Melbourne.

Bureau of Meteorology (BoM) and CSIRO (2019) *State of the Climate 2018*. Bureau of Meteorology and CSIRO, Melbourne.

Carey SP, Hughes MJ (2002) Regolith of the West Victorian Uplands, Victoria, Australia. In GN Phillips and KS Ely (Eds) *Victoria Undercover, Benalla 2002: Collaborative Geoscience in Northern Victoria. Conference Proceedings and Field Guide.* CSIRO Publishing, Melbourne.

Cayley RA, Taylor DH, Vandenberg AHM, Moore DH (2002) Proterozoic–Early Palaeozoic rocks and the Tyennan orogeny in central Victoria: the Selwyn Block and its tectonic implications. *Australian Journal of Earth Sciences* **49**, 225–254. doi:10.1046/j.1440-0952.2002.00921.x

DPI (2010) *Victoria: Databases and Seamless Geology DVD.* Geoscience Victoria, Department of Primary Industries, Melbourne.

Duddy IR (2003) Mesozoic. In WD Birch (Ed.) *Geology of Victoria*. Geological Society of Australia (Victoria Division), Melbourne.

Edwards J, Leonard JG, Pettifer GR, McDonald PA (1996) *Colac 1:250 000 Map Geological Report, Geological Survey Report.* Department of Natural Resources and Environment, Melbourne.

Evans TJ (2006) Geology and groundwater flow systems in the West Moorabool River Catchment and their relation to river salinity. Unpublished MSc (Hydrogeol.) thesis, University of Technology Sydney.

Fergusson CL, Vandenberg AHM (2003). Ordovician. In WD Birch (Ed.) *Geology of Victoria*. Geological Society of Australia (Victoria Division), Melbourne.

Gibson G, Brown A (2003) Earthquakes. In WD Birch (Ed.) *Geology of Victoria*. Geological Society of Australia (Victoria Division), Melbourne.

Gill ED (1948) Geology of the Point Lonsdale–Queenscliff Area, Victoria. *Victorian Naturalist* **65**, 38–46.

Gourley TL, Gallagher SJ (2004) Foraminiferal biofacies of the Miocene warm to cool climatic transition in the Port Phillip Basin, south-eastern Australia. *Journal of Foraminiferal Research* **34**, 294–307. doi:10.2113/34.4.294

Gradstein FM, Ogg JG, Schmitz MD, Ogg GM (Eds) (2012) *The Geologic Time Scale 2012*. Elsevier, Boston.

HCV (2017) *Victorian Heritage Database: Greenstones Axe Quarry.* Heritage Council Victoria, Melbourne. http://vhd.heritagecouncil.vic.gov.au/places/21767 (accessed 9 January 2017).

Holdgate GR, Gallagher SJ (2003) Tertiary. In WD Birch (Ed.) *Geology of Victoria*. Geological Society of Australia (Victoria Division), Melbourne.

Holdgate GR, Geurin B, Wallace MW, Gallagher SJ (2001) Marine geology of Port Phillip. *Australian Journal of Earth Sciences* **48**, 439–455. doi:10.1046/j.1440-0952.2001.00871.x

Holdgate GR, Wallace MW, Gallagher SJ, Witten RB, Stats B, Wagstaff BE (2006) Cenozoic fault control on 'deep lead' palaeoriver systems, Central Highlands, Victoria. *Australian Journal of Earth Sciences* **53**, 445–468. doi:10.1080/08120090600632334

Holdgate GR, Wagstaff B, Gallagher SJ (2011) Did Port Phillip Bay nearly dry up between ~2800 and 1000 cal. yr BP? Bay floor channelling evidence, seismic and core dating. *Australian Journal of Earth Sciences* **58**, 157–175. doi:10.1080/08120099.2011.546429

King RL, Weston KS (1997) *Dimension Stone in Victoria*. Geological Survey of Victoria, Department of Natural Resources and Environment, Melbourne.

Legget RF (1973) *Cities and Geology*. McGraw-Hill, New York.

Marsden MAH (1988) Bellarine Peninsula. In I Clark, B Cook and GC Cochrane (Eds) *Victorian Geology Excursion Guide*. Australian Academy of Science and Geological Society of Australia (Victorian Division), Melbourne.

McHarg IL (1969) *Design with Nature*. American Museum of Natural History, Natural History Press, New York.

McNiven IJ (1998) Aboriginal settlement of the saline lake and volcanic landscapes of Corangamite Basin, western Victoria. *Artefact* **21**, 63–94.

Miner AS, Dahlhaus PG (2000) Preliminary study of seismic risk for the City of Greater Geelong. Unpublished report prepared by PJ Yttrup & Associates Pty Ltd for the City of Greater Geelong.

Mines Department (1949) *Victorian Building Stones*. Victorian Mines Department, Melbourne.

Morand VJ (1995) The Barrabool Hills metagabbro: a curious piece of the Cambrian jig-saw puzzle in Victoria. *Geological Society of Australia Abstracts* **40**, 111–112.

Mulvaney DJ, Kamminga J (1999) *Prehistory of Australia*. Allen & Unwin, Sydney.

Olshina A, Jiricek F (1999) *Geelong Supply Area: Extractive Industry Interest Areas*. Department of Natural Resources and Environment, Melbourne.

Pascoe B, Krishna-Pillay SH (1968–2007) *Wathawoorroong Dictionary*. Wathaurong Aboriginal Co-operative, Geelong.

Price RC, Nicholls IA, Gray CM (2003) Cainozoic igneous activity. In WD Birch (Ed.) *Geology of Victoria*. Geological Society of Australia (Victoria Division), Melbourne.

Rosengren NJ (1973) Lake Connewarre and the Barwon Estuary. *Victoria's Resources*, September–November, 19–22.

Rosengren NJ (1994) *Eruption Points of the Newer Volcanics Province of Victoria: An Inventory and Evaluation of Scientific Significance*. National Trust of Australia (Victoria) and Geological Society of Australia (Victorian Division), Melbourne.

Rosengren NJ (2009) *Murtnaghurt Lagoon, Bellarine Peninsula & Related Landforms: Nature, Origin & Geoscience Significance*. Report prepared by La Trobe University and Environmental Geosurveys Pty Ltd for Save Barwon Heads Alliance, Melbourne.

Sandiford M (2003) Geomorphic constraints on the Late Neogene tectonics of the Otway Range, Victoria. *Australian Journal of Earth Sciences* **50**, 69–80. doi:10.1046/j.1440-0952.2003.00973.x

Taylor DH, Whitehead ML, Olshina A, Leonard JG (1996) *Ballarat 1:100 000 Map Geological Report, Geological Survey of Victoria Report*. Department of Energy and Minerals, Melbourne.

Vandenberg AHM, Willman CE, Maher S, Simons BA, Cayley RA, Taylor DH, Morand VJ, Moore DH, Radojkovic A (2000) *The Tasman Fold Belt System in Victoria*. Geological Survey of Victoria, Melbourne.

Wallace MW, Dickinson JA, Moore DH, Sandiford M (2005) Late Neogene strandlines of southern Victoria: a unique record of eustasy and tectonics in south-east Australia. *Australian Journal of Earth Sciences* **52**, 279–297. doi:10.1080/08120090500139455

Webb JA (1991) Geological history of Victoria. In GW Cochrane, GW Quick and D Spencer-Jones (Eds) *Introducing Victorian Geology*. Geological Society of Australia (Victorian Division), Melbourne.

Williams MAJ, Dunkerley DL, De Deckker P, Kershaw AP, Stokes TJ (1993) *Quaternary Environments*. Edwards Arnold, London.

3

Pre-European vegetation in the Geelong region

Geoff Carr

> *... there is no doubt that the invasion of Australia by humans ... had a profound impact on the ecosystems of the continent. There is doubt about the exact nature of the impact. It is unlikely that this doubt will be resolved in the near future given the paucity of evidence available from 40 000–15 000 years ago.*
>
> – Kirkpatrick (1999, p. 25).

Introduction

The history of vegetation we see around us in Australia goes back to the dawn of the emergence of the terrestrial flora over 400 million years ago (White 1986, 1994; Hill *et al.* 1999; Henderson and Johnson 2016). Similar evolutionary histories and trajectories prevail for the fauna, at least vertebrates (Vickers-Rich and Rich 1993; Rich et al. 2000; Long *et al.* 2002; Henderson and Johnson 2016). These evolutionary developments in the biota and the environments they occupied have been reconstructed from the geological record (Henderson and Johnson 2016) and the associated fossil record (Hill 1994; Hill *et al.* 1999) – microfossils (pollen, spores, diatoms), and macrofossils (roots, stems, wood, leaves, flowers, fruits/seeds, charcoal) preserved in physical environments that often differed dramatically from those now prevailing, geographically, geologically, geomorphologically and climatically (Frakes 1999; Henderson and Johnson 2016).

We have many remarkable insights into the past from the fossil record and research continues to yield many new data, interpretations and refinements. In the Geelong region, for example, the middle to late Eocene sediments associated with Anglesea coal measures, 30–40 million years old, have yielded a rich record of macrofossils and microfossils (pollen) indicating a rainforest flora, similar to that now in north-east Queensland (Christophel *et al.* 1987). Much more recently, Quaternary fossil evidence – pollen, charcoal and fruits – recovered from cores in swamps and lakes allows critical insights into vegetation, climate change, fire history and perhaps some of the impacts of the first human inhabitants (Mills *et al.* 2013).

Pre-European vegetation is generally meant to refer to that encountered by European invaders in the late 18th and early 19th centuries and unmodified by them, though it may have been modified for upwards of 50 000 years ago by Aboriginals who were present at Lake Mungo in south-western New South Wales (NSW) 50 000–46 000 years ago (Bowler *et al.* 2003). It is here taken to be the conditions encountered by the Europeans, commencing in 1836 when Melbourne was established, and 1838 when Geelong was founded by the government-sanctioned freebooters, the squatters (Boyce 2011). Portland was established in 1834 (Boyce 2011).

Crucial to developing a vision of pre-European vegetation and its dependent fauna is the description of the geophysical environments of the biota – geology, geomorphology, soils, climate and fire. For excellent expositions of these factors as they relate to our region see Christophel (1993) for the flora over geological time, Douglas (1993) for geology and geomorphology, BoM and Walsh (1993) for climate and Gibbons and Rowan (1993) for soils.

When the Europeans set up shop in Victoria in the Port Phillip region in 1836 they seized the lands of the *Kulin* Nation, the Aboriginal people, specifically the *Wadawurrung* language group and clan in the Geelong region (Presland 2010). Opinions differ as to how populous the Aboriginals were, but there is general agreement that they numbered not more than 100 000 in Victoria, and highest numbers occurred in the most ecologically productive parts of the country such as the west Port Phillip region and south-west Victoria. The extent to which the Aboriginal hunter-gatherers, as traditionally conceived, modified the vegetation is contentious in the absence of unequivocal evidence. To what extent were these cultural landscapes modified, as accepted for example by Gammage (2012) and Mulvaney and Kamminga (1999), or is the Aboriginal influence exaggerated?

It is not the intention of this chapter to present a reconstruction of the pre-European vegetation communities – their distribution, structure and floristic composition. That important exercise demands much more research and analysis of disparate data. Rather, in this chapter the following is outlined: types of evidence that can be used to reconstruct the vegetation communities and flora (plant species); what we have done to the biota in 180 years; how Aboriginals may have influenced vegetation and fauna; the imperative to document what biota remains; conservation status of the current vegetation and flora; contemporary responses by governments and society to an unprecedented conservation crisis; and what the future may hold.

Why reconstruct the original vegetation?

Approximately 95% of the pre-European vegetation of the Geelong region has been destroyed (Ecology Australia 2001), and what little remains is generally in poor or very poor condition.

Reconstruction (on paper) of the original vegetation may fulfil several objectives:

- to provide detailed information on the structure and floristic composition of the former vegetation, essential to plan and implement revegetation exercises, rather than utilising the often crude one-size-fits-all Ecological Vegetation Class (EVC) revegetation templates generally applied
- to document the substantially lost biological heritage, the products of 400+ million years of evolutionary history, before so much is lost that reconstruction is no longer possible. We owe this to future generations while we still have maximum data at our disposal
- scientific interest, as part of the ripping yarn of the life of Earth, the greatest story that will ever be told (Henderson and Johnson 2016; Lomolino *et al.* 2017), making human affairs seem trivial
- to honour the Aboriginal owners from whom Europeans usurped the land and destroyed much of its biota and ecological function.

The corollary question to 'what was the pre-European vegetation?' is the question, 'what are we doing to protect what remains'?

Evidence that can be used to reconstruct former vegetation

The pre-European, indigenous vegetation and faunal habitats of the region have been reduced, over ~180 years since European invasion, to a minuscule representation of their former selves. We do not have ready access to statistics about percentage loss except the modelling and mapping of EVCs by the Department of Environment, Land, Water and Planning (DELWP 2017), but this is a fairly coarse representation of reality. Further, many remnant patches of recognised indigenous vegetation are highly degraded structurally and floristically, with a very large suite of exotic plants. Additionally, there have been uneven losses of vegetation across the three bioregions straddled by the greater Geelong region (Otway Plain, Victorian Volcanic Plain and Central Victorian Uplands). The Otway Plain between Torquay and Anglesea, unlike the Bellarine Peninsula and Barrabool Hills, is relatively intact because its generally low-nutrient soils are unfavourable for agriculture. The same applies to the dissected terrain and shallow stony soils and low-nutrient sands of the Brisbane Ranges and You Yangs. The low-relief Victorian Volcanic Plain has been resoundingly trashed and little trace of its original vegetation remains.

There are several general lines of enquiry (if available) that may be employed when attempting a reconstruction of the original vegetation.

In situ vegetation remnants: the on-ground evidence

These are often highly fragmentary remnants of single or few trees in agricultural or urban/industrial landscapes and along roads and railway reserves. Remnant vegetation has been almost utterly annihilated in the Barrabool Hills (Anon 2017; Carr, unpubl. data).

Scientific documentation

This includes historic biological collections or observations, botanical research and scientific literature.

- Historic collections of plants lodged in public herbaria (Willis 1956, 1964; Willis and Cohn 1993). Wilson (1889) recorded nine species of plants, including white correa (*Correa alba*) at Buckley Falls on the Barwon River, that are no longer extant (Carr 1987).
- Fossils: microfossils (pollen, spores, diatoms) and macrofossils (leaves, plants, flowers, fruits, wood, charcoal) preserved in dated swamp or lake sediments (Mills *et al.* 2013) or buried under basalt or volcanic ash deposits.

Artefacts of social, political and art history

- Historic parish plans and surveyors' maps. Sinclair and Atchison (2012), for example, in their reconstruction of Werribee Plains vegetation, used four types of information from historic plans: forest boundaries, annotations about vegetation, stylised depictions of tree cover (e.g. stippling), and lot corner-markers where identified tree species occurred.
- Ecology Australia (2013), in attempting to determine the pre-European distribution of the weedy Coast Tea-tree (*Leptospermum laevigatum*) on the Bellarine Peninsula, consulted surveyors' plans with no resolution for that species but with other clearly marked vegetation types, such as Honeysuckle (*Banksia marginata*) Woodland (Fig. 3.4). For the reconstruction of pre-European vegetation in the Wimmera, Morcom and Westbrooke (1998) used four sources of information, including parish plans.

- Journals, diaries, reports and letters of explorers, travellers, naturalists, landowners (especially the squattocracy), government officials and so on. Hateley (2010) cites a broad sample of such sources and the valuable information they contain, as do Sinclair and Atchison (2012) and Gammage (2012). It is striking how frequently in many historic accounts two particular plants are mentioned: 'Honeysuckle' (*Banksia marginata*) (Silver Banksia) and 'Sheoak' (*Allocasuarina verticillata*) (Drooping Sheoak) which were landscape vegetation dominants (Fig. 3.4) but which are now rare (Sinclair and Atchison 2012), extinct or bordering on extinction in many parts of the Geelong region.
- Landscape paintings, lithographs and sketches. Paintings, drawings, lithographs and other artistic representations from colonial times ostensibly provide valuable images indicating vegetation cover before clearing, or extensive clearing, took place. Interpretation of these cultural artefacts needs to be done with caution: they can be useful or, conversely, potentially misleading. Geelong Art Gallery has an important collection of nine von Guérard paintings and lithographs, several depicting scenes around Geelong in the 1850s and 1860s before massive transformation effaced almost all indigenous vegetation. The large oil painting 'View of Geelong' (1856) shows a view (from the present Highton) looking north to the Newtown escarpment at Fyansford, with the distant You Yangs on the horizon; I grew up in this landscape. Instantly recognisable in terms of location we nonetheless see vertical topographic exaggeration (a drama-enhancing artistic device), a much wider Barwon River (water as a painting design element) and a landscape substantially devoid of trees. Trees which are currently visible in the landscape – old-growth Red Gum (*Eucalyptus camaldulensis*) along the river – are nowhere to be seen. These and many more are seen in the painting 'The River Barwon' (1847) by John Skinner Prout, reproduced in Gammage (2012). Trees are prominent throughout the landscape, including on the Newtown escarpment. Some nine years separates these landscape depictions. Gammage (2012) sometimes divorces colonial landscapes from their art-historical context (Sayers 1998, pp. 53–54) in relation to von Guérard, shorn of social and political contexts; he appears to treat paintings as quasi-scientific documents equivalent to colour photographs, as documented by Adam (2017). A literal reading of von Guérard's paintings would suggest that a large part of the Geelong landscape was naturally treeless. Other landscape depictions authentically show the landscape (Hateley 2010), for example those on the Western Plains showing Red Gum (*Eucalyptus camaldulensis*) grassy woodland and grassland by Duncan Cooper (between 1842 and 1854), reproduced in Hateley (2010). A local example, a lithograph, is telling: it shows a landscape on the northern flank of the Barrabool Hills covered in open woodland dominated by Drooping Sheoak, of which no trace remains. This was published in the *Clyde Co. Papers* (Brown 1941) and reproduced and discussed in Carr (1987); it is a crucial source to inform revegetation of the Barrabool Hills (see Anon 2017).

Synthesis and analysis of data

Several kinds of data and analytical techniques are routinely used in reconstructing pre-European vegetation.

- Ecological modelling and mapping by interpretation of physical environmental data, particularly geology, geomorphology, soils, climate and hydrology in relation to structural and floristic vegetation data. This approach (described by

Woodgate *et al.* 1994) has produced the Ecological Vegetation Class (EVC) typology (pre-European and contemporary). An EVC encompasses areas of vegetation with one or more floristic communities that show consistent structure, landscape context and function. Some 556 EVCs are recognised and mapped as pre-1755 Victorian vegetation, and as current remnant vegetation (http://maps.biodiversity.vic.gov.au/viewer/?viewer = NatureKit). The resolution of EVCs and their mapped locations are variable; some are specific and accurate, others very broad and sometimes erroneous in detail. Field surveys confirm that, in many areas, not all EVCs have been recorded. In their reconstruction of the pre-European vegetation of the Werribee Plain (which includes part of the City of Greater Geelong), Sinclair and Atchison (2012) only used EVCs as secondary descriptions of vegetation.

- Extrapolation of distributions of plant species and vegetation reconstruction from remnant vegetation. Remnant vegetation, even if reduced to minute scraps in the landscape, can reveal a great deal about the former vegetation and environments because of its inherent information – its ecological signature. This is because plant species have genetically determined lower and upper limits of tolerance in respect to environmental factors, outlined by Kirkpatrick (1999) and Gibbons and Rowan (1993) (for soil factors), which determine distribution, survival and reproduction of individuals and population growth:
  - water supply
  - temperature
  - light
  - atmosphere (including air supply to roots)
  - edaphic (soil), including nutrient supply, mechanical forces in soil and toxicity, notably pH, salinity and toxic metals
  - fire
  - biotic – the numerous positive and negative ecological interactions of living organisms
  - the regeneration niche and vegetation dynamics
  - dispersal and migration.

Few plants occupy the whole of their potential range, the main reason being competition from other plants (Kirkpatrick 1999): 'species tend to perform best near the middle of their range, with their performance attenuating at the extremes. Different species tend to have different response curves for at least some of the … necessities of life' (Kirkpatrick 1999, p. 10).

In their superb book, Lomolino *et al.* (2017) distinguish two crucial concepts in relation to plant (and animal) distributions: the fundamental ecological niche versus the realised niche, as expressions of distribution.

On the basis of ecological signatures we can usually assign, by extrapolation, plant species to former vegetation communities and by extension the likely structural form and floristic composition of the vegetation, even if many species are extinct locally or regionally. This presupposes a knowledge of the ecology and biology of the flora gleaned from the literature and direct familiarity with environments of *in situ* plant populations or vegetation, aided by geological maps, climatic data and vegetation compositional data if such exist. The latter are floristic quadrat data from sampling vegetation elsewhere (available in the *Atlas of Victorian Wildlife* if not collected by the worker).

Fig. 3.1. An ancient Moonah (*Melaleuca lanceolata*) on limestone at Waurn Ponds. It has been engulfed by African Boxthorn (*Lycium ferocissimum*) (March 2018).

A massive, assumed old-growth Moonah (*Melaleuca lanceolata*) exists at Waurn Ponds (Fig. 3.1). Apart from the few old individuals on coastal cliffs at Corio (Fig. 3.2) and a few at Buckley Falls (Carr 1987; pers. obs.) this is apparently all that remains of what is likely to have been a landscape dominant. The Waurn Ponds example is on Miocene limestone, a characteristic substrate for the species, but not the only type of substrate (Willis 1948, 1964; Moxham and Turner 2009). By extrapolation, it is reasonable to infer that extensive Miocene limestone outcrops in North Geelong, Belmont, Waurn Ponds and elsewhere supported Moonah Woodland (EVC 858 Coast Alkaline Scrub) (Moxham and Turner 2009). If not Moonah Woodland, it is very difficult to conceive what vegetation would have been on this terrain.

The Buloke (*Allocasuarina luehmannii*) forest or woodland remnants (Fig. 3.3), partly in the present study area, documented on the Werribee Plains by Sinclair and Atchison (2012) present a somewhat more complete picture of the original vegetation than the

Fig. 3.2. One of three ancient Moonah (*Melaleuca lanceolata*) trees on the coastal cliffs at Corio (March 2018).

Fig. 3.3. A small stand of Buloke (*Allocasuarina luehmannii*) just north of Little River, indicates a formerly dominant vegetation type of Werribee Plain (March 2018).

forlorn Waurn Ponds Moonah. These authors describe and map Buloke as part of the 'Forest dominated by Grey Box (*Eucalyptus microcarpa*)' but not pure 'Grey Box forest'. Grey Box and other woody and herbaceous vegetation persist, albeit highly degraded, in some road reserves and the floristics, hence structure could be determined from these remnants and much more intact Grey Box forest further north (Sinclair and Atchison 2012). These plains are in a pronounced orogenic rainshadow (interception of rainfall systems by the Otway Ranges, Brisbane Ranges and Central Highlands: BoM and Walsh 1993, Fig. 4.3). Average annual rainfall in this zone is 300–500 mm, which otherwise occurs north of the Great Dividing Range in north-western Victoria and the Riverina. Thus there are strong floristic affinities with vegetation in the Wimmera, as evidenced by Buloke, Grey Box and mistletoes (on Buloke), namely Wire-leaf Mistletoe (*Amyema linophylla*) and Harlequin Mistletoe (*Lysiana exocarpi*). These biogeographic considerations need to be factored in when reconstructing the flora from floristic lists (e.g. Carr 1999) and regional quadrat data.

Photographic evidence

Photographs of landscapes or individual trees or stands of vegetation can be extremely valuable as records of former vegetation. For example, Steve Sinclair (Arthur Rylah Institute, pers. comm. 2017) unearthed some amazing photographs of Silver Banksia (tree form) woodland near Lismore (undated) and at Stoneleigh (undated) on the Victorian Volcanic Plain held in family archives (Fig. 3.4), which indicate the extent and structure of

Fig. 3.4. A scene from a property at Lismore on the Victorian Volcanic Plain in the late 1800s. Silver Banksia (*Banksia marginata*) then formed an open woodland but the species is now all but extinct in the bioregion.

a vegetation type, once widespread but now essentially extinct on the Victorian Volcanic Plain (Sinclair and Atchison 2012). From the evidence of the usually single, rare old-growth Banksia trees scattered on the Volcanic Plain and historic documents (Hateley 2010) we could have inferred that such vegetation formerly existed and was widespread (Steve Sinclair, Arthur Rylah Institute, pers. comm. 2017).

Aerial photography first utilised in the 1940s in Victoria (Sinclair and Atchison 2012) has proved valuable, indicating where vegetation once occurred and gross vegetation structure (e.g. woodland, forest, grassland, wetlands versus drylands), but not floristic composition. The advent of high-resolution colour air photography has changed this situation greatly. In reconstructing the original vegetation of the Werribee Plain, Sinclair and Atchison (2012) used recent digital orth-aerial photographs to identify clues in the landscape indicated by colours: soil, presence of rocks, wetlands and drainage lines or creeks.

Personal recollection

In over 60 years of conscious observation in the Geelong region I have witnessed the destruction of a great deal of indigenous vegetation, directly by clearing or indirectly by weed invasion and stock grazing. Such observations are rarely recorded, but examples include the following.

- A very large Red Gum (*Eucalyptus camaldulensis*) beside a pond survived at the corner of Church Street and Minerva Road, Herne Hill, until the late 1950s.
- A stand of species-rich grassy Red Gum woodland survived in the western part of Western Cemetery, Herne Hill, until the early 1960s or later; it is now a graveyard. Aside from *Eucalyptus camaldulensis* it supported Grey Parrot-pea (*Dillwynia cinerascens*), Milkmaids (*Burchardia umbellata*), Chocolate Lily (*Arthropodium strictum*), Blue Grass-lily (*Caesia calliantha*), Bulbine Lily (*Bulbine bulbosa*) and Yellow Autumn Lily (*Tricoryne elatior*).
- The You Yangs in the 1960s were a relative paradise, before Boneseed (*Chrysanthemoides monilifera* subsp. *monilifera*) (an exotic, introduced species dumped as garden rubbish, probably from Geelong where it was cultivated (Carr, pers. obs.) demolished the landscape (Carr 1993). A huge number of plant and animal species became extinct and the understorey is almost wholly exotic, by no means confined to Boneseed. We expect a continuing downward spiral of degradation, exacerbated by intense rabbit grazing and severe fire. This one of the saddest known spectacles of biological collapse.
- Populations of plant species now extinct in the region but known from personal communications by observers, collectors and photographs, particularly orchids (e.g. the late Eric Bound, Austral Ladies Tresses (*Spiranthes australis*) at Anglesea and the late Andrew Thornley, Metallic Sun-orchid (*Thelymitra epipactoides*) at Jan Juc.

The trick is to record these observations before the opportunity to do so is lost. Personal estate effects, such as the valuable orchid herbarium of Andrew Thornley, may be thrown out or fall into unappreciative hands. A huge number of valuable photographic collections have probably ended in the incinerator or tip.

Influence of Aboriginals

Aboriginal people, the first Australians, inhabited the Australian landscape for 65 000 years, as exciting new research shows (Clarkson *et al.* 2017). There is widespread agreement that they had a major impact on the landscape and biota. Several kinds of influence could have been exerted by Aboriginals: use of fire as an ecological tool, cultivation of food plants rather than the strict hunter-gatherer socio-economic model and (indirectly) the impacts on the grazing and browsing marsupial fauna, specifically the now extinct megafauna.

Very divergent estimates of the size of the Aboriginal population in Australia and in Victoria when Europeans invaded have been made. Broome (2005) estimated the pre-contact Australian population to have been around 60 000, but Taylor (2005) estimated it at approximately 500 000. For western Victoria, Mulvaney (1964) estimated a density of people in the vicinity of one person per 1295 ha, thus conservatively 1800 'might be accepted' as the safe minimum figure. Presland (2010) gives an estimate of 15 000 people for the Port Phillip district.

The catastrophic genocidal impacts of the invaders on the First Australians rapidly manifested: the official census of 1877, only 40 years after invasion, recorded that only 170 Aboriginals lived in the whole of western Victoria (West 1971).

Fire

Much has been written about the use of fire by Aboriginals. In his popularly acclaimed *The Biggest Estate on Earth: How Aborigines made Australia,* Gammage (2012) draws on three lines of evidence: historic documents, colonial artistic landscape depictions, and ecological interpretation of vegetation patterns and 'templates'. Gammage (2012) asserts that Aboriginals essentially burned everywhere across Australia, frequently according to necessity and cultural obligations (Law). Thus Aboriginals hugely modified the vegetation of the landscape, keeping it 'open', as evidenced by the allegedly widespread vegetation 'thickening' post-European invasion in the absence of burning. The nature and extent of the influence of Aboriginal fire practices on the vegetation and fauna is far from settled, however, and is likely to remain so because of a general lack of scientific evidence (rather than inference). Additionally, observations of Aboriginal behaviour made by European explorers and early colonists suffered from cultural prejudice (Kirkpatrick 1999).

The scientific community has strongly questioned Gammage's thesis, and in a detailed critique noted ecologist Paul Adam (2017) concludes that there is danger in Gammage's ideas if they are used to support 'particular approaches to fire management ... and advocacy for high-frequency control burning, returning the country to the Aboriginal fire management regimes'. Parks Victoria, charged with managing 4 million ha of the public estate, has recently instituted 'Aboriginal burning' (https://www.ffm.vic.gov.au/media-releases/aboriginal-traditional-burning-with-barengi-gadjin-land-council), even though Aboriginal burning – part of an ancient oral tradition and practice – was ended approximately 180 years ago.

Hunter-gatherers or agriculturalists?

Recently Pascoe (2014) claimed that Aboriginals across Australia were agriculturalists rather than 'hunter-gatherers' – they used domesticated plants, sourcing, harvesting, irrigating and sowing. If this is the case such practices may have modified the landscape, at least at localised scales in favoured locations. Pascoe (2014) cites few of the plant species cultivated and does not analyse the biological, ecological and environmental attributes that predisposed them to domestication. Whether we can resolve the question of 'agriculture' versus 'hunter-gathering' in Victoria is a moot point because data are and will remain scant.

Megafauna extinctions

Towards the end of the Pleistocene, as revealed by fossil evidence, Australia lost 21 of 23 genera of terrestrial reptiles, birds and marsupials, many of the megafauna with body masses exceeding 44 kg (Koch and Barnosky 2006): all large browsers, most large grazers, the largest predators and all large omnivores, among others *Diprotodon* (six species), Giant Rat Kangaroo (*Dipodomys ingens*), Kangaroos, Short-faced Kangaroos (*Procoptodon goliah*), marsupial 'tapir' (*Palorchestes*), marsupial 'lion' (*Thylacoleo*) and small, large and giant wombats (Johnson 2006). Not all large and medium-sized grazers and browsers became extinct, with three kangaroos and various wallaby species, and several wombat species remaining. Regionally the Eastern Grey Kangaroo (*Macropus giganteus*) (body mass 66 kg: Poole 1995) remains abundant, while the Common Wombat (*Vombatus ursinus*) (body mass 39 kg: McIlroy 1995) became extinct after European invasion (McCarthy and Prescott 1985; Williams and Menkhorst 1995). The Black Wallaby (*Wallabia bicolor*) is increasingly common regionally and the Red-necked Wallaby (*Macropus rufogriseus*) is now only in the Anglesea–Aireys Inlet area (Menkhorst 1995).

There is widespread acceptance that the timing of the megafauna extinctions cluster around 45 000–50 000 years ago (Bowman *et al.* 2012 and references therein) shortly after the arrival of humans in Australia 45 000–55 000 years ago (Roberts *et al.* 1990; Turney *et al.* 2001) but see Clarkson *et al.* (2017). The cause(s) of the megafauna extinctions have been contentious, whether the result of Aboriginal overhunting (Flannery 1994), climate change or other factors, but there is now widespread agreement that human colonisation 'was pivotal in causing these extinctions' (Bowman *et al.* 2012). The data presented by Sandom *et al.* (2014) and Saltré *et al.* (2016) unequivocally point to human agency as the major cause of the extinctions. Flannery (1994) contended that following the extinction by overhunting, and the consequent relaxation of grazing and browsing pressure on vegetation by the megafauna, fire became much more prominent in Australia because of elevated fuel loads coupled with Aboriginal burning. Much debate remains about the impact of Aboriginal burning as a causal agent in changing vegetation and plant species distributions and patterns but little or no support exists for the view that Aboriginal burning was a driver of plant speciation (Bowman *et al.* 2012), as proposed by various authors based on evidence of fossil pollen and charcoal recovered from late Pleistocene and Holocene swamp and lake sediments (references in Bowman *et al.* 2012).

Fire has been an environmental phenomenon influencing vegetation and the evolution of plant taxa since the Tertiary, at least the late Eocene ~35 million years ago, evidenced by fossil charcoal in coal measures and other sediments (Figs 68 and 71 in Hill *et al.* 1999 and references therein; Mooney *et al.* 2012; Sluiter *et al.* 2016). Thus, under an Aboriginal burning regime, spanning 50 000 years, evolution of flora occurred during not more than ~1.5% of the time fire was used – a brief period in geological and evolutionary time.

Whatever the first Australians did by way of impacting flora and fauna, they did not, unlike Europeans, unleash a devastating tide of exotic plants, animals and fungi into the environment.

Causes and mechanisms of vegetation and fauna destruction

Aside from 50 EVCs modelled and mapped (see Appendix 3.1) many studies document vegetation in the region (see Chapter 6 this volume; Ecology Australia 2001). The City of Greater Geelong published *Indigenous Plants to the Geelong Region* recognising 19 biophysical zones, including broad descriptions of vegetation and lists of revegetation species.

There is great need to collate all vegetation data for the region and collect documents in a central, electronically accessible repository.

Over the 180 years of European occupation there have been massive losses of native vegetation, flora, fauna habitats and fauna species. An estimated 5% of native vegetation remains in the City of Greater Geelong, which has a moderate or better semblance of its former structure and floristic composition (Ecology Australia 2001). The prognosis for that remaining, on current trends, is bleak.

Numerous direct and indirect processes have operated over European history to cause loss and degradation of the biota (Hobbs and Hopkins 1990; Lines 1991; Kirkpatrick 1999; Johnson 2006; Keith and Auld 2017). Ecology Australia (2001) outlined the threatening processes and causes of extinction (with many subcategories) for flora and fauna populations which apply in the City of Greater Geelong, such as clearing and vegetation removal, changes to physical environments, and negative plant–plant and animal–plant interactions.

The most severe threats at all geographic scales, aside from the large-scale historic clearing of vegetation, and stock and feral mammal grazing, are:

- weed invasion, the pre-eminent conservation issue for flora and fauna (Carr *et al.* 1992; Carr 1993; Adair and Groves 1998; Low 2001; Groves 2011; Booth 2014)
- incremental vegetation loss and fragmentation contingent on residential and infrastructure development and agriculture
- climate change, which now tends to occupy centre stage as the great threat. It undoubtedly is, but under any climate-change scenario weed and exotic animal invasions will prevail; they will not diminish in significance
- altered fire regimes, involving increased intensity, timing outside the normal fire season (e.g. winter), incremental frequency and large-scale extent.

The *Flora and Fauna Guarantee Act 1988* (Vic.), Schedule 3, lists 23 threatening processes for Victorian flora and fauna, all of which apply, or would apply, to our region of interest. Whatever the notional purposes of such listing, it is an arid gesture if management actions do not follow, as they rarely have. A major threatening process is the 'Spread of *Pittosporum undulatum* [Sweet Pittosporum] in areas outside its natural range'. This plant species alone will destroy most regional and southern Victorian vegetation and fauna habitats over decadal time-scales, yet nothing is being done about it except by a few community groups (e.g. Jan Juc Coast Action Group) and by Parks Victoria in tiny areas of the public estate. The cultivation and sale of Sweet Pittosporum is legal everywhere; all infestations owe their origin to cultivated plants.

Rabbits

So familiar are we with landscapes profoundly influenced by the European Rabbit (*Oryctolagus cuniculus*) – introduced at 'Barwon Park' near Geelong in 1859 – it is almost impossible to conceive what some of these landscapes may have looked like in the absence of rabbits; we will never know. There have been few published studies in Australia on the diet of the rabbit but impacts on threatened plant species by rabbit grazing, preventing recruitment by seed, have been studied, for example *Allocasuarina verticillata* (Cooke 1988) which is of great relevance to the Geelong region. Long (2003), and references therein, list the general environmental impacts of rabbits in Australia and elsewhere as a catastrophic introduction, ecologically and economically. They cause soil erosion, prevent natural regeneration, kill trees by ringbarking (e.g. Sweet Bursaria (*Bursaria spinosa*); Carr,

Fig. 3.5. The escarpment along the Moorabool River floodplain at Fyansford supporting wholly exotic vegetation 'engineered' by rabbit grazing (March 2018).

pers. obs.), drive plant species and fauna habitat to extinction (locally or absolutely), alter floristic composition and destroy vegetation communities directly or indirectly, and promote weeds. Differential grazing pressure between indigenous plants and exotic plant species eliminates palatable indigenous plants (most) and creates a community of exotics, exemplified for example on the Moorabool River and Barwon River escarpments at Fyansford (Fig. 3.5). There, indigenous species are essentially extinct and only toxic exotics form a more-or-less stable, rabbit-generated and -maintained community. Exotic weed species immune to grazing include African Boxthorn (*Lycium ferocissimum*), Soursob (*Oxalis pes-caprae*) (except bulbs) and Blanket Weed (*Galenia pubescens*). Rabbits have had the profoundest irreversible negative influence as landscape engineers, and their impacts will continue into the future.

The future?

As we survey the wreckage of our biota after 180 years of destruction and abuse, deliberate and inadvertent, by the European invaders and their successors (so much of which was unnecessary or avoidable), what and where does it leave us? These questions are crucial in the attempt to take lessons from history and in getting a bearing on what the future might hold so that we can act appropriately, if so motivated. What is clear by any rational assessment is that if we (society) continue along the same path as that taken over the last 180 years, very little of the indigenous biota will exist in much less than half that time. What the Aboriginal custodians left after perhaps 60 000 years of occupation when Europeans usurped their lands, will be largely extinct. There are strong parallels in how the Europeans abused the land and its biota, and the Aboriginal peoples. The issues outlined below are extensively reviewed by Mathews *et al.* (2011) and Booth (2014), and solutions proposed.

What does this leave us with?

A brief summary shows an unprecedented conservation crisis.

- Severe depletion of vegetation communities and faunal habitats and a very parlous conservation status of most. The EVC typology for the City of Greater Geelong (Appendix 3.1) strongly exaggerates the area of remnant native vegetation, notably Plains Grassland.
- Ongoing and accelerating massive regional extinctions of plant and animal species; of those remaining, a large proportion are endangered or critically endangered.
- A very wide array of degradation and threatening processes, and a very significant extinction debt such that processes currently in train (e.g. rural tree decline) will run their course. Foremost among the numerous threats and processes are weed and feral animal invasions, and climate change. There is a huge weed flora – 685 species (38% of the total flora) – in the City of Greater Geelong, and numerous feral fauna species (Ecology Australia 2001).
- Though part and parcel of the biodiversity losses, the plight of trees in rural landscapes (freehold and public land) is particularly worrying. Trees, perhaps most (many old-growth), on grazing and cropping lands, road reserves and other places will senesce and die without recruitment because of stock and rabbit grazing and weed invasions (Figs 3.1–3.3). Dependent biota and landscape amenity will crash.

Where does this leave us?

Even the most casual audit shows that, as a society, we are spectacularly failing to protect and manage our biodiversity and halt its extinction at any geographic scale, local to continental, a sobering global message (Ripple *et al.* 2017). The Geelong region is a microcosm of the wider world.

One thing the recent *Biodiversity Strategy* (DELWP 2017) gets right is its acceptance that the issue involves the whole of society: it involves everyone, all levels of government, all agencies charged with management of biodiversity, the scientific community, industry and business, conservation NGOs and organisations, citizen scientists and others. It is clear that failure so far in the types, quality and quantity of appropriate responses represents the apparent inability of society to effect meaningful management of biodiversity and ecosystem services and functions on the ground, the only measure of achievement (Mathews *et al.* 2011). Formulation and articulation of biodiversity policies, strategies, legislation, international covenants and agreements and recovery plans at all levels of government are never surrogates for progress or success in conservation. They count for nothing without effective on-ground action.

The central state government agency – the Department of Environment, Land, Water and Planning and its forerunners – has not always succeeded in preventing biodiversity destruction and decline. The regulator charged under law, policy, international agreement and covenants to protect, wisely use and manage biodiversity has not always achieved this, through failure of policy, failure of action or actual interventions that damage or degrade our biodiversity. We have witnessed many astounding failures of the planning process under the *Planning and Environment Act 1987* (Vic.), and neglect in condoning and authorising destruction of very highly significant biodiversity. Boon and Prahalad (2017) go a long way in explaining how this state of affairs came about. Particularly egregious local examples include:

- the destruction of many hectares of very high-quality, nationally significant Yellow Gum (*Eucalyptus leucoxylon*) Grassy Woodland at Bannockburn, a unique stand (Ecology Australia 1996) surrounded by many thousands of hectares of cleared land, for a sewage treatment plant
- authorising the destruction of 40 ha of nationally significant wetlands (saltmarsh, Saline Aquatic Meadows) and Moonah Woodland at Point Lonsdale for a proposed canal estate. The flora and fauna values were documented by Ecology Australia (2005) and Mueck and Smales (2007).

Such actions engender little faith that the latest iteration of the DELWP (2017) biodiversity strategy, *Protecting Victoria's Environment – Biodiversity 2037*, will achieve its 20 articulated priorities.

Scientists must accept a measure of blame for the dire conservation status and prognosis for Australian biodiversity, what Recher (2013, 2015) identified as a 'failure of science' and the 'death of nature'. Recher presents a very cogent case, arguing that many ecologists are ignorant of the arts and humanities (essential for understanding how society functions) and are generally unwilling to provide moral direction about how technological and scientific advances align with ecological sustainability and ethics. This allows politicians (and multi-national organisations) to dictate science policy. Also, scientists can be conservative and often fail to challenge authority as well as societal norms, and can be reluctant to advocate on behalf of the environment. These concerns are shared and amplified by Boon and Prahalad (2017) in an important critique of conservation biology, 'ecologists rarely frame their work or its implications in a milieu that is fully cognisant of the broader social, political and economic forces that drive conservation biology'. The central concerns of Boon and Prahalad (2017) are the roles played by neoliberal ideology and its free market assumptions that now guide so much government response to the environmental crisis.

Ecologists richly deserve the criticisms of Recher (2013, 2015) and Boon and Prahalad (2017); however, a major corollary problem is the blind faith in technology of certain scientists and the lay people they influence – that a techno-fix is, or will be, available to get us out of the mire, and a general assumption, at least among non-scientists, that all science is good and self-corrective. Several concerns of the greatest relevance to how conservation and conservation biology guide policy and practice in Victoria and beyond, are outlined below. They are integral to the malaise in which we find ourselves.

The restoration paradigm

'Restoration ecology', currently a vogue discipline, embraces many different activities, some of which may be important for the management of particular communities or species (e.g. habitat for Grey-crowned Babbler (*Pomatostomus temporalis*); Robinson 2006) and other highly desirable measures (e.g. the Barrabool Hills revegetation project; Anon. 2017). But the redemptive scenarios conjured up by many restorationists are often divorced from reality, particularly in regard to 'corridors' which have been researched and discussed for ~30 years (Saunders and Hobbs 1991). But the concept has seized popular imagination.

Such restorationist schemes and their companion idea of achieving biodiversity 'offsets' as 'compensation' for destruction of diversity, sanctioned by the government under 'No

Net Loss' ('sanctioned disposability': Boon and Prahalad 2017) were analysed and strongly criticised by Boon and Prahalad (2017) as fraudulent, bogus and, by implication, morally and ethically wanting. Similar charges may be made against schemes discussed below; collectively, these faith-based restorationist schemes are variants of the 'Field of Dream's' expectation (Palmer et al. 1997): build it (the physical environment) and they (the animals and plants) will come. That is, once the necessary physicochemical conditions have been created, a self-organising community will develop. This paradigm is called 'Cargo Cult Conservation' and its handmaiden, 'Tree-guard Dreaming'.

Connectivity

The community, management agencies and non-government conservation organisations have been led to believe, apparently by academic biologists, that planting vegetation corridors will allow fauna (especially birds and other vertebrates) and flora to move around the landscape, restoring connectivity in fragmented landscapes. It is claimed that this will reverse a central threatening process – habitat fragmentation in cleared landscapes. Such corridors supposedly also allow organisms to move in the landscape to 'choose' their optimum climatic envelope under a climate-change scenario.

Such schemes are highly dubious for various reasons, among which are:

- no evidence is available concerning what species will utilise the corridors, how, when and in what numbers, and what, if any, habitats are available at the ends of corridors
- there is no apparent consideration of the quality of the fragments to be connected, many of which are irreparably degraded (e.g. in the Wimmera with numerous exotic grasses, and in the You Yangs with Boneseed, exotic grasses, etc.)
- except in very rare instances, the existing conservation estate on public and private land receives no management and is itself rapidly and irreparably degrading (mostly because of weed invasion)
- understoreys of installed corridors (formerly agricultural lands) will achieve around 100% herbaceous weed cover within several years
- most corridors will be rapidly colonised by Noisy Miners (*Manorina melanocephala*), which will aggressively exclude almost all bird species (Maron *et al.* 2013)
- there is no apparent consideration of the need to manage the corridors, and in any event many management issues are effectively or definitely intractable
- the reinstatement of multiple ecological functions (so the rhetoric goes) is unlikely.

Plains Grassland 'restoration'

Research has shown that species-rich Plains Grassland communities – formerly a dominant vegetation in the Victorian Volcanic Plain bioregion in the Geelong area – can be created on sites now completely degraded. This can be achieved by scraping off nutrient-rich topsoil and seed banks of exotics, then sowing indigenous grasses, graminoids and forbs (Gibson-Roy and Delpratt 2015). Impressive displays of grassland flora can result. We are assured by Williams *et al.* (2015) that this strategy is the future for grassland conservation; boundless enthusiasm is the order of the day:

> *The restoration of degraded grassland and creation of new areas of native grasslands is of critical importance ... It is only through restoration that we can rapidly expand our grassland areas. Restoration can reverse what has been to date, a story of loss and*

degradation ... diverse grassland communities can now be established in a single sowing.

Direct sowing has released restoration from the tyranny of scale. As confidence and expertise grow there can be no technical limit to the area ... restored (Williams et al. 2015, p. 425).

'Field of Dreams' expectations *sensu* Palmer *et al.* (1997) indeed, with moral and ethical dimensions. Ecologically, technically and economically, management of the 'restored' site is a highly dubious prospect:

- since Stuwe and Parsons (1977) placed grasslands squarely in focus, the great majority of sites have been destroyed outright or irreparably degraded
- the higher-quality natural Plains Grassland, supporting numerous critically endangered species, itself bordering on extinction, is essentially receiving no management to prevent its destruction – the current and rapid trajectory. Weed invasion is the major issue, particularly by the exotic Phalaris (*Phalaris aquatica*) (Ecology Australia 1994). Degradation must be prevented in the first instance by appropriate management
- prior to the 'restoration' effort on a site the indigenous flora and fauna will be close to 100% extinct. Reinstatement of ecological processes and function are likely to be functionally impossible
- the degradation processes which led to the initial community destruction will continue unabated in 'restored' grassland
- management, essential for any 'restored' site, presents extraordinarily difficult, perhaps intractable, technical problems
- if as a society we have proved unable to manage the existing grassland estate, how are we going to find the motivation and resources to 'restore' grasslands?

Environmental and social historians of the future will piece together evidence to try to understand why the current generations went so disastrously wrong in protecting biodiversity, ecological functions and ecosystem services. They may be puzzled. What are we as a society directly or indirectly conveying to posterity? That we failed the biota and future generations because we did not value it and appreciate or acknowledge unarguable duties of management and stewardship; did not care or thought it does not matter; thought it was someone else's responsibility; didn't want to be seen as too 'green'; thought it was acceptable to pass legislation, albeit inadequate and flawed (*Flora and Fauna Guarantee Act 1988* (Vic.), *Catchment and Land Protection Act 1994* (Vic.)), and yet fail to act on it; didn't recognise that failing to find financial resources to manage the environment, in a society which by any measure is among the wealthiest nations globally, represents anything other than a failure of the imagination?

There can be no second chance. In many ways we are consigning to oblivion a biological heritage unique to this place and time, products of hundreds of millions of years of evolutionary history. We are centre-stage players in the sixth great extinction event in geological history, the Anthropocene (Corlett 2015; Kolbert 2014), the only such event caused by *Homo sapiens*, that most unwise of species.

We find ourselves, morally and ethically, on a precarious path and must do better. It is neither too late nor too hard to protect what we have left and to restore, if and where appropriate, with all actions based on scientific data and input from multiple sources and disciplines.

Acknowledgements

The advice and assistance of the following people is gratefully acknowledged: Lisa Crowfoot, John Kershaw, Leila Brook, Lucy Gow and Jamie McMahon (Ecology Australia); Steve Mathews (Mullum Trust); Steve Sinclair and Matt White (Arthur Rylah Institute); and Mark Trengove (Mark Trengove Ecological Services). Comments by Steve Mathews and two anonymous reviewers improved the text.

References

Adair RJ, Groves RH (1998) *Impact of Environmental Weeds on Biodiversity: A Review and Development of a Methodology.* Occasional Paper. National Weeds Program, Environment Australia, Canberra.

Adam R (2017) Can ideas be dangerous? *Australian Zoologist* **38**, 329–374.

Anon (2017) Flora of the Barrabool Hills: project outline. Unpublished paper. A partnership between the local community, Barrabool Hills Landcare Group, Geelong Landcare Network and the Gordon Institute TAFE.

BoM, Walsh NG (1993) Climate of Victoria. In DB Foreman and NG Walsh (Eds) *Flora of Victoria* (Vol. 1, pp. 47–60). Inkata Press, Melbourne.

Boon PI, Prahalad V (2017) Ecologists, economics and politics: problems and contradictions in applying neoliberal ideology to nature conservation in Australia. *Pacific Conservation Biology* **23**, 115–132. doi:10.1071/PC16035

Booth C (2014) *Natural Victoria. Conservation Priorities for Victoria's Natural Heritage.* Victorian National Parks Association, Melbourne.

Bowler JM, Johnston H, Oley JM, Prescott JR, Roberts RG, Shawcross W, Spooner NA (2003) New ages for human occupation and climate change at Lake Mungo, Australia. *Nature* **421**, 837–840. doi:10.1038/nature01383

Bowman DMJS, Murphy BP, Burrows GE, Crisp MD (2012) Fire regimes and the evolution of the Australian biota. In RA Bradstock, AM Gill and RJ Williams (Eds) *Flammable Australia: Fire Regimes, Biodiversity and Ecosystems in a Changing World* (pp. 27–47). CSIRO Publishing, Melbourne.

Boyce J (2011) *1835, The Founding of Melbourne and the Conquest of Australia.* Black Inc., Melbourne.

Broome R (2005) *Aboriginal Victorians: A History Since 1800.* Allen & Unwin, Sydney.

Brown PL (Ed.) (1941) *Clyde Company Papers.* Oxford University Press, London.

Carr GW (1987) The vegetation and management of Buckley Falls on the Barwon River near Geelong, Victoria. Unpublished report for Friends of Buckley Falls. Ecological Horticulture, Melbourne.

Carr GW (1993) Exotic flora of Victoria and its impact on indigenous biota. In DB Foreman and NG Walsh (Eds) *Flora of Victoria* (Vol. 1, pp. 256–297). Inkata Press, Melbourne.

Carr GW (1999) Checklist of the vascular flora of the Victorian Volcanic Plain (1997). In R Jones (Ed.) *The Great Plains Crash: Proceedings of a Conference on the Grasslands and Grassy Woodlands of Victoria* (pp. 91–119). Appendix 1. Indigenous Flora and Fauna Association and Victorian National Parks Association, Melbourne.

Carr GW, Yugovic JV, Robinson K (1992) *Environmental Weed Invasions in Victoria: Conservation and Management Implications.* Department of Conservation and Natural Resources and Ecological Horticulture, Melbourne.

Christophel DC (1993) Prehistory of the flora of Victoria. In DB Foreman and NG Walsh (Eds) *Flora of Victoria* (Vol. 1, pp. 3–23). Inkata Press, Melbourne.

Christophel DC, Harris WK, Syber AK (1987) The Eocene flora of the Anglesea locality, Victoria. *Alcheringa* **11**, 303–323. doi:10.1080/03115518708619139

Clarkson C, Jacobs Z, Marwich B, *et al.* (2017) Human occupation of northern Australia by 65,000 years ago. *Nature* **547**, 306–310. doi:10.1038/nature22968

Cooke BD (1988) The effects of rabbit grazing on regeneration of sheoak, *Allocasuarina verticillata* and saltwater ti-tree, *Melaleuca halmaturorum,* in the Coorong National Park, South Australia. *Austral Ecology* **13**, 11–20. doi:10.1111/j.1442-9993.1988.tb01414.x

Corlett RT (2015) The Anthropocene concept in ecology and conservation. *Trends in Ecology & Evolution* **30**, 36–41. doi:10.1016/j.tree.2014.10.007

DELWP (2017) *Protecting Victoria's Environment: Biodiversity 2037.* Department of Environment, Land, Water and Planning, Melbourne.

Douglas JG (1993) Geology and geomorphology of Victoria. In DB Foreman and NG Walsh (Eds) *Flora of Victoria* (Vol. 1, pp. 24–46). Inkata Press, Melbourne.

Ecology Australia (1994) Environmental implications of large-scale conversion of exotic perennial pasture. Unpublished report prepared by ARG McMahon, P Peake, J McRobert and GW Carr for Office of the Environment, Department of Natural Resources and Environment, Melbourne.

Ecology Australia (1996) Flora and fauna values of Harvey's property, Bannockburn, Victoria. Unpublished report prepared by AM Muir and MH Collinson for Fisher Stewart Pty Ltd.

Ecology Australia (2001) Biodiversity management plan: an inventory of sites of biodiversity significance in the City of Greater Geelong, Vol. 1. Unpublished report prepared by GW Carr, SE McMillian, CE Silviera, M Trengove and LP Smith for City of Greater Geelong.

Ecology Australia (2005) Proposed Stockland Lonsdale Lakes residential development, Point Lonsdale: flora, fauna and geomorphological values and issues. Unpublished report prepared by GW Carr, LE Conole, N Rosengren and C Campbell for Point Lonsdale Coastal Space Group.

Ecology Australia (2013) Bellarine Peninsula: delineation of coast tea-tree and coast wattle. Unpublished report prepared by GW Carr, ARG McMahon, FM Sutton and JS Kershaw for City of Greater Geelong.

Flannery TF (1994) *The Future Eaters.* Reed Books, Melbourne.

Frakes LA (1999) Evolution of the Australian environments. In AE Orchard (Ed.) *Flora of Australia,* 2nd edn (Vol. 1, pp. 163–203). Australian Biological Resources Study and CSIRO Publishing, Melbourne.

Gammage B (2012) *The Biggest Estate on Earth: How Aborigines made Australia.* Allen & Unwin, Sydney.

Gibbons F, Rowan J (1993) Soils in relation to vegetation in Victoria. In DB Foreman and NG Walsh (Eds) *Flora of Victoria* (Vol. 1, pp. 159–194). Inkata Press, Melbourne.

Gibson-Roy P, Delpratt J (2015) The restoration of native grasslands. In A Marshall, NSG Williams and JW Morgan (Eds) *Land of Sweeping Plains: Managing and Restoring the Native Grasslands of South-eastern Australia* (pp. 332–387). CSIRO Publishing, Melbourne.

Groves RH (2011) The impact of alien plants in Australia. In D Pimental (Ed.) *Biological Invasions: Economic and Environmental Cost of Alien Plant, Animal and Microbe Species* (pp. 1–10). CRC Press: Boca Raton.

Hateley R (2010) *The Victorian Bush, Its 'Original and Natural' Condition.* Polybractea Press, Melbourne.

Henderson R, Johnson D (2016) *The Geology of Australia,* 3rd edn. Cambridge University Press, Melbourne.

Hill R (Ed.) (1994) *History of the Australian Vegetation: Cretaceous to Recent.* Cambridge University Press, Melbourne.

Hill RS, Truswell E, McLoughlin S, Dettmann ME (1999) Evolution of the Australian flora: fossil evidence. In AE Orchard (Ed.) *Flora of Australia*, 2nd edn (Vol. 1, pp. 251–320). Australian Biological Resources Study and CSIRO Publishing, Melbourne.

Hobbs RJ, Hopkins AJM (1990) From frontier to fragments: European impact on Australia's vegetation. In DA Saunders, AJM Hopkins and RA How (Eds) *Australian Ecosystems: 200 Years of Degradation, Utilisation and Reconstruction* (pp. 93–114). Surrey Beatty & Sons, Sydney.

Johnson C (2006) *Australia's Mammal Extinctions: A 50,000 Year History.* Cambridge University Press, Cambridge.
Keith DA, Auld TD (2017) Conservation of Australian vegetation. In DA Keith (Ed.) *Australian Vegetation*, 3rd edn (pp. 677–710). Cambridge University Press, Cambridge.
Kirkpatrick JB (1999) *A Continent Transformed: Human Impact on the Natural Vegetation of Australia*, 2nd edn. Oxford University Press, Melbourne.
Koch PL, Barnosky AD (2006) Late quaternary extinctions: state of the debate. *Annual Review of Ecology Evolution and Systematics* **37**, 215–250. doi:10.1146/annurev.ecolsys.34.011802.132415
Kolbert E (2014) *The Sixth Extinction. An Unnatural History.* Bloomsbury, London.
Lines WJ (1991) *Taming the Great South Land.* Allen & Unwin, Sydney.
Lomolino MV, Riddle BR, Whittaker RJ (2017) *Biogeography. Biological Diversity across Space and Time*, 5th edn. Sinauer Associates, Sunderland.
Long J (2003) *Introduced Mammals of the World: Their History, Distribution and Influence.* CSIRO Publishing, Melbourne.
Long JA, Archer M, Flannery T, Hand S (2002) *Prehistoric Mammals of Australia and New Guinea: One Hundred Million Years of Evolution.* University of New South Wales Press, Sydney/Johns Hopkins University Press, Baltimore.
Low T (2001) *Feral Future: The Untold Story of Australia's Exotic Invaders.* Penguin Books, Melbourne.
Maron M, Grey MJ, Catterall CP, Major RE, Oliver DL, Clarke MF, Loyn RH, McNally R, Davidson I, Thomson J (2013) Avifaunal disarray due to a single despotic species. *Diversity and Distributions* **19**, 1468–1479. doi:10.1111/ddi.12128
Mathews S, Carr GW, McMahon ARG (2011) Victorian National Parks Association Nature Conservation Review. Unpublished paper prepared by Ecology Australia for Victorian National Parks Association, Melbourne.
McCarthy G, Pescott T (1985) The common wombat in the Geelong region. *Victorian Naturalist* **22**, 58–61.
McIlroy JC (1995) Common wombat. In R Strachan (Ed.) *The Mammals of Australia* (pp. 204–205). Australian Museum and Reed Books, Sydney.
Menkhorst PW (Ed.) (1995) *Mammals of Victoria: Distribution, Ecology and Conservation.* Oxford University Press, Melbourne and Department of Conservation and Natural Resources.
Mills K, Gell P, Kershaw P (2013) *The Recent Victorian Drought and its Impacts: Without Precedent?* RIRDC Publication No 12.040. Rural Industries Research and Development Corporation.
Mooney S, Harrison SP, Bartlein PJ, Stevenson J (2012) The prehistory of fire in Australia. In RA Bradstock, AM Gill and RJ Williams (Eds) *Flammable Australia: Fire Regimes, Biodiversity and Ecosystems in a Changing World* (pp. 3–25). CSIRO Publishing, Melbourne.
Morcom LA, Westbrooke ME (1998) The pre-settlement vegetation of the western and central Wimmera Plains of Victoria, Australia. *Australian Geographical Studies* **36**, 273–288. doi:10.1111/1467-8470.00057
Moxham C, Turner V (2009) A journey through coastal Moonah woodland in Victoria. *Victorian Naturalist* **126**, 169–178.
Mueck S, Smales I (2007) Flora and fauna of the Point Lonsdale Residential and Waterways Project, Point Lonsdale, Victoria. Unpublished report prepared by Biosis Research for Stockland Developments Pty Ltd.
Mulvaney DJ (1964) Prehistory of the basalt plains. *Proceedings of the Royal Society of Victoria* **77**, 427–432.
Mulvaney DJ, Kamminga J (1999) *Prehistory of Australia.* Allen & Unwin, Sydney.
Palmer MA, Ambrose RJ, Poff NL (1997) Ecological theory and community restoration ecology. *Restoration Ecology* **5**, 291–300. doi:10.1046/j.1526-100X.1997.00543.x
Pascoe B (2014) *Dark Emu: Black Seeds: Agriculture or Accident.* Magabala Press, Broome.

Poole WE (1995) Eastern grey kangaroo. In R Strachan (Ed.) *The Mammals of Australia* (pp. 335–338). Australian Museum and Reed Books, Sydney.

Presland G (2010) *First People: The Eastern Kulin of Melbourne, Port Phillip and Central Victoria.* Museum Victoria Publishing, Melbourne.

Recher HF (2013) What makes this old scientist grumpy? In D Lunney, P Hutchings and HF Recher (Eds) *Grumpy Scientists: The Ecological Conscience of a Nation* (pp. 1–8). Royal Zoological Society of New South Wales, Sydney.

Recher HF (2015) Failure of science, death of nature. *Pacific Conservation Biology* **21**, 2–4. doi:10.1071/PC14907

Rich PV, van Tets GF, Knight F (2000) *Kadimakara: Extinct Vertebrates of Australia.* Princeton University Press, Princeton.

Ripple WJ, Wolf C, Newsome TM and 15 359 signatories from 184 countries (2017). World scientists' warning to humanity: a second notice. *Bioscience,* bix 125, https://doi.org/10.1093/biosci/bix125

Roberts RG, Jones R, Smith MA (1990) Thermoluminescence dating of a 50,000 year old human occupation site in northern Australia. *Nature* **345**, 153–156. doi:10.1038/345153a0

Robinson D (2006) Is revegetation in the Sheep Pen Creek area, Victoria, improving grey-crowned babbler habitat? *Ecological Management & Restoration* **7**, 93–104. doi:10.1111/j.1442-8903.2006.00263.x

Saltré F, Rodríguez-Rey M, Brook BW, Johnson CN, Turnery CSM, Alroy J, Cooper A, Beeton N, Bird MI, Fordham DA, Gillespie R, Herrando-Pérez S, Jacobs Z, Miller GH, Nogués-Bravo D, Prideaux GJ, Roberts RG, Bradshaw CJA (2016) Climate change not to blame for the late Quaternary megafauna extinctions in Australia. *Nature Communications* **7**, 10511. doi:10.1038/ncomms10511

Sandom C, Faurby S, Sandel B, Svenning J-C (2014) Global late Quaternary megafauna extinctions linked to humans, not climate change. *Proceedings. Biological Sciences* **281**, 20133254. doi:10.1098/rspb.2013.3254

Saunders DA, Hobbs RJ (Eds) (1991) *Nature Conservation 2: The Role of Corridors.* Surrey Beatty & Sons, Sydney.

Sayers A (1998) The shaping of Australia landscape painting. In E John, A Sayers, E. Kornhauser and A Ellis (Eds) *New Worlds from Old: 19th Century Australian and American Landscapes* (pp. 52–69). National Gallery of Australia, Canberra.

Sinclair SJ, Atchison K (2012) The pre-colonial distribution of grassland, woodland and forests on the Werribee Plains, Victoria. *Cunninghamia* **12**, 213–227. doi:10.7751/cunninghamia.2012.12.017

Sluiter IRK, Blackburn DJ, Holdgate GR (2016) Fire and Late Oligocene to Mid-Miocene peat megaswamps of south-eastern Australia: a floristic and palaeoclimatic interpretation. *Australian Journal of Botany* **64**, 609–625. doi:10.1071/BT16165

Stuwe J, Parsons RF (1977) *Themeda australis* grasslands on the basalt plains, Victoria: floristics and management effects. *Australian Journal of Ecology* **2**, 467–476. doi:10.1111/j.1442-9993.1977.tb01162.x

Taylor J (2005) *Population and Patterns of Residence.* Macquarie/Macmillan, Sydney.

Turney CSM, Bird MI, Fifield LK, Roberts RG, Smith M, Dortch CE, *et al.* (2001) Early human occupation at Devil's Lair, southwestern Australia. *Quaternary Research* **55**, 3–13. doi:10.1006/qres.2000.2195

Vickers-Rich P, Rich TH (1993) *Wildlife of Gondwana.* Reed Books, Sydney.

West A (1971) Aborigines of western Victoria. In MH and L O'Brien (Eds) *The Natural History of Western Victoria* (pp. 62–73). Australian Institute of Agricultural Science, Horsham, Victoria.

White ME (1986) *The Greening of Gondwana: The 400 Million Year Story of Australia's Plants.* Reed Books, Sydney.

White ME (1994) *After the Greening: The Browning of Australia.* Kangaroo Press, Sydney.

Williams LM, Menkhorst PW (1995) Common wombat. In PW Menkhorst (Ed.) *Mammals of Victoria: Distribution, Ecology and Conservation* (pp. 82–83). Oxford University Press, Melbourne.

Williams NSG, Marshall A, Morgan JW, Delpratt J, Gibson-Roy P, Wong N (2015) The future of south-eastern Australia's native temperate grasslands. In A Marshall, NSG Williams and JW Morgan (Eds) *Land of Sweeping Plains: Managing and Restoring the Native Grasslands of South-eastern Australia* (pp. 420–451). CSIRO Publishing, Melbourne.

Willis JH (1956) Robert Brown's collectings in Victoria. *Muelleria* **1**, 51–54.

Willis JH (1964) Vegetation of the basalt plains in western Victoria. *Proceedings of the Royal Society of Victoria* **77**, 397–418.

Willis JH, Cohn HM (1993) Botanical explorations of Victoria. In DB Foreman and NG Walsh (Eds) *Flora of Victoria* (Vol. 1, pp. 61–78). Inkata Press, Melbourne.

Wilson JB (1889) *Excursions near Geelong in Search of Plants, or Florula corioensis: With a List of those Collected.* G. Mercer, Geelong.

Woodgate PW, Peel WD, Ritman KT, Corman JE, Brady A, Rule AJ, Banks JCG (1994) *A Study of the Old-growth Forests of East Gippsland.* Department of Conservation and National Resources, Melbourne.

Appendix 3.1. Ecological Vegetation Classes (EVCs) modelled for the City of Greater Geelong, June 2017

EVC number	EVC name
EVC 1	Coastal Dune Scrub/Coastal Dune Grassland Mosaic
EVC 3	Damp Sands Herb-rich Woodland
EVC 6	Sand Heathland
EVC 9	Coastal Saltmarsh
EVC 16	Lowland Forest
EVC 20	Heathy Dry Forest
EVC 21	Shrubby Dry Forest
EVC 22	Grassy Dry Forest
EVC 47	Valley Grassy Forest
EVC 48	Heathy Woodland
EVC 55	Plains Grassy Woodland
EVC 56	Floodplain Riparian Woodland
EVC 67	Alluvial Terraces Herb-rich Woodland
EVC 68	Creekline Grassy Woodland
EVC 71	Hills Herb-rich Woodland
EVC 72	Granitic Hills Woodland
EVC 74	Wetland Formation
EVC 83	Swampy Riparian Woodland
EVC 104	Lignum Swamp
EVC 125	Plains Grassy Woodland
EVC 132	Plains Grassland
EVC 134	Sand Forest

EVC 136	Sedge Wetland
EVC 140	Mangrove Shrubland
EVC 161	Coastal Headland Scrub
EVC 163	Coastal Tussock Grassland
EVC 164	Creekline Herb-rich Woodland
EVC 175	Grassy Woodland
EVC 177	Valley Slopes Dry Forest
EVC 196	Seasonally Inundated Sub-saline Herbland
EVC 198	Sedgy Riparian Woodland
EVC 200	Shallow Freshwater Marsh
EVC 282	Shrubby Woodland
EVC 291	Cane Grass Wetland
EVC 292	Red Gum Swamp
EVC 300	Reed Swamp
EVC 302	Coastal Saltmarsh/Mangrove Shrubland Mosaic
EVC 641	Riparian Woodland
EVC 647	Plains Sedgy Woodland
EVC 691	Aquatic Herbland/Plains Sedgy Wetland Mosaic
EVC 851	Stream Bank Shrubland
EVC 858	Coastal Alkaline Scrub
EVC 891	Plains Brackish Sedge Wetland
EVC 892	Heathy Woodland/Sand Heathland Mosaic
EVC 894	Scoria Cone Woodland
EVC 895	Escarpment Shrubland
EVC 898	Cane Grass-Lignum Halophytic Herbland
EVC 899	Plains Freshwater Sedge Wetland
EVC 991	Water body – salt
EVC 992	Water body – fresh

4

Welcome to *Wadawurrung* Country

Uncle Bryon Powell and Tandop David Tournier with David S. Jones and Phillip B. Roös

Acknowledgement of Country

We wish to acknowledge the Elders, families and forebears of the *Wadawurrung* peoples who remain the spiritual and cultural custodians of our lands and waters and who continue to practise their values, languages, beliefs and customs.

Welcome ...

Nyoorra woorreeyn,
Koonyaba morrgalyoo/Kardeeneeyoo,
Woorreeng woorreng; Nedock; Tandop; Babarang; Wanoong;
Wairnga knettuks; Peetyarr; Ngardang; Tatee baa Yanabeel.
Yoongga nyatne-ngitj karreengalabeel-a wela meerree-munya barre-ngak
Yoongga nyatne-ngitj karreengalabeel-a wela meerree-munya barre-ngak dyarreeka-ngitj
Yoongga nyatne-ngitj karreengalabeel-a wela meerree-munya barre-ngak yanabeel getyaweel
Bengalak mok-borreeyn kappa kandor moorroop keen keen beel Wada nyanbo bengadak.
Keem barne barre Wadda Wurrung
Kitjarra-ngitj, bitjarra ngala, mok barra tjarra ngitj
Keen keen beel baa yoowang ngitj
Nyatne
How do you do
Good morning/afternoon
Grand Mother; Grand Father; Uncle; Aunty; Brother;
Sister Father; Mother; Friend and Visitor
To give Thankyou we Creator – for the who to watch-he land-our
To give Thankyou we Creator – for the who to watch-he land-our to stand we
To give Thankyou we Creator – for the who to watch-he land-our visitor many
We proud to follow track spirit Aborigine to go before us.
This is Wathaurong Country

Let us talk together, let us not fight, let us have peace,
Black and White together
Thankyou
Tournier (2013, p. 1).

As an Elder of the *Wadawurrung* community, I welcome you to my Country.

I pay my respects to co-author *Tandop* David Tournier (1955–2016), author of the above 'Welcome to Country' text in *Wathaurong* and English tongues, who entered the Dreaming on 31 December 2016 while co-authoring this chapter (Anon 2016, 2017).

This chapter offers an introduction to past, present and future *Wadawurrung* Country, culture and knowledge (Powell 2019; Powell and Jones 2019; Pascoe 2003, 2012). The *Wadawurrung* people ('the red soil people') are part of the greater *Kulin* Nation (Anon 2014; Massola 1968, 1969, 1971a, b). In *Wadawurrung* language, *Wada* means 'no lip', referring to a linguistic and not a physical characteristic, and *Wurrung* means 'people' (Powell 2019, pers. comm.). The narration voices in this chapter are primarily Uncle Bryon and *Tandop* David talking about their Country that is their community's Country.

The *Kulin* Nation includes the peoples of the *Taungerong*, *Dja Dja Wurrung*, *Boon Wurrung* (Briggs 2008, 2014; Eidelson 2014) and *Wurundjeri* peoples. To the south were the *Gadubanud* of the Otway Ranges (Niewójt 2009) and the *Gulidjan* of the Lake Colac catchment, who were part of the *Mara* Nation. Written for the Geelong–Bellarine Peninsula portion of *Wadawurrung* Country (Surf Coast Shire 2014) that sweeps up to Ballarat and Ararat (Nicholson *et al.* in press), this chapter offers several stories, explains six key principles of our culture, introduces 12 places that are important to our culture, and provides a learning resource about our living culture and Country. It is narrated as a learning resource, so while the text may be cogent and focused, the referencing is extensive, allowing readers to scope more information (Powell 2015a, b, 2019; Powell and Jones 2019; WAC 2018b).

In our beginning ...

In the beginning, our Country was created by our Creation Being who left their mark on this landscape. The following story demonstrates the link that Aboriginal people maintain with the land, sentient and non-sentient beings, and the cosmos. My story of this beginning starts when ...

Bunjil (Wedge-tailed Eagle (*Aquila audax*)) summoned six men to assist him in the creation of the land, the people and all living things and to pass on his teachings and knowledge to all men and women.

The six men were *Djurt Djurt*, the Nankeen Kestrel (*Falco cenchroides*); *Thara*, the Black-shouldered Kite (*Elanus axillaris*); *Dantun*, the Blue Mountain Parrot (*Platycercus elegans*); *Tadjeri*, the Brush-tailed Phascogale (*Phascogale tapoatafa*); *Turnun*, the Feather-tailed Glider (*Acrobates pygmaeus*); and *Yukope*, the Great Parakeet (*Melopsittacus undulatus*).

All were capable of mighty deeds in the name of *Bunjil*.

After *Bunjil* made the Country and all the living things in it, he taught the people how to use their tools and the rules of social behaviour.

When he had finished, *Bunjil* gathered his wives (one of whom was *Ganawarra*, the Black Swan (*Cygnus atratus*)) and his children (one of whom was *Binbeal*, the Rainbow) and instructed *Belin-Belin* (the Musk Crow (*Corvus coronoides*)) who was in charge of the winds: '*Belin–Belin*, open your bag and let out some wind.'

Belin-Belin opened his bag and released a wind so strong it carried big trees into the air, roots and all.

Bunjil wanted more wind, so *Belin-Belin* opened all his bags and released a wind so great that it carried *Bunjil* and his family to *Tharangalk-Bek* (the heavens).

Bunjil and his family now live in the heavens and are the stars looking down on the world.

You can see them in the night sky: *Bunjil* is the star Altair in the Eagle Constellation; the two stars beside *Bunjil* are *Ganawarra* and her spirit; *Bunjil*'s brother *Palyang* is the star Antares in the Scorpion Constellation; *Turnung* is Venus; *Tadjeri* is the star Achernar, in the Eridanus Constellation; *Dantun* is the star β Crucis, the Southern Cross; *Djurt-Djurt* is the star Beta Centauri, one of the pointers; and *Thara* is the star Alpha Centauri, the other pointer (Powell 2015c; see also Massola 1968).

Once the landscape was created, *Bunjil* turned his attention to human creation at *Kareet Bareet* (Black Hill) near Gordon. To the *Wadawurrung*, *Kareet Bareet*:

> *... is very special place. It's special for my ancestors and me. That's where Bunjil helped Creation Being. He gathered up the* [red] *clay from the mountain. He divided it into two on two sheets of bark and formed them into the shapes of men. He got his brother Palyang – Palyang has control of all the creeks and rivers and billabongs – and he began to thump the water with his hands. The water became thicker, and thicker, and took on the shape of two woman. So each man was presented with a hunting spear, and the woman were presented with digging sticks so they could dig the edible roots and yams* [Murnong, Microseris lanceolata]. *And, from that, this where we all came from: Kareet Bareet* (Powell 2015d; Smyth 1878, p. 423).

Importantly, in the latter creation of our ancestors, the *Wadawurrung* believe that they were formed and created at *Kareet Bareet* and did not travel across the land bridge to Australia some 120 000–60 000 years ago as advanced by western science (Powell 2015d).

Our 'history' is told orally in our stories – the Western text-printed version is but one version of the oral narrative (Powell 2015a, b, c, d, e, f, g, h, 2019; Tournier 2010a, 2011a, b, c, 2013, 2014 a, b; Whitehead with Tournier 2014).

In the western science beginning ...

Some 60 000 years ago our human forebears travelled along the Indonesian land bridge from Africa and southern Asia to reach the present Australian land mass (Nunn and Reid 2016). This was a time when the sea level was lower, glacial melting had not occurred, and land temperatures and vegetation were markedly different from today, offering a rich moist temperate and tropical environment. Megafauna roamed the land mass.

The first Aboriginal arrivals in today's Northern Territory, our human forebears, dispersed across the then unoccupied landscape devoid of humans but rich in animals, potable water and plants, eventually arriving in south-eastern Australia some 40 000 years ago.

On *Wadawurrung* Country, a settlement pattern was established accordingly to food resource activity and the nature of the landscape. At the time, *Nerm* (Port Phillip Bay) and Western Port were extensive grassland plains, much of the landscape was covered in wet sclerophyll forests and lush open woodlands, misty rivers circuitously trickled from the northern Great Dividing Ranges and the Otway Ranges, one could walk the land bridge to the Tasmanian Countries and the sea level was much lower, enabling food resource-luxuriant and shallow plains along the Bass Strait edges (Briggs 2010; Holdgate *et al.* 2011; Nunn and Reid 2016).

Some 60 000 to 12 000 years ago climatic conditions in south-eastern Australia were considerably drier and cooler (Markgraf *et al.* 1992; Pickett *et al.* 2004). The lower sea level enabled a land bridge to exist from southern Victoria across Bass Strait to Tasmania (Lambeck and Chappell 2001), and there were extensive wetland-rich plains along the Bass Strait shoreline. From 12 000 to 9000 years ago changes occurred. These shifted the landscape from a moist tropical environment towards an arid Australian environment, resulting in a significant transformation and reduction of wet forests and an expansion of open and dispersed fire-reliant dry woodland and grasslands, termed today as savannah vegetation (Gammage 2011) Archaeological evidence points to major forest loss in temperate Australia except along the sinuous band of temperate forest of the eastern and south-eastern coastlines (Dodson 1989, 2001), with enclaves of tropical woody vegetation in localised favourable microsites such as deep sheltered and well-watered river valleys (Thom and Roy 1985; Thom *et al.* 1994).

There is an extensive geological history of volcanic eruptions across the region, with eruptions every 25 000 years, at a frequency of eruptions now estimated as 1:10 800 years (Boyce 2013). The most recent eruptions occurred at Mount Gambier and Mount Schank in South Australia ~5000 years ago.

Contemporary archaeological evidence of *Nerm* (Port Phillip Bay) and its catchment records human occupation dating back at least 40 000 years. The oldest dated archaeological site in Victoria occurs at Keilor in Melbourne, on *Wurundjeri* Country. Its evidence includes charcoal from a hearth dated to 31 000 years BP (Flood 1995). Archaeological research of the Melbourne metropolitan region has identified evidence of artefact scatters, scarred trees and stone arrangements documenting evidence of Aboriginal life in the localities from the Pleistocene (~2 588 000 to 11 700 years ago) through to the immediate pre-European past. Most of these sites also evidence relationship connections between Aboriginals, sites, landscapes and food resources within the immediate area.

Some 10 000–20 000 years ago we witnessed volcanic eruptions by our ancestors that resulted in the creation of *Bonan Youang* (Mt Buninyong) (Powell 2015f), *Warrenyeep* (Mt Warrenheip) (Powell 2015i), *Djerrinallum* (Mt Elephant) (Powell 2015i), The Anakies (Tournier 2010, 2011b, 2014b), in the shadows of the Devonian era (60 million years to 419.2 million years ago) created *Wurdi Youang* (You Yangs) (Powell 2015h) on our Country (Powell 2019; Powell and Jones 2019). As long-term residents we had to modify our resource-gathering and residency patterns while these explosions and associated lava flows wreaked havoc across our Country. But, importantly, we witnessed these geological incidents and our stories tell the history of the events (Powell 2015i).

Thus, our peoples watched major episodes of climate change transformation and, demonstrating resilience, we discovered adaptive strategies to live with these changes including changing our food resource diets and seasonal movements. Our ancestors experienced major climate change impacts and know first-hand how to be resilient and adaptive to continue to look after and live on our Country (Jones *et al.* 2018; Low Choy *et al.* 2013; Powell 2019; Powell and Jones 2019).

While we can date tangible European settler and 'explorer' uninvited invasion of our Country around 1800, it was the intangible invasions that wreaked the most destruction upon our culture and our Country. Country dispossession, clan people herding and resettlement, attempts at cultural nomenclature and language erasure and religious and philosophical re-indoctrination of language, knowledge and law were prolific during this time. Additionally, land and its waters were fenced, dammed, drained and fired and vegetation was extensively felled (Gammage 2011; Pascoe 2014). These were very tangible social engineering strategies by Europeans to control the landscape. Unconscious was the rampant spread of European disease including smallpox and plagues, and sexually transmitted diseases carried by European settlers and their animals on lands and waters and by Aboriginals using traditional trade routes and songlines (Kerwin 2012). There were new human and animal-borne diseases and weed seeds carried on animals and ships, invasions by noxious weeds and exotic animals (rabbits, foxes, cats, dogs, etc.) and birds that vehemently preyed upon our traditional animals and our traditional food vegetation resources (Jones and Clarke 2018; Tournier 2015a). Furthermore, there was an extensive recasting of the landscape with northern hemisphere pasture grasses and shelterbelt corridors to incorporate images of their homeland landscapes (Gammage 2011).

Individually and collectively, these intangible invasive agents decimated and weakened *Kulin* Nation populations, animals, birds and fish as well as the *Wadawurrung* community and its Country. What was once a very healthy human (and animal) community historically reliant upon a starchy, lean, non-artificial sugar-based (and preservative-free) diet and a solid exercise regime, never exposed to the dormant disease ravages of Europe, was suddenly engulfed by the transformation of traditional diet and exercise regimes (Gott 1982, 1983, 2005; Gott and Conran 1991; Zola and Gott 1992). Our transitionary seasonal resource-harvesting encampments were replaced with permanent European food-dependent encampments. Exercise and nutritious diets were suddenly substituted with preservative-rich and artificial sugar-laden foods, heightening overweigh and diabetes-related deaths. A human community that had been genetically insulated for over 60 000 years from the northern hemisphere plagues and diseases was suddenly enveloped in this onslaught, with no plant and medical solutions and therefore a lack of capacity to adapt.

Between 12 000 and 15 000 years ago, our ancestors and other Aboriginal peoples along Australia's coastlines witnessed a major physical transformation of our Country, resulting in a 'time of chaos' (Briggs 2008, 2010, 2014; Nunn and Reid 2016). The sea level rose quickly, flooding our popular seaside encampments and destroying popular crustacean harvesting places. The Tasmanian land bridge disappeared under sea waters.

Nerm (Port Phillip Bay), long a source of grassland plants and animals and characterised by its twisting central *Birrarung* freshwater river, was suddenly engulfed in salt water as the seas entered into the depression (Holdgate *et al.* 2011; Briggs 2010; Massola 1968). Riverine corridors like the present Barwon, Moorabool and Werribee rivers changed segments of their alignments. Extensive freshwater bodies like Lake Corangamite changed to saline water and also changed their exit flow routes. And the last of the volcanoes (7000–30 000 years ago) far distant in *Gunditjmara* (*Budj Bim*, Mt Napier, Tower Hill) and *Gulidjan* Country's (Mt Noorat) erupted (Gunditjmara People with Wettenhall 2004). The mixed lush moist open temperate grassy woodlands and forests quickly changed to dry open forests and extensive grasslands plains, and the tropical forests of the Otways receded into its sheltered gullies.

As noted by Bird (1993, pp. 8–9):

> *... about 18 000 years ago when the glaciers and ice sheets reached their maximum extent, the sea was 120 to 140 metres below its present level, and the continental shelves were exposed as wide coastal plains ... The warming of the climate and the diminution of the glaciers and ice sheets caused a rise in the sea level which continued until the mid Holocene about 6000 years ago; an event known as the late Quaternary marine transgression.*

Both the Australian landscape and *Wadawurrung* Country experienced these major geographical and environmental transformations that resulted in today's landscapes and environments. Importantly, we witnessed these environmental transformations, and our stories (and those of neighbouring communities) recall these historical events.

In the 'whitefella' beginning ...

The first non-European 'explorers' to *Wadawurrung* Country were the Chinese in 1421 as part of a larger maritime exploration venture captained by Vice-Admiral Zhao Man under the command of Admiral Zheng He (1371–1433/1435) during the Ming era. One of their mahogany ships was wrecked on *Gunditjmara* Country shores at Warrnambool (Menzies 2003), although some hold that the wrecked ship was a Portuguese vessel venturing through the Bass Strait in 1522 (Shaw 1996). The crew's knowledges, ideas and genetics were disseminated into the adjacent *Gunditjmara* Country.

Europeans first sighted and set foot on *Wadawurrung* Country in the early 1800s. Matthew Flinders (1774–1814), with George Bass, sailed past in Bass Strait in 1798 on the sloop *Norfolk*, mapping the coastal edge (Bass 1895; Flinders 1978). John Murray (*c*.1775–*c*.1807) on the white-sailed survey vessel *HMS Lady Nelson* entered *Nerm* (Port Phillip Bay) on 14 February 1802. Murray's party walked through our Country, around the shores of Corio Bay. Murray records that on the 'south shore lies a cluster of small rocky islands and all round them a shoal of sand; plenty of swans and pelicans were found on them when the boat was down, from which I named them Swan Isles [Swan Island]' (Murray in Keble 1947, p. 131).

Flinders returned again on the white-sailed survey ship *HMS Investigator* in April–May 1802 and moored off Indented Head on 30 April 1802, recording the receptiveness and attempts of a 'Welcome to our Country' by members of the *Wadawurrung Bengalat balug* clan:

> *In the morning a fire was perceived two hundred yards from the tent; and the Indians appeared to have decamped from thence on our landing ... a party of the inhabitants showed themselves about a mile from us; and on landing there we found a hut with a fire in it, but the people had disappeared, and carried off their effects. I left some strips of cloth, of their favourite red colour, hanging about the hut, and proceeded westward along the shore to examine the arm of the port running in that direction.*
>
> *Three natives having made their appearance abreast of the boat, we again landed. They came to us without hesitation, received a shag and some trifling presents with pleasure, and parted with such of their arms as we wished to possess without reluctance. They afterwards followed us along the shore; and when I shot another bird, which hovered over the boat, and held it up to them, they ran down to the water-side and received it without expressing either surprise or distrust. Their knowledge of the effect of fire-arms I then attributed to their having seen me shoot birds when unconscious of being observed; but it had probably been learned from Mr. Murray* (Flinders 1966, 26 April–2 May 1802).

On the next day, 1 May 1802, Flinders climbed to the top of *Wurdi Youang* (You Yangs; Station Peak) then returned to the *Investigator*, recording again the receptiveness of members of the *Wadawurrung Bengalat balug* clan:

> *Fires had been seen moving along the shore, but the people seemed to have fled; though we found two newly erected huts with fires in them, and utensils, which must have*

> *belonged to some of the people before seen, since there was boiled rice [plant seeds being detoxified] in one of the baskets. We took up our quarters here for the night, keeping a good watch; but nothing was seen of the Indians till we pushed off from the shore in the morning* [Sunday 2 May 1802], *when seven showed themselves upon a hill behind the huts. They ran down to examine their habitations, and finding every thing as they had left it, a little water excepted of which we were in want, they seemed satisfied; and for a short time three of them followed the boat* (Flinders 1966, 26 April–2 May 1802).

In December 1802, Acting Lieutenant Charles Robbins arrived in Port Phillip Bay and set foot at Sorrento on *Boon Wurrung* Country (Shillingshaw 1972). For the next five weeks the Robbins party ventured around the shores of Port Phillip Bay. On 18 January 1803, near Point Cook, Robbins met a party of 11 *Wadawurrung* people who were seeking to initiate a meeting:

> *As we went out of the boat eleven natives met us; they were civil; I gave one of them a biscuit; he looked at it; I took it again, eat of it, when he did the same; whatever we said they said after us. There was one who appeared to be their chief. They handed us their spears to look at ... They followed us as we went on ... when the boat came up we went on board. I made signs for them to come into the boat, but they would not venture* (Shillingshaw 1972, p. 32).

The Van Diemen's Land (Tasmania) colonial administrators sent a military and convict contingent northward to evaluate the economic pastoral capacity of the *Nerm* (Port Phillip Bay) region in October 1803. Establishing an outport at Sorrento, in *Boon Wurrung* Country, they concluded that the lack of potable water and saline alkaline calcareous soils negated agricultural suitability. They quickly returned to Launceston in January 1804, but not before several European convicts escaped in December 1803.

One of the escapees, who survived, was the infamous convict William Buckley (1780–1856) who walked from Sorrento through *Boon Wurrung* Country around the Bay into *Wadawurrung* Country to live along the Barwon River corridor and on the Bellarine Peninsula (Tudehope 1962a; Morgan 1980; Massola 1969; Brownhill 1955). Buckley related his first meeting with a *Wadawurrung* family group to colonialist George Langhorne in 1835, telling of their generosity with assistance, food and language sharing, before he parted from their company. Morgan, in 1852, published Buckley's recollections including his travels around *Wadawurrung* Country, staying at Painkalac Creek, *Mangowak* (Aireys Inlet) and *Nooraki* (Mount Defiance), and his first meeting with a group of *Wadawurrung* women of the *Bengalat balug* clan. At this meeting, the women recognised a stick Buckley was using for walking as a spear belonging to a relative who had recently died, and invited him back to their camp where he was welcomed and adopted by the group. For the next 32 years he lived with these *Bengalat balug* people on the Bellarine Peninsula, engendering their trust and respect. Morgan wrote, 'By virtue of his age and peaceful ways, Buckley ... became a *Ngurungaeta* [a *Woi wurrung* word of the *Wurundjeri*], a person of considerable respect among his people and his voice was influential in deciding matters of war and peace' (Flannery 2002, pp. xii, 192; Morgan 1852).

While Buckley's travels, and 'Buckley's chance', has been narrated into whitefella myth for Australia, little is known of the activities of the independent sealers and whalers who visited and camped on our shores along Bass Strait from the 1790s to the 1840s. These entrepreneurs harvested Humpback Whale (*Megaptera novaeangliae*) and Australian Fur Seal (*Arctocephalus pusillus*) for their oil, meat and bone (Hainsworth 1971). Sealers and whalers, often involving crews of up to 200, as well as official parties like the *Mary Ann* and *William and Ann* dispatched by Governor Arthur Phillip in 1791, prompted long-term industry viability concerns by the colonial governance authorities to 'restrain individuals from resorting there in too great numbers, and to fix certain times for their visiting these places to prevent the destruction of that commercial advantage' (King in Townrow 1997, p. 7).

Many of these sealers and whalers met, sought travel and potable water guidance from, traded with, shot, partook of food exchanges and engaged in intercourse with, and purchased or acquired 'indentured' *Wadawurrung* peoples to support their encampments and living requirements (Boys 1935; Hainsworth 1971; Ellis 1981; Labilliere 1878; Micco 1971; Murray-Smith 1973; Todd 1835, 28 June 1835; Townrow 1997; Tuckey 1805).

Thus, non-*Wadawurrung* artefacts, foodstuffs, guns and other weapons, coloured fabrics (especially red), English language and mannerisms and so on were 'unofficially' introduced to our *Wadawurrung* coastal clan peoples, which explains why European explorers expressed surprise about the familiarity of our peoples with some of these non-*Wadawurrung* technologies including the sound and use of the gun.

Unannounced, on 16 December 1824, Englishmen Hamilton Hume (1797–1873) and William Hovell (1786–1875) entered our Country across the Werribee River. They observed a:

> *... harbour or bay* [that] *consisted of an immense sheet of water... The bay too is literally covered with black swans, and various other aquatic birds. Caught some black bream in the creek. Messrs. Hovell and Hume each marked his initials on a tree with an iron tomahawk, at some distance from the left bank of the creek, about two miles from the beach* (Hume and Hovell in Anon 2018).

Hume and Hovell proceeded southwards to the northern shores of Corio Bay over the next two days, before returning on the same route back into *Wurundjeri* Country. Their party met members of the *Wadawurrung Neerer balug* clan who also offered a 'Welcome to Country' greeting:

> *... Friday, December 17th. — ... This morning, one of the men, James Fitzpatrick, having proceeded a short distance up the creek, to shoot wild fowl, was suddenly surprised by a couple of natives who were lurking behind some reeds; the man no sooner perceived them, than he begun to retreat, and they to advance, throwing off their cloaks, and with their arms in their hands; perceiving this, he turned and snapped his piece at one of them; but as it missed fire, he had no resource left, except flight, and which also would have been unavailing, had not his shouts for assistance, brought him timely aid. About two hours after this occurrence, as two of the people were employed in procuring firewood, in a small clump of trees, not far from the tent, two natives sprung towards them from behind the trees. These, however, on the men presenting their muskets at them, made signs of peace. Mr. Hume who was at hand now approached, when laying down his arms, and beckoning to the men to do the same, the natives followed the example, and after much conversation, but of which not a word was understood by either party, they proceeded with Mr. Hume to the tent. These people by degrees began to be a little better understood, when they seemed to wish to describe that a vessel had been in that bay, and that the people had landed;*

> *and to imply that both the master and the people were continually in a hurry. They also appeared to point out where the vessel lay, and suiting 'the action to the word,' endeavoured to explain that they had seen men felling trees in that direction, and this was all done with a gesture and grimace, evincing that these people were at least not bad mimics* (Andrews 1981, pp. 221–222, 265, 296; Hume and Hovell in Anon 2018).

Importantly, our people 'did not appear to be astonished at the sight of Hume and Hovell's horses and bullocks', clearly having sighted them previously – at least 80 feral cattle had been released at Sorrento 20 years earlier (Andrews 1981; Boys 1935). Our people also offered knowledge, disclosing to Hume that 'the harbour they called Geelong [*Djilang* – the downs, *Iramoo* [Werribee Plains], - and a remarkable high hill, on the downs, they informed me was called *Wilanmarnatar* [You Yangs]' (Andrews 1981, p. 219), a visit corroborated by Hovell in his diary (Andrews 1981).

On 20 May 1835 John Batman and his party sailed into Port Phillip Bay, recording that it was 'one of the finest basins of water I ever saw ... Just called upon the deck to see ~100 [Magpie] geese flying near the vessel' (Batman 1835a, 29 May; Billott 1979, pp. 101–102).

On 29 May 1834 a small band of Europeans, led by John Batman, a childhood friend of Hume, arrived on our shores in the *Rebecca* at (and named) Indented Head; this is the *Wadawurrung* 'Invasion Day'. Ironically, this is very close to the designated National Sorry Day on 26 May, observed annually in Australia by Aboriginal peoples. Uninvited, the newcomers assessed the western-perceived agricultural economic potential of the land while their dogs (*Canis lupus familiaris*) frolicked on the beach with our dingoes (*Canis lupus dingo*), before shooting one of our dingoes without our consent and without sensitive re-use of its carcass properties. Dingoes were not our pets, and we were not masters to them. Rather, they co-existed with us mutually respectfully as partners in the landscape.

Batman, through Bonwick's editorial, wrote of this encounter:

> *'The wind was light, and after some difficulty we got in [to Port Phillip Bay]; width about a mile and a-quarter; the depth five and a-half to seven fathoms of water. We got well into the port about ten o'clock, where the water is very smooth, and one of the finest basins of water I ever saw, and most extensive.'*
>
> *'As we were sailing up the port* [Phillip Bay], *heard a dog* [dingo] *on the shore howling; cannot think what brought it there.' Going twelve miles within the port to anchor they saw another dog on the shore. The fate of this animal was a sad omen of the fate of the native dog's aboriginal master from the coming of the white man. 'We put off',' says the journal, 'and came up to the dog, which proved to be a native dog of New Holland, which had surely left the natives within a day or two, he came quite close to my natives, and did not appear at all afraid, but would not suffer them to take hold of him. Our dogs, after some time, took after him, and run him into the water, where we shot him.*
>
> *Now they are after the poor dog's native friends, tracks of whom had been seen with the huts by the beach, where they had been eating mussels. This is the description of the country: 'Beautiful land, and all good sheep country, rather sandy, but the sand black and rich, covered with kangaroo grass about ten inches high, and as green as a field of wheat, beautiful plains, excellent land, &c.' Can we wonder that Mr. Batman's report of such a place should have sent the island settlers mad about it? Sealers and sailors had talked before, but now they had the evidence of such a bushman and farmer as John Batman himself.*

> *On the Saturday he has more to say in praise. 'Good hay could be made, and in any quantity. I never saw anything equal to the land in my life. I was never so astonished in my life.' On the Friday they had walked but twelve miles, but on the Saturday twenty. They had anchored about St. Leonard's, wandered over the Bellerine hills, and beheld the magnificent Geelong plains. Not astonishing that he said: 'From what I have seen I am quite delighted with Port Phillip'* [sic.] (Batman 1835a; Billott 1979, pp. 87–88; Bonwick 1868).

In the following week Batman and his party ventured along the ridges and shores of the Bellarine Peninsula and skirted around parts of the present Corio Bay and Geelong, before sailing northwards to present-day Melbourne into *Wurundjeri* and *Boon Wurrung* Countries, our neighbours to the north-east and east. Near Werribee River the party found 'seven large huts' and a group of 20 *Wadawurrung* women and 24 children. 'They had four native dogs and every woman has a load of 60 or 70 lbs. [27–31 kg] on her back … Each had two or three baskets, net bags, native tomahawks, bones, etc' (Batman 1835a, 31 May). Batman also observed an aquaculture system in Hovell's Creek, the width of 'fifty or sixty yards [45–55 m] … built of stones about four feet [1.2 m] high, and well done and well placed out. Two or three of these places following each other down the stream with gates to them, which they appear to stop with a bundle of rushes' (Batman 1835b, 1 June). Both diary entries demonstrate our residency in this Country, our fabrication skills, our aquaculture technology expertise and our buildings.

This visit was propitious because Batman (1835a, b; Attwood 2009; Blackman 1885–86; Brownhill 1955; City of Yarra 2018) entered into a treaty with the *Wurundjeri*, our neighbours, to purchase tracts of their Country and, more significantly, our Country unbeknownst to us, for:

> *… 40 pairs of blankets, 42 tomahawks, 130 knives, 62 pairs of scissors, 40 looking glasses, 250 handkerchiefs, 18 shirts, 4 flannel jackets, 4 suits of clothes and 150lb. of flour …*
>
> *Be it Remembered That on the day and Year within written possession and delivery of the Tract of Land within mentioned was made by the within named Jagajaga Jagajaga Jagajaga, Cooloolock Bungarie Yanyan Moowhip Mommarmalar Chiefs of the Tribe of Natives called Dutigallar Geelong to the within named John Batman by the said Chiefs taking up part of the Soil of the said tract of Land and delivering the same to the said John Batman in the name of the whole* (Batman 1835a, b; Attwood 2009; Blackman 1885-86; Brownhill 1955, pp. 4-7; City of Yarra 2018).

The *Wadawurrung* were not invited, we were not consulted, we were not even advised of this event and treaty-making – yet it was our Werribee Plains, Geelong regional landscape and our Bellarine Peninsula lands and waters that were 'bought' by this land grab.

In August 1835 surveyor John Helder Wedge also arrived at Indented Head. He met several members of the Batman party who were encamped there as well as members of the *Wadawurrung Bengalat balug* clan (Anderson 1970; Billott 1979). Wedge (1835, 7 August) recorded that:

> *I soon learnt the most friendly understanding exists with the natives … Although they brought home with them plenty of provisions, consisting of various edible roots, kangaroos, bats, and calkiet (young ants, in a fly state, taken from decayed hollow*

trees), they soon began to importune us for bread and other things, not excepting the cutlery. From this I inferred at once that, to satisfy their newly acquired taste for our food and other things which we brought with us, such as knives, tomahawks, and blankets, was a sure way to conciliate them.

Major Thomas Mitchell (1792–1855), reputedly the 'discoverer' of inland south-eastern Australia, sighted our Country while standing on *Wurundjeri* Country on top of *Gaboor* (Mount Macedon) on 30 September 1836 (Mitchell 1839; Stapylton 1971). He obtained this view after returning from the western 'pastures' that he had named 'Australia Felix' that comprised the Countries of the *Gulidjan*, *Tchap Wurrung* and *Gunditjmara* peoples.

As evidenced above, in each instance of *Wadawurrung* first encounters with Robbins, Flinders, Buckley, Wedge, Hume and Hovell and Batman, the *Wadawurrung* offered a 'Welcome' to these visitors, an opportunity for discussion and mutual understanding. Sometimes these communications were misinterpreted and treated with suspicion, evidenced in our *Wadawurrung* lack of English language fluency and the absence of any 'houses' or 'structures' or 'fences' as tangible representations of 'ownership', 'technological advancement' or 'culture', atypical of similar representations on European shores. But sometimes the communications were understood as offering food, knowledge and directional guidance.

Suddenly, in 1835, the shores and waters of *Nerm* (Port Phillip Bay) were enveloped in Europeans arriving *en masse* accompanied by hordes of guns, wagons, sheep, cattle, seeds, pet animals and the uncontrolled use of fire and axe. They came with a predilection to claim, conquer, fell, drain, fence and burn any western-perceived economically prosperous land in the belief that all was *terra nullius*. This European notion of *terra nullius* was, under English law, interpreted as land vested in the Queen of Great Britain, termed the 'Crown'. The definition was applied because the Crown's governance representatives interpreted that the Australian (and *Wadawurrung* Country) landscape possessed no distinct laws. Thus, a 'first in, first served' Western colonisation policy was applied. This Western policy assumed that because, in their eyes, no civilisation existed on these lands and waters there was no need to introduce, discuss, negotiate or purchase lands, waters and their resources from us, the *Wadawurrung* who had resided as a civilisation here under our own system of laws and custodial arrangements for over 40 000 years.

Clark claims that between 1837 and 1852 the *Wadawurrung* population fell from 275 to 30 people; a decline of 90% due to disease, conflict and the gun (Clark 1990). Many of the *Wadawurrung* peoples were also dispossessed of their Country and relocated from the suddenly European-transformed landscapes of Geelong to Countries to the north and west. Widespread conflict about 'whose land' led to the establishment of Aboriginal Protectorates (1835–49), Reserves (e.g. Mt Duneed (Turnbull 2009), Kargun, Steiglitz), Methodist, Presbyterian, Lutheran and Catholic church-led Missions (Buntingdale), and ration outstations at Stockyard Hill (1860–74) intended to offer protection and supplies to displaced *Wadawurrung* people (Clark 1990; Felton 1981). This Protectorate Reserve system was disbanded in 1849, after being considered 'unsuccessful' by colonial administrators. The formerly Aboriginal-dedicated lands and waters reserves were sold between 1901 and 1906, not re-vested to the *Wadawurrung* peoples. The *Wadawurrung* people living on the reserves were again relocated, this time to the Coranderrk Reserve on *Wurundjeri* Country on the upper edges of *Birrarung Marr* (Clark 1990).

Many of the *Wadawurrung* people were relocated to and lived on these Mission Stations, Reserves and government stations. A significant number of *Wadawurrung* successfully sought work on pastoral stations, or worked in industries such as fishing, goldmining and logging. *Wadawurrung* people also continued to quietly gather at camp sites on the outskirts of towns, and engaged in the new European sports of cricket, football and athletics.

While most of our *Wadawurrung* community died as part of this tangible and intangible colonisation process, several quietly merged into colonial pastoral society through employment or marriage, on-Country and off-Country, or were Country-displaced to the Protectorates like Coranderrk. Many secretly held onto and maintained language, narrative story-telling, nomenclature and environmental knowledge. These peoples are today's *Wadawurrung* descendants who are actively reviving language, culture, sovereignty and Country obligations, albeit modified within today's Western system of laws and land relationships.

Living in and on *Wadawurrung* Country

Before European arrival in our Country, our lifestyle and relationship to Country was totally different from the culture that we collectively live in today, as well as the way *Wadawurrung* peoples live and work today. Our present life and work patterns have been reshaped mediate by Australian culture together with its laws and protocols. That is not to say that we have cast aside our *Wadawurrung* laws and cultural protocols, but rather that our laws and cultural protocols have been substantially modified to work with the new culture that we co-exist in today.

Our *Wadawurrung* people had and continue to have a close connection to the coast and the sea. *Wadawurrung* Country includes the Sea Country, which is the sea and the coastal strip along the Port Phillip Bay and Bass Strait coastline that includes adjoining estuaries, beaches, coastal waters (up to ~148 km offshore), rivers and ocean. The *Wadawurrung* peoples make no distinction between lands, waters and seas. They see themselves as having responsibilities and custodianship across the lands, waters and sea boundaries that have been vested in their care for many thousands of years. The relationship between the *Wadawurrung* people and the Sea Country that goes back to Creation. The wetlands, estuaries, rivers, lakes and seas have sustained our people culturally and economically for tens of thousands of years. Responsibility to both Country and Sea Country form part of our society that embraces all aspects of social, legal, economic and religious life (Roös 2017).

The area of land over which the *Wadawurrung* peoples hunted and gathered food is today described by archaeologists as a 'stretch of country identifiable as the economic range, normally included the estate and was thus owned by clans [of the *Wadawurrung*]. The band seasonally occupied and utilised various parts of the range in a settlement pattern that was a response to the group's habitat' (Clark 1990, pp. 4–5).

The *Wadawurrung* travelled extensively in Country, moving with the seasons and following food resources while adhering to cultural laws and protocols. They moved from inland and the mountains across the plains to the seas, then back inland again. Temporary bark shelters were used for camping, but our people always returned to their permanent settlements with houses made of rock walls and mud roofs covered with bower spinach and other herbs (Tournier, pers. comm. 2014, recorded in Roös 2017, p. 103).

The *Wadawurrung* settlement patterns followed and respected food sources. Clans occupied a specific range over which they moved for gathering food, doing trading and fulfilling their social obligations. As indicated by *Tandop* Tournier at the introduction of this Chapter, movement of the clans was very much connected to the seasons, and the associated food. The *Wadawurrung* Country was a very important area that included gathering places for various Aboriginal groups from Victoria and beyond. Special gatherings and meetings occurred at sites near the towns now called Queenscliff and Geelong. Clans from the *Yari Yari*, *Bangerang*, *Baraba Baraba*, *Wiradjuri* and other tribes of the *Kulin* Nation were often invited via message sticks (the older version of mobile phones). They navigated over lands by foot and on canoes along the rivers down to *Nerm* (Port Phillip Bay), crossing the Bay before eventually meeting for weeks at designated sites to settle disputes, trade, arrange marriages, heal the sick and conclude with song and dance (Tournier, pers.comm. 2014, recorded in Roös 2017, p. 112).

Presland (2008) has observed the extensive burning regimes practised in the grasslands west of Melbourne, in adjacent *Wurundjeri* Country, 'in grassland areas, firing led to new growth, which attracted game and, carried out at the right time of year, promoted higher yields in some tuber-bearing plants' (Presland 2008, p. 119):

> *Regular burning by Aborigines almost certainly kept large parts of the Melbourne area, particularly to the west of the Maribyrnong River and in the north, clothed with a structure of grassland ... Similar results almost certainly defined smaller-scale changes such as the assisted growth of particular plants. The range of plants such as Murnong may have been extended to some degree but it still only occurred within the areas that provided its preferred habitat. It remained a plant of the plains and dry foothills; the actions of Aboriginal people did not lead to it growing in damp river valleys or swampy locations.*

Our landscapes were both our 'supermarkets' of food resources and our laws, cultural protocols and custodial relationships (Jones and Clarke 2018), and *Wadawurrung* language names for the most common species used are set out in Table 4.1 (WAC 2018b). The freshwater creek and wetland environs were home to animal species including the Growling Grass Frog (*Litoria raniformis*) and the Striped Legless Lizard (*Delma impar*), as well as permanent and migratory birds. Native riverine and wetlands vegetation along the creeks and around wetlands, including Red River Gum (*Eucalyptus camaldulensis*) and White Mangroves (*Avicennia marina*), provided important timber artefact and fire resources, and Short-finned Eel (*Anguilla australis*) habitat (McNiven and Bell 2010). Reptiles, including the Tiger Snake (*Notechis scutatus*), Eastern Blue-tongued Lizard (*Tiliqua scincoides scincoides*), Eastern Long-necked Turtle (*Chelodina longicollis*), Rakali (*Hydromys chrysogaster*) and Eastern Brown Snake (*Pseudonaja textilis*), were commonly observed and respected along creeklines and at wetlands. Native birds were prolific at these waterbodies, offering food, eggs and warnings. Particularly favoured and respected were the Pacific Black Duck (*Anas superciliosa*), Australian Pelican (*Pelecanus conspicillatus*), Australian White Ibis (*Threskiornis molucca*), Great Egret (*Ardea alba*), Native Companion (now known as the Brolga) (*Antigone rubicunda*) and Sulphur-crested Cockatoo (*Cacatua galerita*).

Inland, Eastern Grey Kangaroos (*Macropus giganteus*) and Swamp Wallaby (*Wallabia bicolor*) were abundant and prolific on the grassland plains and open woodlands, seeking shelter in the forests and water at waterholes and creeks. The rich riparian habitat and vegetation of creek valleys and wetlands provided bountiful food, fabrication and tool resources for *Wadawurrung* people. Vegetation, fresh water, fish, shellfish like oysters (*Saccostrea glomerata*) (Gill and Lane 1985) and mussels (*Mytilus planulatus*), Short-finned Eel (*Anguilla australis*), water birds and small mammals provided ready resources within this landscape. Accordingly, the spatial intersection of these plants and animals offered

Table 4.1. *Wadawurrung* language names for birds, mammals, fish, reptiles and food plants for the Geelong region

	Wadawurrung name	Common name	Scientific name
Birds	*Bundjil*	Wedge-tailed Eagle	*Aquila audax*
	Djirnap	Sulphur-crested Cockatoo	*Cacatua galerita*
	Gherin, Gherang or *Willam*	Yellow-tailed Black Cockatoo	*Calyptorhynchus funereus*
	Kawirr	Emu	*Dromaius novaehollandiae*
	Kunuwarra	Black Swan	*Cygnus atratus*
	Kuwarrk	Kookaburra	*Dacelo novaeguineae*
	Moodiwarr	Musk Duck	*Biziura lobata*
	Parwan	Magpie	*Gymnorhina tibicen*
	Porronggiti	Brolga	*Antigone rubicunda*
	Toolim	Pacific Black Duck	*Anas superciliosa*
	Waa	Crow/Raven	*Corvus coronoides*
Mammals	*Barnong*	Ringtail Possum	*Pseudocheirus peregrinus*
	Bu	Southern Brown Bandicoot	*Isoodon obesulus obesulus*
	Goim	Eastern Grey Kangaroo	*Macropus giganteus*
	Go-yin	Swamp Wallaby	*Wallabia bicolor*
	Mon.ngarrk	Echidna	*Tachyglossus aculeatus*
	Ngaambulmum	Koala	*Phascolarctos cinereus*
	Ngurrr-ngurr	Wombat	*Vombatus ursinus*
	Wad-dirring or *Perridak*	Platypus	*Ornithorhynchus anatinus*
	Walert	Brushtail Possum	*Trichosurus vulpecula*
	Yurn	Eastern Quoll	*Dasyurus viverrinus*
Fish	*Buniya*	Short-finned Eel	*Anguilla australis*
	Turrpurt	A small Trout	
	Ware-rap	Blackfish	
Reptiles	*Karn Kadak*	Snake	
Food plants	*Balout*	Cherry Ballart	*Exocarpos cupressiformis*
	Bal-yan	Bullrush or Cumbungi	*Typha latifolia*
	Banganga	Fringe lilies	*Thysanotus* spp.
	Bom	Grass lilies	*Caesia* spp.
	Bom	Chocolate Lily	*Dichopogon strictus*
	Burn-naa-look or *Mooee-yung*	Blackwood	*Acacia melanoxylon*
	Garra or *Jerring-up*	Golden Wattle	*Acacia pycnantha*
	Koon-yang or Gunyang	Kangaroo Apple	*Solanum laciniatum*
	Murnang	Murnong, Yam-daisy	*Microseris lanceolata*

Table 4.1. *(Continued)*

	Name not known	Spiny-headed Mat Rush	*Lomandra longifolia*
	Name not known	Austral Bugle	*Ajuga australis*
	Name not known	Orchids	*Orchidaceae* spp.
	Name not known, most lily-roots called *Bom*	Pale Vanilla-lily	*Arthropodium milleflorum*
	Nareem	Small-leaved Bramble or Native Raspberry	*Rubus parvifolius*
	Nareta	Ruby Saltbush	*Enchylaena tomentosa*
	Nurm-bal	Flax-lily	*Dianella* sp.
	Pike	Yellow Bulbine-lily	*Bulbine bulbosa*
	Polango Warangare	Water Ribbons	*Triglochin procera*
	Tark	Common Reed	*Phragmites australis*
	Tarook or *Dintini*	Small-leaved Clematis	*Clematis microphylla*
	Terraat	Native Geranium	*Geranium solanderi*
	Tharook or *Min nam-berung*	Pink Bindweed	*Convolvulus arvensis*
	Toolim	Pale Rush	*Juncus pallidus*
	Warour re rup or *Toolain*	Late Black Wattle	*Acacia mearnsii*
	Wurrak or *Wooriki*	Silver Banksia	*Banksia marginata*
	Yep-eurt or *Bom*	Milkmaids	*Burchardia umbellata*
Trees and plants	*Bar-rang* or *Woo-loitj*	Kangaroo Grass	*Themeda triandra*
	Bi-et-mai	Yellow Gum	*Eucalyptus leucoxylon*
	Biyal or *Bull*	River Red Gum	*Eucalyptus camaldulensis*
	Kokibainang	Swamp Wallaby Grass	*Amphibromus nervosus*
	Larrap	Manna Gum	*Eucalyptus viminalis*
	Ngarri	Drooping Sheoaks	*Allocasuarina verticilliata*
	Wangarra	Messmate	*Eucalyptus obliqua*

Source: Powell (pers. comm., 2018); WAC (2018b, pp. 5–20).

multiple 'supermarkets' and 'hardware stores' as well as sheltered, protected and defensive locations for the *Wadawurrung* to camp for seasonal or extended periods.

The coastal and riverine *Wadawurrung* clans had access to a wide variety of fish, including Mullet (*Mugilidae*), Whiting (*Sillaginodes punctatus*), Flounder, Flathead (*Platycephalus fuscus*), Trevally, Tommy-rough, Australian Grayling (*Prototroctes maraena*). These and many other species were seasonally speared and netted, particularly in the tidal flats and in saline estuaries. We loved our shellfish and seafood, including abalone (*Haliotis* sp.), Turbot (*Scophthalmus maximus*) and sand-loving pipi shells (*Paphies australis*). While all mammals were target species for food-resource hunting and artefact fabrication, there were cultural limits and protocols to harvesting so that resources were managed and sustained, not overharvested. There were also cultural protocols about which species, where, when and how it was harvested. Thus, birds and their eggs were harvested, along with lizards and insects, but in a considered manner to enable their successional growth and longevity, often indirectly ensuring population control, but also the continuance of their availability for our 'supermarket'.

The core *Wadawurrung* diet included a mix of terrestrial and aquatic plants and roots (Gott 1982, 1983, 2005; Gott and Conran 1991; Tournier 2015a; Zola and Gott 1992). The most important food included the Murnong (*Microseris lanceolata*), or 'Yam-daisy', a tuber that resembled a dandelion that possessed a rich starchy root, and was reliant upon active cultivation (agriculture) for continued growth (Gott 1983; Pascoe 2014). Thus, we actively cultivated favoured tracts of the lands to ensure the continued flowering and sensitive harvesting of this species. Other roots that we also harvested and cooked, often by boiling them in hot water, included potatoes, Milkmaids (*Burchardia umbellata*), Tarook (Blushing bindweed (*Convolvulus erubescens*)) and *Puewon* (Bulbine Lily (*Bulbine bulbosa*)). Our key agricultural tool was a sharp stick (*karni*) that was used to dig roots out, separate bulbs and aerate the soil. Other favoured food plants and fruits included watercress (*Boyungkaal* (*Nasturtium officinal*)), the Warrigal Cabbage (New Zealand Spinach, Captain Cook's Lettuce (*Terragonia tetragoniodes*)) and the Native Raspberry (*Rubus parvifolius*), and aquatic plants such as Water Ribbons (*Triglochin procera*) (Gott and Conran 1991). All of these you can still eat today, either fresh or as a flavour enhancer. Many of these species were commonly used, eaten and harvested by both *Wadawurrung* and Aboriginal peoples residing in or travelling through an area for food, and in the creation of weapons, clothing, shelter, and carrying vessels. In addition to terrestrial plants, animal and bird foods, freshwater and saline plants, crustaceans, fish and kelp provided a resource for food, medicines and artefacts, often resulting in extensive midden complexes that were regular 'food festival' locations (Clarke 2008; Pascoe 2003; Thurstan *et al.* 2018; Zola and Gott 1992).

Other plant species with resource potential included the Spiny-headed Mat-rush (*Lomandra longifolia*) and Coastal Sword Sedge (*Lepidosperma gladiatum*) whose strap-like foliage were harvested to make baskets and build fish-traps (Gott and Conran 1991). Kangaroo Grass (*Themeda triandra*) seeds were harvested and ripened in summer before being ground into a flour for the preparation of damper. Riceflower (*Pimelea spicata*) bark was stripped and threaded to fabricate fine nets for summer catches of Bogong Moths (*Agrotis infusa*) (Flood 1980). The starchy roots of Austral Bracken (*Pteridium esculentum*) were eaten raw or roasted, and prepared into a medicinal paste (Flood 1980; Gott and Conran 1991; Zola and Gott 1992).

Vegetation was a source of food, artefact fabrication materials and medicines (Clarke 2008). Blackwood (*Acacia melanoxylon*), a very hard wood, was highly favoured for spear-throwers and shields. Additionally, its bark was heated and infused in water to bathe rheumatic joints (Gott and Conran 1991). The top 50 cm of the stem of the Soft Tree-fern (*Dicksonia antarctica*), found in the moist Otway and Ballarat mountain forests and in riparian zones, was cut to scoop out its starchy pith for food, raw or cooked, and fern regrowth (Gott and Conran 1991). Many other plants and fungi also offered valuable food and medicine sources. For example, gum (*Eucalyptus* sp.) and tea-tree (*Melaleuca* sp.) leaves were often crushed and soaked in water to prepare medicinal ointments. Bowls and dishes were carved from tree bark and old gnarled growths, for food and water transportation. Canoes were fashioned from the bark of River Red Gum (*Eucalyptus camaldulensis*) trees, resulting in the characteristic visible modification of the trees that make them identifiable today as scarred or culturally modified. Other items such as boomerangs and spears were fabricated from the timber of *Eucalyptus* sp., Grasstree (*Xanthorrhoea australis*) and *Melaleuca* sp. species.

In terms of vegetation relevant to landscape design purposes for the Geelong region, that possesses *Wadawurrung* edible, medicinal, artefact fabrication or function purposes, *Tandop* David was often asked for guidance. Table 4.2 is his previously unpublished list that was used in the planting design for the REACH-Epworth Hospital Walkway project issued to Deakin University (Tournier with Jones 2015).

There were several semi-permanent settlement nodes within *Wadawurrung* Country, especially where food sources were readily available or trade gatherings could happen between *Kulin* Nation and or other tribes. There were special places for ceremonies, and routes that formed songlines. The *Wadawurrung* would commute to these nodes cyclically every second or third year, or from a message stick announcement or a unique cosmological or environmental event. The many layers of middens, full of shells, at these sites indicate that people were there for a very long time, coming back to and using the same place for thousands of years (Thurstan *et al.* 2018; Tournier, pers. comm. 2014, recorded in Roös 2017, p. 112). More permanently used campsites were established inland in areas such as

Table 4.2. *Wadawurrung* medicinal and food plants relevant for the Geelong region

	Common name	Scientific name	Relevance
Upper storey	Cherry Ballart	*Exocarpos cupressiformis*	Edible seeds; materials; medicinal
	Silver Banksia	*Banksia marginata*	Edible; artefacts; drink
	Lilly-pilly	*Acmena smithii*	Edible
	Blackwood	*Acacia melanoxylon*	Medicinal; materials
Middle storey	Kangaroo Apple	*Solanum laciniatum*	Edible seeds; medicinal
	Grass-trees	*Xanthorrhoea* sp.	Edible; materials
Groundcovers	Murnong	*Microseris lanceolata*	Edible roots
	Pale Vanilla-lily	*Arthropodium milleflorum*	Edible roots
	Milkmaids	*Burcharia umbellata*	Edible roots
	Chocolate Lily	*Dichopogon strictus*	Edible roots
	Native Geranium	*Geranium retrorsum*	Edible; medicinal
	Coast Wattle	*Acacia sophorae*	Edible seeds; medicinal
	Pigface	*Carpobrotus* sp.	Edible; medicinal
	Warrigal Cabbage	*Tetragonia tetragoniodes*	Edible leaves; medicinal
	Carex sp.	*Carex tereticaulis*	Materials
	Spreading Flax-lily	*Dianella revoluta*	Materials
	Sea-berry Saltbush	*Rhagodia candolleana*	Edible
	Ruby Salt-bush	*Enchylaena tomentosa*	Edible
	Common Sneezewood	*Centripeda cunninghamii*	Medicinal
	River mint	*Mentha australis*	Medicinal
	Small-leaved Clematis	*Clematis microphylla*	Edible; medicinal
	Australian Hollyhock	*Lavatera plebeia*	Medicinal; materials
	Austral Bracken	*Pteridium esculentum*	Edible

Source: Tournier with Jones (2015).

the You Yangs or along the Barwon or Moorabool river corridors. These campsites often included the construction of structures of stone and mud, with sticks on top woven together, closed with bark and mud on top, but still permitting the use of fire within and the release of smoke. Mud roofs allowed for the planting of herbs and plants to provide additional food, such as Bower Spinach (*Tetragonia implexicoma*). Along the coastlines, dwellings were not permanent. Small campsites included shelters made of bark, sticks and seaweed (Thurstan *et al.* 2018), or just the physical formation of a half-circle of brushes sourced from close-by bush and shrubs, to protect against the wind. Movement of *Wadawurrung* along the coastline happened during summer months and, because it was warm with very little rain, it was not necessary to build shelters (Su 2014; Tournier pers. comm. 2014, recorded in Roös 2017, p. 103).

Six principles of *Wadawurrung* culture

The following are six principles for understanding *Wadawurrung* culture. They are not all relevant principles, but they are the ones that we are commonly asked to explain.

Nation and clan

Around the Geelong region, the *Bengalat balug* clan looked after much of the Bellarine Peninsula, while the *Wada wurrung balug* clan looked after much of the metropolitan Geelong region and the Barrabool Hills.

The *Wadawurrung* are part of the *Eastern Kulin* Nation that includes the *Wurundjeri* (northern Melbourne and Yarra Valley region), *Boon Wurrung* (southern Melbourne to Wilsons Promontory including the Mornington Peninsula), the *Wadawurrung* themselves, the *Dja Dja Wurrung* (of the Central Highlands–Bendigo region) and the *Taungurong* (of the Broken, Delatite, Coliban, Goulburn and Campaspe rivers upper catchments) peoples. These communities are tied together as a Nation (or confederation) by way of related language dialects, common Dreaming stories and songlines, moiety and pre-colonisation martial arrangements, and resource-sharing arrangements and protocols (Doolan 1979).

Before European invasion, it has been estimated that these Countries hosted some 60 000–100 000 Aboriginals, but by the late 1830s the population was estimated to be under 20 000. The population decline was due to disagreements, massacres and killings (Clark 1995), physical displacement to Mission Stations and Protectorates, starvation, removal of traditional food sources and ability to care for and harvest resources, and incarnation.

Wadawurrung Country, within the *Kulin* Nation, stretched from the edges of Fiery Creek and Beaufort, along the Great Dividing Range down to the Werribee River corridor, through the waters of Port Phillip Bay and Bass Straits, from Aireys Inlet inland to Winchelsea and across the grasslands and saline lakes to Cressy and Skipton. Figure 4.1 depicts this overall *Wadawurrung* Country, as well as the predominant residence locations of our clans within Country. Clan nomenclature and their approximate primary custodial tracts of lands and waters are listed in Table 4.3 (revised from Clark 1990; CoGG 2018).

The basic unit of *Wadawurrung* social organisation was the clan, a group based on kinship through the male line with a shared religious identity (Barwick 1984). A clan was like a family group within the overall Country and often comprised ~20–100 individuals. These clans today make up the extended *Wadawurrung* family united in common Country responsibilities, language, moieties and laws. The clan was a lands/waters-caring unit of people, not a land-owning unit in the western definitions, whose territory (clan-Country) was defined and determined by ritual responsibilities (thus laws and protocols) not by fences

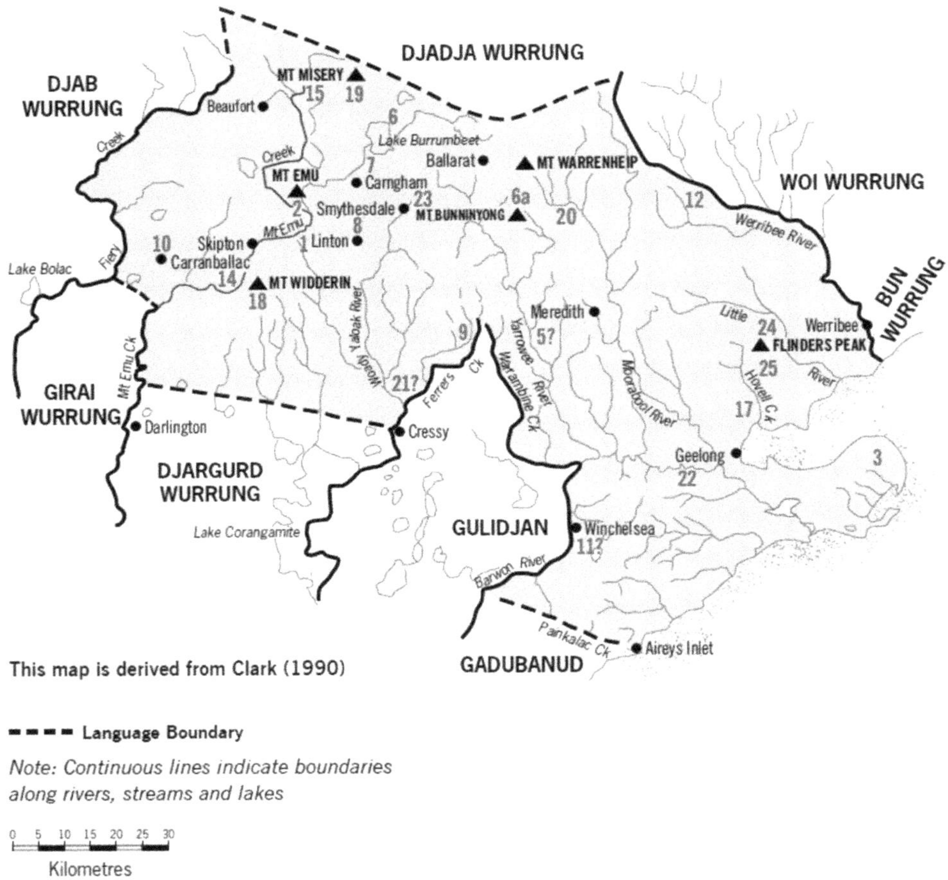

Fig. 4.1. *Wadawurrung* Country. Source: http://www.geelongaustralia.com.au/common/public/images/inpage/kaap02.jpg.

or defined landmarks (Barwick 1984). These common spiritual identities resulted in larger tribal groups also being intimately interconnected. The basis of the *Wadawurrung* religious identity is totemism, which can likened to the Dreamtime – the time of creation when our Creation Beings being roamed the land, creating people and naming the animals and plants.

Country

Non-Indigenous anthropologist Deborah Bird Rose (1996) has suggested that:

> *In Aboriginal English, the word 'Country' is both a common noun and a proper noun. People talk about Country in the same way that they would talk about a person: they speak to Country, sing to Country, visit Country, worry about Country, grieve for Country and long for Country. People say that Country knows, hears, smells, takes notice, takes care, and feels sorry or happy. Country is a living entity with a yesterday, a today and tomorrow, with consciousness, action, and a will toward life. Because of this richness of meaning, Country is home and peace: nourishment for body, mind and spirit; and heart's ease.*

Table 4.3. *Wadawurrung* Country clans and nomenclature

	Clan name	Approximate location
1	*Barere barere balug*	Skipton region
2	*Beerekwart balug*	Mount Emu region
3	*Bengalat balug*	Indented Head region
4	*Berrejin balug*	Unknown
5	*Boro gundidj*	Yarrowee River region and corridor
6	*Burrumbeet balug*	Lakes Burrumbeet and Learmonth region
6a	*Keyeet balug*	Mount Buninyong region
7	*Carringum balug*	Carngham region
8	*Carininje balug*	Linton region
9	*Corac balug*	Rokewood and Kuruc-a-ruc Creek region
10	*Corrin corrinjer balug*	Carranballac region
11	*Gerarlture balug*	West of Lake Modewarre region
12	*Marpeang balug*	Blackwood, Myrniong and Bacchus Marsh region
13	*Mear balug*	Unknown
14	*Moijerre balug*	Mount Emu Creek region and corridor
15	*Moner balug*	Trawalla and Mount Emu Creek region
16	*Monmart balug*	Unknown
17	*Neerer balug*	Between Geelong and the You Yangs (Hovells Creek) region
18	*Pakeheneek balug*	Mount Widderin region
19	*Peerickelmoon balug*	Near Mount Misery
20	*Tooloora balug*	Mount Warrenheip, Lal-Lal Creek, west branch of Moorabool River region
21	*Woodealloke balug*	Woady Yaloak River, south of Kuruc-a-ruc Creek region
22	*Wada wurrung balug*	Barrabool Hills region
23	*Wongerrer balug*	Headwaters of Woady Yaloak River region
24	*Worinyaloke balug*	West side of Little River region
25	*Yaawangi balug*	You Yang Hills locality

Source: Clark (1990), revised by Powell and Jones in 2018.

The Dreaming is the intellectual Country of Indigenous peoples and their communities. The Dreaming is both the ancient time of creation and the present-day reality of Dreaming. The word 'Dreaming' is not a universally agreed word for this content across Australia. The physical manifestation of that Country is the Australian landscape – night sky, day sky, land surface, below land, its waters, and the air.

Country holds special meaning for the *Wadawurrung* people. Everything in our Country is alive and everything is embodied in relationships. Thus, the past, present and future are one, and the spiritual and physical worlds of Country interact. The Dreaming is an ongoing celebration of and reverence for past events: the creation of the land, the creation of law, the creation of people. Stories have been given to and vested in Aboriginal peoples from the Dreaming, everything comes into being through the expression or journey through story or stories, and the Dreaming is the story of our Creation Beings. All things exist eternally in the Dreaming; the Dreaming is alive. An individual is born to

Country, not just in Country but from Country, and his or her identity is inextricably and eternally linked to the Dreaming (Powell 2019; Powell and Jones 2019).

Wadawurrung Country, and our sovereignty of this Country, continues today, mirrored in the recent *Uluru Statement from the Heart* (Referendum Council 2017):

> *Our Aboriginal and Torres Strait Islander tribes were the first sovereign Nations of the Australian continent and its adjacent islands, and possessed it under our own laws and customs. This our ancestors did, according to the reckoning of our culture, from the Creation, according to the common law from 'time immemorial', and according to science more than 60 000 years ago.*
>
> *This sovereignty is* ***a spiritual notion: the ancestral tie between the land, or 'mother nature', and the Aboriginal and Torres Strait Islander peoples who were born therefrom, remain attached thereto, and must one day return thither to be united with our ancestors*** [Creation Beings]. ***This link is the basis of the ownership of the soil, or better, of sovereignty.*** *It has never been ceded or extinguished, and co-exists with the sovereignty of the Crown.*
>
> *How could it be otherwise? That peoples possessed a land for sixty millennia and this sacred link disappears from world history in merely the last two hundred years?*
>
> *We stress the above text in bold.*

Within and interconnecting Country are songlines. Songlines are the pathways that our ancestors travelled along in creating the overall landscape and Country, and are recalled and celebrated in our stories and songs (Chatwin 1987). The pathways included numerous points connected to these stories and songs. The songlines can be site- or Country-specific, or Nation-level traversing across our Country and one or more neighbouring Countries, or regional-level traversing across several Countries across (principally) south-eastern Australia (Neale 2017).

Language and name

The *Wadawurrung* language is dynamic. In our eyes '*Wadawurrung*' is English-written as '*Wadawurrung*' and not '*Wada wurrung*' or '*Wathaurong*' (Powell 2019). It is our meaning, it is our library of law and 'history', it explains or identifies place(s) or resources, and enables daily communication (Clark and Heydon 2002). It is, however, not a written static language that can be bound into a single dictionary. Numerous colonial and ethnographic attempts to transcribe and record our words and language, like most attempts around Australia, were influenced by sound and phonetic interpretations of what recorders though they heard as well as a lack of clarity as to the actual meaning of words and terms (Blake *et al.* 1998; Cary 1899; Dawson 1881; Mathews 1904; Smyth 1878, 1972; Thomas 2014, Jones and Clarke 2018).

Wadawurrung language is intertwined with our Country, and culturally they cannot be separated (Tournier 2015b). To speak or sing our language is to respect and nurture and celebrate our Country.

Because our language, like other Aboriginal languages around Australia, is dynamic, it has a fluidity and changeability. For example, while words may be common, our conversations may vary the use and scope of words. Thus, one 'Welcome to Country' text, like the *Tandop* David text above, is but one version at one point in time for one audience. It is not generically used but can evolve and be modified according to the context and audience.

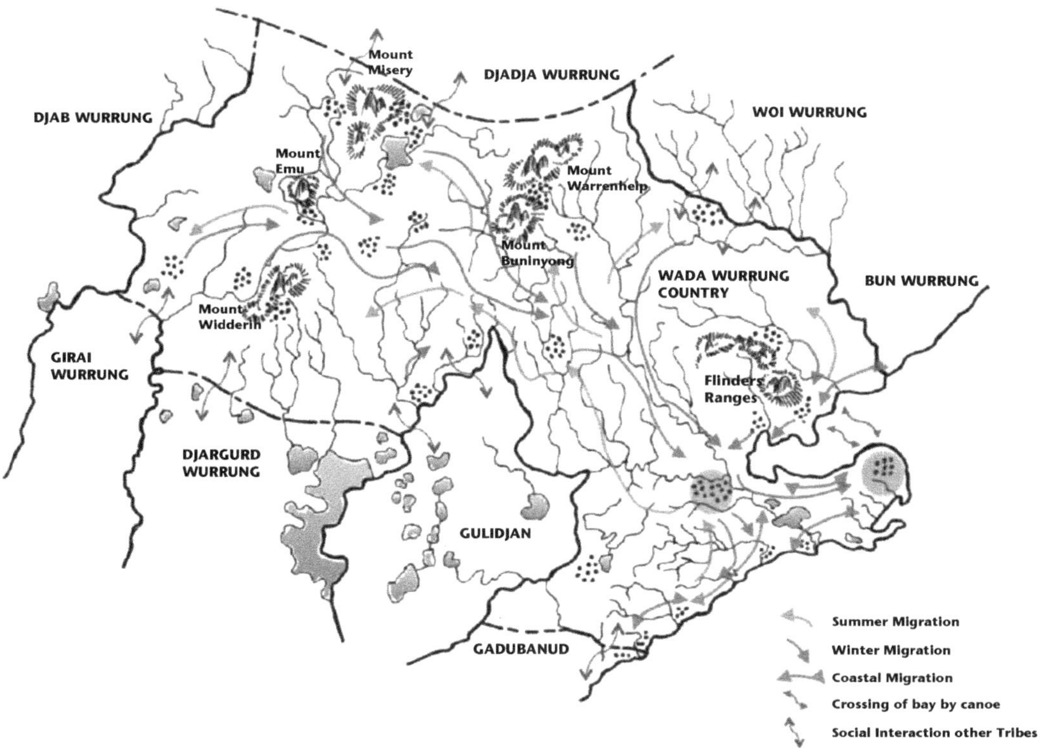

Fig. 4.2. Possible *Wadawurrung* seasonal movement patterns. Source: Roös (2017, p. 109).

Thus, language, word(s), 'Welcome to Country' text and 'Acknowledgement to Country' text are not fixed or static but are dynamic, changeable and malleable to the particular context and circumstance. Specifically, it is and was oral and thus amenable to western linguistic text conformity and encasement and Country-specific.

Seasons

Historically, as a people we seasonally moved across our Country accordingly to specific environmental patterns and clues, and established camps according to the cycles of nature and resource available according to our seasonal calendar (Clarke 2009).

While the European calendar possesses four seasons (Summer, Winter, Spring, Autumn), these involve static northern hemisphere time sectors (Roös 2015). In contrast, an Aboriginal season calendar is dynamic, and can consist of five to nine obvious seasons characterised by weather and environmental attributes.

These seasonal cycles and our *Wadawurrung* calendar signalled various human movement changes including the construction of temporary and semi-permanent shelters, the management of the landscape through fire, the harvesting of plant roots, and the migration along coastal areas and inland. Figure 4.2 indicates the potential migration of some clan *Wadawurrung* people across the *Wadawurrung* Country, following the seasons in a cultural landscape.

For us, in this part of the *Wadawurrung* Country, there were most likely six seasons understood and respected in the pre-invasion days (Fig. 4.3). *Tandop* David has

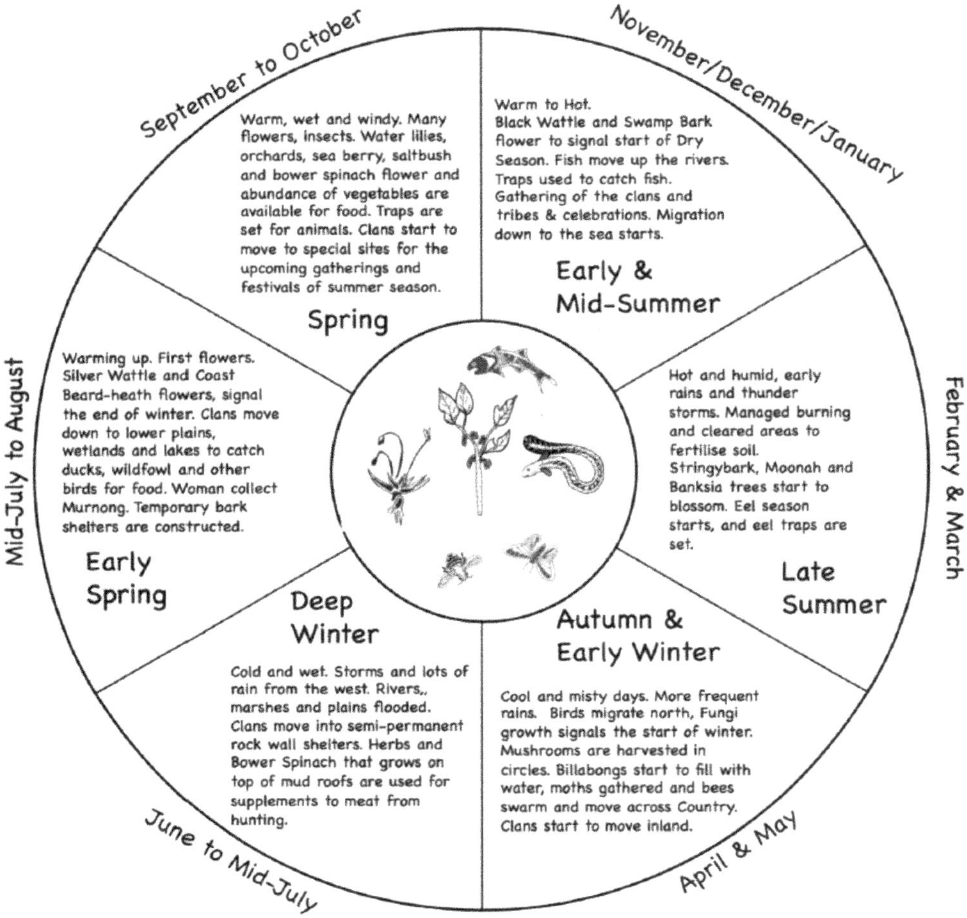

Fig. 4.3. Seasons of the Geelong region *Wadawurrung* peoples. Source: *Tandop* David gathered and synthesised in Roös (2017, p. 112) revised by Powell and Jones in 2018.

explained his version of these six seasons as possessing the characteristics set out in Table 4.4.

The main *Wadawurrung* settlement nodes and coastal migration followed the seasons and food sources (bush tucker/bush tukka), as summarised in Table 4.5.

Law, moiety and 'ownership'

Wadawurrung society is divided into two groups, called moieties, each with specific Totemic Beings. Each *Wadawurrung* person belongs to one or the other of the two moieties: *Bunjil* the Eaglehawk (*Aquila audax*) or *Waa* the Crow (*Corvus coronoides*) (Barwick 1984). Clan moiety in *Wadawurrung* society is patrilineal, determined by male ancestry lineage. For *Wadawurrung* individuals and their clan, the totem is a symbol of membership of that clan and their overall community. Serving notionally like a familial surname, it is a societal structure recognised by members of this clan and those of other clans in that

Table 4.4. Seasons of the Geelong region *Wadawurrung* peoples

Time period	Environmental patterns and characteristics
Early and mid Summer (November, December, January)	The Black Wattle tree (*Acacia mearnsii*) and Swamp Paperbark (*Melaleuca ericifolia*) flower in late November, signalling the beginning of the dry season, prompting people to set up camps around reliable water sources at creeks, rivers and billabongs. Due to the dry season, waterbodies attract wildlife and Eastern Grey Kangaroos (*Macropus giganteus*), Emus (*Dromaius novaehollandiae*), Common Wombats (*Vombatus ursinus*) and Ring-tailed Possums (*Pseudocheirus peregrinus*) that come to drink and can be caught. Grasses flower, fruits ripen and manna is collected from the coast Manna-Gum (*Eucalyptus viminalis*). Fish move up the rivers, and fish traps are used. During this time of abundance, large gatherings of the clans and tribes take place. With permission from the Elders of the coastal clans, people move down to the sea and coastal wetlands and river mouths to gather shellfish, to catch Flathead (*Platycephalus fuscus*), Flathead Galaxias (*Galaxias rostratus*) and Black fish (*Gadopsis marmoratus*). When male Short-finned Eels (*Anguilla australis*) start to move down the rivers to the sea, it signals the start of the next season, late Summer.
Late Summer (February to March)	With the early Autumn rains, the days become cooler. Various parts of the land are densely covered in scrub and grass and, by planned and carefully managed burning, areas are cleared to fertilise the soil before Autumn rains. During this time the Messmate Stringybark (*Eucalyptus obliqua*), Moonah (*Melaleuca lanceolata*) and Coast Banksia (*Banksia integrifolia*) trees come into blossom, providing sweet nectar that attracts many insects and birds. The Coast Saltbush (*Atriplex cinerea*) on the coastal dunes provides fruits for mixing into meals. The Short-finned Eel (*Anguilla australis*) season starts with the migration of large quantities of female eels down the rivers to the sea, to begin their long journey up the east coast of Australia to the South Pacific. Eel traps are set, and the *Wadawurrung* clans set up camps along the rivers. The roots of water plants are part of a vegetable and fruit diet that includes the berries from late Summer fruits. The start of early Winter is marked by birds starting to migrate to the warmer north, and more frequent heavy rains.
Autumn to early Winter (April to May)	The soil is still warm after the rains, and the mushrooms sprout in circles at Freshwater Creek. Large groups of women, young and the Elders sit inside the circles and harvest the mushrooms, while talking about their travels, confirming between them the Dreaming stories of the past and future, endorsing the importance of caring for Country. The billabongs fill with water after more frequent rains. Moths emerge and native bees swarm across Country collecting nectar from the Flax-lily Grass (*Dianella caerulea*) and Cranberry Heath (*Astroloma humifusum*). The clans start to move inland to higher ground, as the rain of Winter starts to settle in.
Deep Winter (June to mid July)	Winter rains result in frequent flooding of the flats near the rivers and creeks, while low lands and salt marshes are mostly wet and unsuitable for camping. Therefore, people move uplands to higher sheltered spots. Ring-tailed Possums (*Pseudocheirus peregrinus*), Koalas (*Phascolarctos cinereus*), Common Wombats (*Vombatus ursinus*) and Eastern Grey Kangaroos (*Macropus giganteus*) are also seeking shelter on higher ground, and people can catch them close to the campsites. Herbs and the Bower Spinach (*Tetragonia implexicoma*) growing on top of the mud roofs of the semi-permanent rock-wall shelters are convenient for food supply, while people stay inside the shelters and keep fires burning for warmth. Possum skins stitched together are wrapped around family members to protect against the intense cold spells that occur as the westerly winds rip through the Country ahead of the cold fronts.

Table 4.4. *(Continued)*

Early Spring (mid July to August)	The Golden Wattle (*Acacia pycnantha*), Silver Wattle (*Acacia dealbata*) and Coast Beard-heath (*Leucopogon parviflorus*) start to flower, the first signs of a change in season. Winter is coming to an end and warmer weather is on its way. People move towards the lower lands and plains, wetlands and lakes where they catch ducks, wildfowl and other birds. Eggs are in abundance and are collected from nests, including from turtles along the beach. While the men hunt, the woman collect large quantities of Murnong (*Microseris lanceolata*) in string bags to take back to the camps. Temporary bark shelters are constructed in camp clusters along the watercourses, dry wood is collected and fires are kept burning during the cold nights to keep warm.
Spring (September to October)	The Murnong (*Microseris lanceolata*) continues to flower, and provides an abundance of root vegetables. Water lilies (*Nymphaeaceae* sp.), orchards, Sea Berry (*Hippophae rhamnoides*), Coastal Saltbush (*Atriplex cinerea*) and Bower Spinach (*Tetragonia implexicoma*) flower and large amounts of greens are used in food. This is a time of abundance; trees, grasses and shrubs flower, and bees and insects collect nectar. Snakes and lizards become active, and young kangaroos come out of the pouch. The floodplains are flooded and tadpoles appear in ponds, billabongs, rivers and lakes. Traps are set and tadpoles, snakes and lizards are roasted on the coals of the circle ovens in the ground. The different clans start to move towards the special sacred places for the annual gatherings of the tribes that will happen during the coming Summer season.

Source: *Tandop* David gathered and synthesised in Roös (2017, pp. 109–111) revised by Powell and Jones in 2018.

Country. The totem has strong territorial and mythological associations and obligations, and it is believed that the totem can warn an individual of approaching danger.

The *Wadawurrung* and *Kulin* Nation share the patrilineal form of moiety system. The *Kulin* Nation social world was divided into the *Waa* (Crow) or *Bunjil* (Eaglehawk) moieties (Clark 1990). Marriage was always across the moieties, with a *Waa* person having to marry a *Bunjil* person, preferably from a distant clan group (Barwick 1984). In some instances, members of the *Wadawurrung* clans intermarried with the matrilineal clans of the *Gulidjan, Djab wurrung* and the *Djargurd wurrung* (*Gunditjmara*), although in many instances these practices ended in inter-clan hostility.

The *Wadawurrung* peoples were divided into 25–26 clans with group sizes of 40–60 people, each of which was responsible for a specific area of land within the wider *Wadawurrung* territory. Each clan occupied a distinctive geographical area and belonged to one of the two moieties. *Wadawurrung* groups that belong to the *Waa* moiety include the *Beerekwart balug* (Mount Emu), *Bengalat balug* (Indented Head region), *Carninje balug* (Linton region), *Corrin corringer balug* (Carranbullac region), *Moner balug* (Trawalla and Mount Emu Creek region), and the *Tooloora balug* (Mount Warrenheip, Lal Lal Creek, west branch of Moorabool River region) (Clark (1990; Presland 2010).

Clans of the *Bunjil* moiety include the *Burrumbeet balug* (Lakes Burrumbeet and Learmonth region), *Keyeet balug* (Mount Buninyong), *Marpeang balug* (Blackwood, Myrniong, Bacchus Marsh region), *Moijerre balug* (Mount Emu Creek region), *Peerickelmoon balug* (near Mount Misery), *Wadawurrung balug* (Geelong–Barrabool Hills region) and the *Wongerrer balug* (head of Woady Yallock Creek). Clans of unknown moiety association

Table 4.5. *Wadawurrung* settlement areas and bush tucker (food sources)

Landscape type / vegetation community	Indicator species (*Wathaurong* name in brackets)	Season
High ground and mountains	Wombat (*Gnor-Gnor*)	All year
	Wattle grubs (*Wooeekam*)	Late Summer
	Tree sap running	All year
	Fungi/mushrooms (*Tubar*)	Autumn/Winter
	Manna (*Laap*) – gum tree	Autumn/Winter
Open areas and plains	Kangaroo (*Goim*)	All year
	Fungi/mushrooms (*Tubar*)	Autumn/Winter/Spring
	Grass flowers	All year
	Tree sap running	Late Summer
	Lerp and manna (*Laap*)	After rain
	Honey ants	All year
	Salt bush	
Coast and river mouths, sandhills	Lizards (*Gaan*)	Summer
	Snakes (*Kangalang*)	Summer
	Wattle grubs (*Wooeekam*)	Spring/Summer
	Eggs (including turtles)	Early Spring
	Eels (*Kooyang*)	Late Summer
	Salt bush	All year
	Oyster (*Barrabool*)	Spring/Summer
	Mussel (*Moorabool*)	Spring/Summer
	Flathead fish	Spring/Summer
	Black fish	Spring/Summer
Waterways, riverbanks, wetlands	Frogs (*Djerrm*)	Summer
	Snakes (*Kangalang*)	Summer
	Lizards (*Gaan*)	Spring/Summer
	Ducks and swans	All year
	Wattle grubs (*Wooeekam*)	Spring/Summer
	Tree sap running	All year
	Eggs (including turtles)	Early Spring
	Eels (*Kooyang*)	Late Summer
	Grass flowers	Spring/all year
	Yam-daisy (*Murnong*)	Spring/Summer
Mallee	Emu (*Karwir*)	All year
	Wattle grubs (*Wooeekam*)	Spring/Summer
	Tree sap running	All year
	Lerp and manna (*Laap*)	Summer/Winter
	Mallee fowls/turkey	All year
	Grass flowers	Spring/all year
Bush and scrub	Possums (*Wallart*)	All year
	Lizards (*Gaan*)	Summer
	Snakes (*Kangalang*)	Summer
	Wattle grubs (*Wooeekam*)	Spring/Summer
	Tree sap running	All year
	Lerp and manna (*Laap*)	Summer/Winter
	Galahs/pigeons/birds	All year

Source: *Tandop* David, pers. comm. 2014. The language included is the version used by *Tandop* David. Revised by Powell and Jones in 2018.

include the *Barere barere balug* (Skipton region), *Borogundidj* (Yarrowee River region), *Carringum balug* (Carngham region), *Corrac balug* (Rokewood and Kuruc-a-ruc Creek region), *Gerarlture* (west of Lake Modewarre), *Neerer balug* (between Geelong and the You Yangs), *Pakeheneek balug* (Mount Widderin region), *Woodealloke gundidj* (Woady Yallock River, south of Kuruc-a-ruc Creek), *Worinyaloke balug* (west side of Little River) and the *Yaawangi* (You Yang Hills) (Clark 1990, but names and places revised by Powell and Jones in 2018).

'Ownership' is a western term associated with land and or objects. Today one obtains 'ownership' of land or object by economic purchase to gain legal rights to development and or to use the land or object. Thus, under western legal interpretation, one can 'own' land and have control over that asset.

In contrast, the *Wadawurrung* believe that 'land' (= Country) 'owns' the people and all animals and resources thereon, below and above. Additionally, specific responsibilities associated with Country (including lands, waters, objects, animals, fish, stories, laws etc.) obligate the *Wadawurrung* people to manage these lands, waters and their resources for and on behalf of our Creation Beings in anticipation of their return.

In this sense, 'ownership' is not vested in us as an individual, as a clan or as a people. Rather, it is vested in the land or object which we (as an individual, clan or people) are culturally responsible and obligated to care for according to the relevant cultural laws and moiety arrangements. That care places a cultural law obligation to monitor, care for, look after, respect (a place, an animal, a songline, a story or song) upon an individual, clan and/or community. In this sense, there is no certificate of title, contract of law or purchase invoice that shows 'ownership'. Instead, there are inherited bloodline stories or songs, birthright vested role(s) or Eldership(s).

Respect

A key attribute in *Wadawurrung* culture is respect. Respect relates to people, places, animals and vegetation, as well as to our Elders. Remember, we have been the managers of this Country for over 40 000 years, and we hold the longevity of historical memory, records and environmental knowledge of place within our Country.

We ask for respect for this knowledge, and understanding that it exists but may not be comprehensive or accessible knowledge. We ask to be consulted, listened to and respected about place and this knowledge.

Twelve places on *Wadawurrung Country*

The following are 12 places that embody and or explain *Wadawurrung* culture and Country. They are not all relevant places, but they are the ones that we are commonly asked to explain.

Murnong

The Murnong (*Microseris scapigera*) plant was extensively present throughout *Wadawurrung* Country (Frankel 1982; Gott 1983, 1982; Johnson 2017; Walsh 2016). The word '*Murnong*' means 'thumb' in *Wadawurrung* language, referring to the plant's series of bulbous root tubers (Powell, pers. comm. 2019). It is graphically embodied in the *Wadawurrung* (Wathaurung Aboriginal Corporation) logo because it was a key staple vegetation food source across our Country. Generations of women and children passively cultivated this species, together with many other food-source tuberous and flowering plants species to provide daily and or seasonal food supplies. This passive cultivation, or agriculture, included plant splitting and sowing, soil tilling and aerating, vegetation litter burning and selective harvesting mediated by selective herbivores such as Eastern Grey Kangaroo (*Macropus giganteus*), Swamp Wallaby (*Wallabia bicolor*) and Emu (*Dromaius novaehollandiae*) which browsed on foliage and flower tips, pruning them to stimulate seasonal plant growth.

The image by colonial surveyor John Helder Wedge (1793–1872) of *Wadawurrung* women harvesting Murnong (*Microseris scapigera*) depicts a practice that occurred historically, continuously and regularly across our Country.

With the advent of European pastoralism, this agricultural regime and the health of these plant species declined. Cessation and prevention of traditional agricultural practices stifled plant species growth, uncontrolled firing killed swathes of plant patches, and rampant sheep and cattle grazing of the sweet flowers, foliage and stalks to the ground surface negated their regeneration. Dispossession of *Wadawurrung* people from their Country meant removal from our farmlands and the cessation of the regular plant-raising practices on which the species depended for maintaining its growth and succession.

Indented Head

Indented Head is the first place of invasion of our Country. We associate 29 May 1835 – not 26 January – as our 'Invasion Day' because this was the day that Europeans set foot, uninvited, on our Country, wandered on our Country, let their dogs (*Canis lupus familiaris*) frolic on our beaches, and shot one of our dingoes (*Canis lupus dingo*).

In Western histories, the day is celebrated as the day on which John Batman and his band of explorers 'discovered' the Bellarine Peninsula, and ventured across its shores and meadows before setting northwards to the present site of Melbourne on *Wurundjeri* Country.

The present stone cairn at Indented Head 'celebrates' that event. It offers one interpretation of the event, one version of history – but the *Wadawurrung* see the memorial and the date differently.

Yollinko Park

Nestled on a deep bend on the Barwon River near Highton is the presently named Yollinko Park. To the *Wadawurrung* this was a gathering place, a plant and aquatic food resource sharing point, a crossing point on a trail from Corio Bay to the Barrabool Hills, and place where middens from food sharing were constructed and burial mounds located. Willem Baa Nip ('King Billy') was the last survivor of the *Wadawurrung balug* clan who would have lived with his family at this place before European settlement (Dearnaley 2015).

This Aboriginal living station, on the confluence of Kardinia Creek and the Barwon River, was a major Winter *Wadawurrung balug* and *Wadawurrung* camping place. There is archaeological evidence of at least 5000 years of occupation (Roberts 1993) – 780 artefacts were recovered from the mounds. Lane observes that:

> *This camping station had been the winter resort for the …* [*Wadawurrung balug* clan] *when they returned from their summertime bivouacs along the Bass Strait … The Wadawurrung retired to Kardinia Creek in late autumn, the bulk of the protein needs would have been satisfied by the wood grubs from under the bark of the river red gum [Eucalyptus camaldulensis] and or the possums* [Common Ring-tailed Possum *(Trichosurus vulpecula)*] *in the hollows of their branches, along with all the*

birds which frequented the environs. The Wadawurrung would stay in this location all winter or at least until the migratory snipe [Latham's Snipe *(Gallinago hardwickii)*] *which came jinking at full moon in August / September, in such vast flocks they could be heard approaching. Such a glut of food would be provided by the snipe that a meeting of clans would convene for a corroborree* (Lane 1991b, p. 7).

Johnstone Park

Notionally, Johnstone Park in Geelong today comprises an area of ~2.4 ha bounded by Railway Terrace, Gheringhap Street, Fenwick Street and Little Malop Street. The Park was 'developed' in the late 1860s and opened officially on 11 March 1872. The railway extension from Geelong to Colac in 1875 divided the Park into two segments, of which only the eastern segment remains today as the Park. The Park was renovated to a Percy Everett design in 1916 although the design was only partially realised, and has been subject to several recent conservation assessments (Jones and Whitehead 2005; Lovell Chen 2016; O'Connor 2016).

On 27 April 1872 the *Geelong Advertiser* reported on the origin of 'Maiden's Walk' in Johnstone Park. The Walk creation involved three Aborigines of the 'Barrabool tribe' – 'Harry, Billy [Willem Baa Nip] and Jenny' – who each planted a Sheoak (*Allocasuarina* sp.) in the park (Clark 1990, p. 305; Jones and Whitehead 2005; Billot 1969).

Before invasion this was a waterhole juncture for two rivulets that feed along today's Gordon Avenue into Manifold Heights, and the *Worrowing Willam* rivulet that drained southwards before exiting into Corio Bay. Superintendent of the Port Phillip District, Charles La Trobe, selected the site for 'reserves for Ornamental and Public Garden Grounds' when he visited Geelong in February 1848. La Trobe also ordered the creek to be earthen levy-dammed as a water supply for horse and bullock teams in the area of the Mercer, Malop and Gheringhap Streets intersection, and provided a level roadway where the gully intersected the main road to Melbourne and the west. The dam leaked and was abandoned; the area was known as the Dam Reserve for some time.

Willem Baa Nip

Today, Willem Baa Nip (1836–85) (also known as 'William Gore' and 'King Billy'), erroneously called the 'last of the Barrabool tribe' (*Wada wurrung balug* clan), is buried in the Western Cemetery in Manifold Heights (Brownhill 1955; Massola 1969; Ryan 2016). A memorial plaque in Little Malop Street recalls him, a painted wall mural celebrates him, a cemetery headstone on his grave paid for and erected by then Mayor of Geelong Robert de Bruce Johnstone (1884–1944) recalls his memory (Billot 1969), and the recently named Baanip Boulevard in Waurn Ponds/Armstrong Creek respects his original clan Country (Milligan 2005a, b; Wardle 2008a, b).

Willem Baa Nip was born in 1836 on the banks of a lagoon, probably the former wetland that now forms Johnstone Park, on the edge of the *Worrowing Willam* rivulet, and died in Geelong Hospital in 1885. Baa Nip means 'Bunyip' in *Wadawurrung* language. On the day Willem Baa Nip was born his father saw a Bunyip in a waterhole, and called Willem Baa Nip after this apparition and place. This 'naming' was a normal *Wadawurrung* human naming process – *Wadawurrung* human names derive from a place, an event, a sighting at a point in time. They are not European patrilineal names. Willem Baa Nip was well respected and a learned man, often wore a coat with australite buttons (Clarke 2018), and was a recognised spokesperson and negotiator for his clan. After Geelong's settlement Willem Baa Nip could be seen wandering daily from a reserve at Mount Duneed to Corio Bay to North Shore and back.

You Yangs

To the *Wadawurrung*, the name *You Yangs* derives from an abridgement of *Wurdi Youang* that means 'big hills' (Rowe and Jacobs 2013; Tournier 2014a). The highest peak, Station Peak (or Flinders Peak or *Wooloomanata*), was known as *Wurdi-yawang*, meaning 'large hill'. Clark (1990) perceived that the You Yangs are within the clan territory of the *Yawangi baluk* centred on the You Yangs, and the *Worinyaluk baluk* on the west side of the Little River (*Worinyaluk*) (Clark 1990; Doce and Clark 2014; Smyth 1878). Clark also records that there is a lack of Western clarity as to the origins and meanings of the name You Yangs; that You Yangs is a variant of *Yawang*; the variant spellings of *Yaughon* (Todd 1835) and *Youang* (Smyth 1878), that were translated as 'big or large hill' or 'bald' or simply 'hill' (Clark and Heydon, 2002). Robinson recorded '*yow wong*' as the *Woi wurrung* word for 'granite stone'.

In 1969 Aldo Massola published a narrative of a car journey to Victoria's primary Aboriginal heritage sites. He included the following discussion of the You Yangs:

> *In his Journal, Flinders ...* [he] *could they find any water. Although he did not know it, water was quite close to him on the Peak. It was in one of those strange water wells hollowed in granite, a cavity developed over the centuries, through the impurities in the hard rock being first softened and then washed away by the action of the elements ... The basin measures forty-nine inches by forty-two inches and is twenty-eight inches deep. Consequently it holds rain water well into the dry season ... they must have been used by the Aborigines, since a number of microliths and other small stone implements have been picked up near them. Given time, these shallow depressions would all have become water storage basins. The Aborigines that roamed over and around the You-Yangs were the Yawangi ... Upon the plains, in the early days, there was an abundance of game and edible roots of many kinds and everywhere there was surface water ... Like all natives, however, they preferred to camp on a sandy spot ... It is a sandy patch, or 'blow,' with an area of between four and five acres, situated on the south-west of one of the low undulations so prevalent on these plains, ... the entire surface of the blow is almost covered with chips, flakes, and broken stones of all description, mixed with hundreds of implements, thus proving the site's suitability ... It is known that Frederick Armytage and his wife, who then owned* [Wooloomanata] *station, took a kindly interest in the natives, especially in Billy Leigh, who was 'king' of the Yawangi at the time. Billy was baptized and confirmed in the Church of England at Lara, and when he died, in August 1912, the Armytages erected an upright cross on his grave in the Eastern Cemetery* (Massola 1969, pp. 11–13).

Waarden's (1986) archaeological report details the presence of the *Yawangi baluk* over much of the park's area (Tudehope 1962b; Waarden 1986). The research documents four rock well sites used by the *Yawangi baluk*, and at each well site archaeological evidence points to regular habitation there.

Dog Rocks

Dog Rocks, in Batesford, is a special place to the *Wadawurrung* for quarrying greenstone for implements (Brumm 2010; McBryde 1978, 1984a, b; Massola 1969; Roberts 2017). The *Wada wurrung balug* clan of the Barwon River riverine corridor and Barrabool Hills was responsible for its management.

While Mount William near Lancefield is today nationally recognised (and before European colonisation was recognised by the wider *Kulin* Nation) as a key greenstone quarry for south-eastern Australia, the Dog Rocks quarry provided a similar resource for the *Wadawurrung* and a different type of greenstone stone. The Dog Rocks stone was quarried, cut, traded across the *Kulin* Nation and fabricated into cutting and hunting tools. It was a common artefact for any *Wadawurrung* family because of its sharp cutting properties.

The Anakies

The Anakies, or the Three Sisters, is a place of cultural significance to the *Wadawurrung* (Ryan 2016; Tournier 2010a; 2011b; 2014b). The thee scoria hills, culturally strategically positioned between the Anakie Gorge pathway and the You Yangs, are both a place-specific songline as well as part of a longer songline story.

Anakie Youang consists of the three peaks known as the Three Sisters, being *Coranguilook*, *Baccheriburt* and *Woollerbeen*. *Coranguilook* is Mount Anakie itself. *Coranguilook*, *Baccheriburt* and *Woollerbeen* collectively comprise *Anakie Youang*, meaning the 'little hills' (the *Wurdi Youang* [You Yangs] are 'big hills'). Archaeoastronomers studying *Wurdi Youang* have concluded that three stones at *Wurdi Youang* mimic the crest of each sister, *Coranguilook*, *Baccheriburt* and *Woollerbeen* (Norris *et al.* 2013).

Of '*Anikai Youwan*', the field journals of John Helder Wedge record on 29 August 1835 that:

> *We passed over extensive open plains about a mile, a slight rise on our right for the 1st six mile which turn to the left and run to the foot of Anikai Youwam – at the foot of which rise I went for about 4 miles intending to ascend one of the eminences which for the cluster of Hills [Anikai Youwan] but learning from the two native youths that were with me that there was no water in that direction. I altered my course and passed over a ridge of hills which run from Mt Youwan to open plains continue nearly to the ridge they are quite destitute of timber and covered with grass but not of that luxuriance I should have expected from the soil. The hills are thinly timbered, some are gum trees* (Wedge in Ryan 2017, pp. 19-20; Wedge 1835–36).

Wurdi Youang

Wurdi Youang, on the flank of the You Yangs, remains an enigmatic site of cultural significance to the *Wadawurrung* as well as the larger *Kulin* Nation peoples. The name *Wurdi Youang* for the stone arrangement is not its original *Wadawurrung* name, but has been designated that by contemporary Aboriginals because of its geographical proximity to *Wurdi Youang* (the 'big hills' of the You Yangs). Comprising an oval-shaped stone arrangement, aligned to certain topographical and astronomical features and events, it is today a designated State Heritage Place. It has attracted considerable attention from the international and national archaeoastronomical discipline as being a unique place because of its significant astronomical values (Hamacher 2012; Norris *et al.* 2013; WAC 2018a).

To the *Wadawurrung* it is also a unique culturally special place, which we know is very sensitive to human, animal and fire modification. It is thus a place to which we wish to restrict visitor access.

McLeod's Waterholes

McLeod's Waterholes near Drysdale was an important meeting and food resource gathering and sharing place for the *Bengalat balug* clan of the *Wadawurrung*, and for the community as a whole (O'Connor 2017). There are significant burial sites of the *Bengalat balug* clan surrounding the permanent potable waterholes. This location was unique because of its sandy soils and sand dunes, which meant that its reliable food resources (grasses, tubers, etc.) were different from elsewhere on the Bellarine Peninsula. Long forgotten until anthropologist Louis Lane 'rediscovered' it in the 1950s, it is a rich wildlife haven as well as hosting several middens and grave mounds.

Armstrong Creek Valley

One of the favoured encampment areas of Willem Baa Nip and other members of the *Wada wurrung balug* clan following European colonisation was the vales of Armstrong Creek. While they could wander into the dusty streets of early Geelong, they often preferred Living Stations (semi-permanent encampments rich with terrestrial and aquatic food and potable water resources) like Jerringot, Yellingko (Dearnaley 2015), Buckleys Falls, and the vales of Armstrong Creek, Freshwater Creek, Breamlea Creek and the riverine corridors of the Barwon and Moorabool rivers. Armstrong Creek was a regular travelling corridor for the *Wadawurrung* commuting from inland Living Stations to coastal venues. The vales comprised an expansive set of *Eucalyptus* spp.-dominated grassy meadows often with favoured Kangaroo Grass (*Themeda australis*), Murnong (*Microseris scapigera*), Drooping Sheoak (*Allocasuarina verticillata*), River Red Gum (*Eucalyptus camaldulensis*), Small-leaved Clematis (*Clematis microphylla*), Black-anther Flax-lily (*Dianella revoluta*), Golden Wattle (*Acacia pycnantha*) and Black Wattle (*Acacia mearnsii*), that were shared with kangaroos (*Macropus* sp.), possums, emus, quolls and a wide spectrum of birds, reptiles, insects, eels (*Anguilla australis*) and fish (Lane 1991b; Rhoads 1986; Turnbull 2009).

Clark (1990) concluded that around 1839 there were some 300 members of the *Wada wurrung balug* in the region, but by 1853 the clan had been reduced to some nine women, seven men including Willem Baa Nip, and one child (Davenport in Sayers 1969). In the 1850s they regularly camped among the grassy River Red Gum woodlands of 'Dooliebeal' (now Stewarts Reserve in Warralily) on allotments tenured by settler John Armstrong. Reputedly, John Stewart either purchased or leased this tract of land and 'donated' it to the *Wada wurrung* where our 'dispossessed could camp unmolested (CoGG 2018; Rowe and Huddle 2000; Turnbull 2009, p. 26). The *Wada wurrung balug* peoples were also 'in the habit of camping in the Feehan's Lane area, in the lee of Mount Duneed before and after the land was sold' in 1858 (Lane 1991b).

On 10 June 1861, after 'Dooliebeal' was 'closed, the colonial government official reserved 1 acre (0.4 ha) for the 'Mount Duneed Aboriginal Reserve as a site for shelter hut for Aborigines in the Geelong District' to facilitate the regulation that all Aboriginals 'had to leave town every evening at sundown for their camps at Mount Duneed' (Barker 2014; Bullers 2014; Clark 1990; Felton 1981; Clark 1990; Milligan 2005a; Turnbull 2009). This reservation was revoked in 21 February 1907 (Turnbull 2009).

This Reserve was the last 'residence' of Dan Dan Nook (*d.*1870) ('King Jerry') and Willum Baa Nip (*d.*1885). Robert Johnstone 'at his own expense had several huts built on the southern slope of Mt Duneed, and sent them regular supplies of food, baskets, etc and paid a local doctor to go out and see this camp to do what he could to help the sick one' (Lane 1991b). Of Waurn Ponds and these vales, Lane observes that:

> *This place the Native TRIBES-People called 'EUROKE' because of the sandstone through which WAURRN Chain-of-Ponds had carved a gorge. The abrasive rock they used for grinding the greenstone, epidiorite, from Ceres into hatchet-blades. Country*

people who lived on the land, there, at the turn of the century still call the place WURRN PONDS: and 'WURRN' means a cluster of Aboriginal shelter-humpies. Traces of occupation are to be found along the margins of the stream (Lane 1991b, *p. 2).*

Lake Modewarre

This place is called Lake Modewarre today. *Modewarre* means 'home of the musk duck' (*Biziura lobata*) (Powell 2015g). There is no water in it today because farming practices have dammed the creeks that once fed the lake, stopping the water from flowing. It was once a place of great activity and an abundant source of food for the *Wadawurrung* people.

In Uncle Bryon's words:

> *Before settlement, before whitefellas came here this place was full of Musk Ducks [Biziura lobata]. Plenty of meat, plenty of eggs, plenty of feathers to trade, but it was also a place where there was plenty of* [Short-finned] *eels* [*Anguilla australis*]. *So once a year the Wadawurrung baluk and the Gerarlture balug people would come here, and they would catch eels, smoke them, and trade the smoked eels with their neighbours. And those eels would be traded across to the Bellarine, across to the Barrabool Hills, and down as far as Coogoorah* [meaning 'swampy reed creek'] *down past Anglesea. It was a good industry. There's no water there today* (Powell 2015g).

Between 1803 and 1832 William Buckley described the lake as being perfectly fresh and abundant in Short-finned Eels (*Anguilla australis*), with the lake surrounds having well-established and luxuriant foliage. River Red Gums (*Eucalyptus camaldulensis*) and other trees provided excellent shelter (Bonwick 1856). A journal entry by squatter Dr Thomson in 1846 recorded his observation that this was the first time he had seen the lake dry (Wynd 1971, 1992). During the summer of 2004–05 fish (carp) and eels died *en masse* in Lake Modewarre, although no dead or ill trout or minnow were observed. On 31 January 2005 a resident saw dead eels between the north rocks and the boat harbour, and it was estimated that there were 50 dead eels for every 100 m of shoreline. The eels were described as large, with blood flush around the jaw. In mid February 2006 a similar but larger eel death incident occurred, with numerous sizes of eels observed to be dead or dying. It appeared as if they were trying to escape from the water, travelling about 1 m up the shoreline then returning to the water's edge. Many of the eels were observed to be bleeding from the gills and had large white lesions on their bodies (EPA 2007).

A final thought

As a final thought, we acknowledge that Europeans attempted to purchase our Country in 1837, that we have experienced the ravages and effects of the biological and military invasion and colonisation of our lands and waters, and that we have consistently sought to engage in and offer the opportunity for discussions and negotiations towards a mutually beneficial scenario in which both our worlds can occupy and care for our Country (CCMA 2018).

What we wish for is recognition of our past, existing and future sovereignty of this landscape we call *Wadawurrung* Country, and to entertain constructive and equal discussions as to how to manage and curate these lands and waters so that their qualities are not compromised, that past and current degradation activities are redressed and mediated, and that we collectively move forward in crafting a society that respects the qualities, narratives and cultural richness of these lands and waters we call *Wadawurrung* Country.

'We all bleed red, we all have hearts. It's just a matter of talking and working things out,' *Tandop* David told the *Geelong Advertiser* at a reconciliation event in 2015.

References

Anderson H (1970) The place for a township: John Helder Wedge at Port Phillip in 1835. *Victorian Historical Magazine* **41**(160), 351–359.

Andrews A (Ed.) (1981) *Hume and Hovell: 1824*. Blubberhead Press, Hobar.

Anon (2014) *Nyernila: Listen Continuously – Aboriginal Creation Stories of Victoria*. Arts Victoria, Melbourne.

Anon (2016) Geelong mourns the passing of Wathaurong Elder. *Geelong Advertiser*. http://www.geelongadvertiser.com.au/news/geelong/geelong-mourns-passing-of-wathaurong-elder/news-story/434d7d0ad626d981bca1974f323d33ad (accessed 1 March 2018).

Anon (2017) In memory of Unc, *Wathaurong Aboriginal Co-operative Community Newsletter* February–March, 22–23.

Anon (2018) *Hume & Hovell: Expedition to Port Phillip c1824*. https://poi-australia.com.au/hume-hovell-expedition-to-port-phillip/ (accessed 1 March 2018).

Attwood B (2009) *Batman's Treaty and the Matter of History*. Miegunyah Press, Melbourne.

Barker M (2014) *Proposed Rezoning at 35 Hams Road and 151–229 Anglesea Road, Waurn Ponds: Cultural Heritage Assessment*. Benchmark Heritage Management, Melbourne.

Barwick D (1984) Mapping the past: an atlas of Victorian clans 1835–1904. *Aboriginal History* **8**, 100–131.

Bass W (1895) Bass's voyage in a whaleboat. *Historical Records of New South Wales* **3**. https://archive.org/stream/historicalrecor05walegoog/historicalrecor05walegoog_djvu.txt

Batman J (1835a) *John Batman's Journal*, 31 May 1835. State Library of Victoria, Melbourne.

Batman J (1835b) *The Batman Geelong Treaty*, 6 June 1835. State Library of Victoria, MS13485. http://ergo.slv.vic.gov.au/explore-history/colonial-melbourne/pioneers/batmans-treaty (accessed 1 March 2018).

Billot CP (1969) *The Life of our Years: In and Around Geelong – A Chronological History*. Lothian Publishing, Melbourne.

Billot CP (1979) *John Batman: The Story of John Batman and the Founding of Melbourne*. Hyland House, Melbourne.

Bird ECF (1993) *The Coast of Victoria: The Shaping of Scenery*. Melbourne University Press, Melbourne.

Blackman J (1885–86) The locality of Batman's treaty with the Port Phillip natives. *Transactions of the Historical Society of Australasia* **1**, 45–57.

Blake BJ, Clark ID, Krishna-Pillay SN (1998) *Wathawurrung*: the language of the Geelong–Ballarat area. In BJ Blake (Ed.) *Wathawurrung and the Colac Language of Southern Victoria* (pp. 59–154). Research School of Pacific and Asian Studies, Australian National University, Canberra.

Bonwick J (1856) William Buckley: the wild white man and his Port Phillip black friends 1817–1906. Geo Nichols, Melbourne. www.lib.monash.edu.au/exhibitions/history/xaustcat.html (accessed 1 March 2018).

Bonwick J (1868) *John Batman: The Founder of Victoria*. Samuel Mullen, Melbourne. Reproduced in http://gutenberg.net.au/ebooks13/1303761h.html (accessed 1 March 2018).

Boyce J (2013) The Newer Volcanics Province of south-eastern Australia: a new classification scheme and distribution map for eruption centres. *Australian Journal of Earth Sciences* **60**(4), 449–462. doi:10.1080/08120099.2013.806954.

Boys RD (1935) *First Years at Port Phillip: Preceded by a Summary of Historical Events from 1788*. Robertson & Mullens, Melbourne.

Briggs C (2008) *The Journey Cycles of the Boonwurrung: Stories with Boonwurrung Language*. Victorian Aboriginal Corporation for Languages, Melbourne.

Briggs C (2010) *Boonwurrung*: the filling of the bay – the time of chaos. In *Indigenous Creation Stories of the Kulin Nation* (pp. 8–11). Arts Victoria, Melbourne.

Briggs C (2014) *The Journey Cycles of the Boonwurrung: Stories with Boonwurrung Language,* 2nd edn. Victorian Aboriginal Corporation for Languages, Melbourne.

Brownhill WR (1955) *The History of Geelong and Corio Bay. Geelong Advertiser*, Geelong.

Brumm AR (2010) The falling sky: symbolic and cosmological associations of the Mt William greenstone axe quarry, Central Victoria, Australia. *Cambridge Archaeological Journal* **20**(2), 179–196. doi:10.1017/S0959774310000223

Bullers R (2014) *Spring Creek Precinct Structure Plan, Torquay, Victoria: Aboriginal Heritage Impact Assessment.* Ecology and Heritage Partners Pty Ltd, Geelong.

Cary JJ (1899) *Vocabularies of the Geelong and Colac Tribes, Collected in 1840* (pp. 1–36). Australian Association for the Advancement of Science.

CCMA (2018) *Aboriginal Heritage Handbook.* Corangamite Catchment Management Authority, Colac, Vic.

Chatwin B (1987) *The Songlines.* Jonathan Cape, London.

CoGG (2018) *Stewarts Reserve.* City of Greater Geelong. https://www.geelongaustralia.com.au/parks/item/stewarts.aspx (accessed 1 March 2018).

City of Yarra (2018) *The Aboriginal History of Yarra: 4. Treaty, The 'Batman Land Deed', between John Batman and Aboriginal 'Chiefs' of the Kulin Nation, Port Phillip Area, 1835.* Museum Australia, Accession no. MS6832. http://aboriginalhistoryofyarra.com.au/4-treaty/ and http://ergo.slv.vic.gov.au/image/batman-geelong-treaty (accessed 1 March 2018).

Clark ID (1990) *Aboriginal Languages and Clans: An Historical Atlas of Western and Central Victoria, 1800–1900.* Department of Geography and Environmental Science, Monash University, Melbourne.

Clark ID (1995) *Scars in the Landscape: A Register of Massacre Sites in Western Victoria, 1803–1859.* Australian Institute of Aboriginal and Torres Strait Islander Studies, Canberra.

Clark ID, Heydon T (2002) *Dictionary of Aboriginal Placenames of Victoria*. Victorian Aboriginal Corporation for Languages, Melbourne.

Clarke PA (2008) Aboriginal healing practices and Australian bush medicine. *Journal of the Anthropological Society of South Australia* **33**, 3–38.

Clarke PA (2009) Australian aboriginal ethnometeorology and seasonal calendars. *History and Anthropology* **20**(2), 79–106. doi:10.1080/02757200902867677

Clarke PA (2018) Australites. Part 1: Aboriginal involvement in their discovery, *Journal of Astronomical History and Heritage* **21**(2/3), 116.

Dawson J (1881) *Australian Aborigines: The Languages and Customs of Several Tribes of Aborigines in the Western District of Victoria, Australia*. George Robertson, Melbourne.

Dearnaley J (2015) Restoration of Yollinko Park. Unpublished SRD768 Landscape Masterclass project. School of Architecture and Built Environment, Deakin University, Geelong.

Doce E, Clark ID (2014) The You Yangs Regional Park. In ID Clark (Ed.) *An Historical Geography of Tourism in Victoria, Australia: Case Studies* (pp. 181–205). Walter de Gruyter GmbH & Co KG. https://www.degruyter.com/viewbooktoc/product/430898 (accessed 1 March 2018).

Dodson JR (1989) Late Pleistocene vegetation and environmental shifts in Australia and their bearing on faunal extinctions. *Journal of Archaeological Science* **16**, 207–217. doi:10.1016/0305-4403(89)90066-6

Dodson JR (2001) Holocene vegetation change in the Mediterranean-type climate regions of Australia. *The Holocene* **11**(6), 673–680. doi:10.1191/09596830195690

Doolan JK (1979) Aboriginal concept of boundary: how do Aboriginals conceive 'easements'. How do they grant them? *Oceania* **49**(3), 161–168.

Eidelson M (2014) *Yalukit Willam: The River People of Port Phillip*. City of Port Phillip, Melbourne.

Ellis VR (1981) *Trucanini: Queen or Traitor?* Australian Institute of Aboriginal Studies, Canberra.

EPA (2007) *A Review of Historic Western Victorian Lake Conditions in Relation to Fish Deaths*. Environmental Protection Authority, Melbourne. https://www.epa.vic.gov.au/our-work/publications/publication/2007/march/1108 (accessed 1 March 2018).

Felton P (1981) Victoria: lands reserved for the benefit of Aborigines 1835–1971. In N Peterson (Ed.) *Aboriginal Land Rights: A Handbook* (pp. 168–228). Australian Institute of Aboriginal Studies, Canberra.

Flannery T (Ed.) (2002) *The Life and Adventures of William Buckley; 32 Years a Wanderer amongst the Aborigines of the then Unexplored Country around Port Phillip, now the Province of Victoria*. Text Publishing, Melbourne.

Flinders M (1966) *Voyage to Terra Australis: April 26–May 2, 1802*. Libraries Board of South Australia, Adelaide.

Flinders M (1978) *Observations on the Coasts of Van Diemen's Land, Bass Strait and New South Wales*. Libraries Board of South Australia, Adelaide.

Flood JM (1980) *The Moth Hunters: Aboriginal Prehistory of the Australian Alps*. Australian Institute of Aboriginal Studies, Canberra.

Flood JM (1995) *Archaeology of the Dreamtime*. Angus & Robertson, Sydney.

Frankel D (1982) An account of aboriginal use of the yam-daisy. *Artefact* 7, 43–45.

Gammage B (2011) *The Biggest Estate on Earth: How Aborigines made Australia*. Allen & Unwin, Sydney.

Gill ED, Lane LN (1985) Sea levels and Aboriginal oyster middens at Lake Connewarre, Victoria, Australia. *Proceedings of the Royal Society of Victoria* **97**(2), 95–100.

Gott B (1982) Root use by aborigines of southern Australia. *Archaeology in Oceania* **17**(1), 59–67. doi:10.1002/j.1834-4453.1982.tb00039.x

Gott B (1983) Murnong – *Microseris scapigera:* a study of a staple food of Victorian Aborigines. *Australian Aboriginal Studies* **2**, 2–18.

Gott B (2005) Aboriginal fire management in south-eastern Australia: aims and frequency. *Journal of Biogeography* **32**(7), 1203–1208. doi:10.1111/j.1365-2699.2004.01233.x

Gott B, Conran J (1991) *Victorian Koorie Plants: Some Plants used by Victorian Koories for Food, Fibre, Medicines and Implements*. Yangennanock Women's Group, Aboriginal Keeping Place, Hamilton, Victoria.

Gunditjmara People with Wettenhall G (2004) *The People of Budj Bim: Engineers or Aquaculture, Builders of Stone House Settlements and Warriors defending Country*. EmPress, Ballarat, Victoria.

Hainsworth D (1971) *The Sydney Traders*. Cassell, Melbourne.

Hamacher D (2012) Wurdi Youang: the latest research. *Australian Indigenous Astronomy*. http://aboriginalastronomy.blogspot.com/2012/11/wurdi-youang-latest-research.html (accessed 1 March 2018).

Holdgate GR, Wagstaff B, Gallagher SJ (2011) Did Port Phillip Bay nearly dry up between ~2800 and 1000 cal. yr BP? Bay floor channelling evidence, seismic and core dating. *Australian Journal of Earth Sciences* **58**(2), 157–175. doi:10.1080/08120099.2011.546429

Johnson D (2017) *Plants in Focus: Microseris walteri Murnong*. Friends of the Geelong Botanic Gardens, Geelong. http://www.friendsgbg.org.au/uploads/images/Plants-in-Focus/Microseris%20walteri.Murnong.Sep%202017.1.9.a.pdf (accessed 1 March 2018).

Jones DS, Clarke PA (2018) Aboriginal culture and food–landscape relationships in Australia: Indigenous knowledge for Country and landscape. In J Zeunert and T Waterman (Eds) *Routledge Handbook to Landscape and Food* (pp. 41–60). Routledge, London.

Jones DS, Roös PB, Dearnaley J, Threadgold H, Nicholson M, Wissing R, Berghofer D, Buggy R, Low Choy D, Clarke PA, Serrao-Neumann S, Kitson G, Ryan S, Powell B, Powell G, Kennedy MG (2018) ReCrafting urban climate change resilience understandings: learning from Australian Indigenous cultures. In *Biophilia Smart Resilience: Proceedings of the 55th International Federation of Landscape Architects World Congress 2018*, 18–21 July 2018, Marina Bay, Singapore. http://www.ifla2018.com/

Jones & Whitehead (2005) *Johnstone Park Conservation Management Plan*. Jones & Whitehead Pty Ltd, Melbourne.

Keble RA (1947) History and geology of mud islands. *Memoirs of the Natural Museum of Victoria* **15**, 131–138. https://museumsvictoria.com.au/media/5709/jmmv19471509.pdf (accessed 1 March 2018).

Kerwin D (2012) *Aboriginal Dreaming Paths and Trading Routes: The Colonisation of the Australian Economic Landscape*. Sussex Academic Press, Brighton, UK.

Labilliere FP (1878) *The Early History of the Colony of Victoria*. Sampson & Low, London.

Lambeck K, Chappell J (2001) Sea level change through the last glacial cycle. *Science* **292**(5517), 679–686. http://science.sciencemag.org/content/292/5517/679 (accessed 1March 2018.

Lane LN (1991a) Yollinko Park. Unpublished paper.

Lane LN (1991b) Education in Geelong & District, Victoria – History. Unpublished paper.

Lovell Chen (2016) *Heritage Impact Statement: Johnstone Park, 24–28 Gheringhap Street, Geelong*. Lovell Chen Architects & Heritage Consultants, Melbourne.

Low Choy D, Clarke PA, Jones DS, Serrao-Neumann S, Hales R, Koschade O (2013) *Aboriginal Reconnections: Understanding Coastal Urban and Peri-urban Indigenous People's Vulnerability and Adaptive Capacity to climate Change*. National Climate Change Adaptation Research Facility, Gold Coast, Queensland.

Markgraf V, Dodson JR, Kershaw AP, McGlone MS, Nicholls N (1992) Evolution of late Pleistocene and Holocene climates in the circum-South Pacific land areas. *Climate Dynamics* **6**, 193–211. doi:10.1007/BF00193532

Massola A (1968) *Aboriginal Place Names of South-east Australia and their Meanings*. Lansdowne Press, Melbourne.

Massola A (1969) *Journey to Aboriginal Victoria*. Rigby, Melbourne.

Massola A (1971a) *Bibliography of the Victorian Aborigines: From the Earliest Manuscripts to 31 December 1970*. Hawthorn Press, Melbourne.

Massola A (1971b) *The Aborigines of South-eastern Australia: As They Were*. William Heinemann, Melbourne.

Mathews RH (1904) Ethnological notes on the Aboriginal tribes of New South Wales and Victoria, Part 1. *Journal and Proceedings of the Royal Society of New South Wales* **38**, 71–106.

McBryde I (1978) *Wil-im-ee Moor-ring*: or, where do axes come from? *Mankind* **11**, 354–382.

McBryde I (1984a) Exchange in south-eastern Australia: an ethnohistorical perspective. *Aboriginal History* **8**, 132–153.

McBryde I (1984b) Kulin greenstone quarries: the social contexts of production and distribution for the Mt William site. *World Archaeology* **16**(2), 267–285. doi:10.1080/00438243.1984.9979932

McNiven IJ, Bell D (2010) Fishers and farmers: historicising the Gunditjmara freshwater fishery, western Victoria. *La Trobe Journal* **85**, 83–91.

Menzies G (2003) *1421: The Year China discovered the World*. Bantam Books, London.

Micco H (1971) *King Island and the Sealing Trade, 1802*. Roebuck Society, Sydney.

Milligan R (2005a) *What became of the Wathaurong People of the Geelong District?* http://www.geelongonefire.org.au/images/Ron_Milligan_-_What_became_of_the_Wathaurong_People_of_the_Geelong_District.pdf (accessed 1 March 2018).

Milligan R (2005b) King Billy: the last of a proud people. The Wathaurung lived on a beautiful and bountiful land. *Geelong Advertiser,* 11 November, p. 8. http://www.geelongonefire.org.au/images/King_Billy_in_The_Geelong_Advert._story_Nov_2005.pdf (accessed 1 March 2018).

Mitchell TL (1839) *Three Expeditions into the Interior of Eastern Australia, with Descriptions of the recently explored Region of Australia Felix, and the Present Colony of New South Wales*. T & W Boone, London (facsimile edn, Libraries Board of South Australia, Adelaide, 1965).

Morgan J (1852) *Life and Adventures of William Buckley, Thirty-two Years a Wanderer amongst the Aborigines of the then unexplored Country round Port Phillip now the Province of Victoria*. Archibald MacDougall, Hobart.

Murray-Smith S (1973) Beyond the pale: the islander communities of Bass Strait in the 19th century. *Tasmanian Historical Research Association Papers and Proceedings* **20**(4), 167–200.

Neale M (2017) *Songlines: Tracking the Seven Sisters*. National Museum of Australia Press, Canberra.

Nicholson M, Romanis G, Paton I, Jones DS, Gerritsen K, Powell G (in press) 'Unnamed as yet': putting *Wadawurrung* meaning into the North Gardens landscape of Ballarat. *UNESCO Observatory E-Journal Multi-disciplinary Research in the Arts*.

Niewójt L (2009) *Gadubanud* society in the Otway Ranges, Victoria: an environmental history. *Aboriginal History* **33**, 175–199.

Norris RP, Norris C, Hamacher D, Abrahams RA (2013) *Wurdi Youang*: an Australian Aboriginal stone arrangement with possible solar indications. *Rock Art Research* **30**(1), 55–65 http://aboriginalastronomy.blogspot.com/2012/11/wurdi-youang-latest-research.html (accessed 1 March 2018).

Nunn PD, Reid NI (2016) Aboriginal memories of inundation of the Australian coast dating from more than 7000 years ago. *Australian Geographer* **47**(1), 11–47. doi:10.1080/00049182.2015.1077539

O'Connor A (2016) *Preliminary Aboriginal Cultural Heritage Survey: Johnstone Park, Geelong, Victoria*. Ecology and Heritage Partners Pty Ltd, Geelong.

O'Connor A (2017) *Drysdale Bypass, Jetty Road to Whitcombes Road, Drysdale, Victoria: Aboriginal Cultural Heritage Management Plan*. Ecology & Heritage Partners, Geelong. http://dsewebapps.dse.vic.gov.au/Shared/ATSAttachment1.nsf/(attachmentopen)/DE70EB2EDA96DA22CA258115007E3D01/$File/Greater+Geelong+C369+Supporting+Doc+-+Aboriginal+Cultural+Heritage+Management+Plan+Exhibition+Gazetted.PDF (accessed 1 March 2018).

Pascoe B (2003) *Wathaurong: The People who said No*. Wathaurong Aboriginal Co-operative and Coast Action, Geelong.

Pascoe B (2012) *Convincing Ground: Learning to Fall in Love with your Country*. Aboriginal Studies Press, Canberra.

Pascoe B (2014) *Dark Emu: Black Seeds: Agriculture or Accident*? Magabala Books, Broome, Western Australia.

Pickett EJ, Harrison SP, Hope G, Harle K, Dodson JR, Kershaw AP, Prentice IC, Backhouse J, Colhoun EA, D'Costa D, Flenley J, Grindrod J, Haberle S, Hassell C, Kenyon C, Macphail M, Martin H, Martin AH, McKenzie M, Newsome JC, Penny D, Powell J, Raine JI, Southern W, Stevenson J, Sutra JP, Thomas I, Van Der Kaars S, Ward J (2004) Pollen-based reconstructions of biome distributions for Australia, south-east Asia and the Pacific (SEAPAC region) at 0, 6000 and 18 000 14C yr BP. *Journal of Biogeography* **31**(9), 1381–1444. doi:10.1111/j.1365-2699.2004.01001.x

Powell B (2015a) *Uncle Bryon Powell Interview Part 1 on the Barwon River, Wadawurrung Country*. Culture Victoria. https://cv.vic.gov.au/stories/aboriginal-culture/seeing-the-land-from-an-aboriginal-canoe/uncle-bryon-powell-interview-part-1-on-the-barwon-river-wadawurrung-country/ (accessed 1 March 2018).

Powell B (2015b) *Uncle Bryon Powell Interview Part 2 on the Barwon River, Wadawurrung Country*. Culture Victoria. https://cv.vic.gov.au/stories/aboriginal-culture/seeing-the-land-from-an-aboriginal-canoe/uncle-bryon-powell-interview-part-2-on-the-barwon-river-wadawurrung-country/ (accessed 1 March 2018.

Powell B (2015c) The Bunjil Creation story. In M Satchell (Ed.) *Maude's Bunjil Lookout. Meredith and District News*, April, 16.

Powell B (2015d) *Wadawurrung Language: Kareet Bareet*. Open ABC. https://open.abc.net.au/explore/86715 (accessed 1 January 2018).

Powell B (2015e) *Lake Modewarre*. Open ABC. https://open.abc.net.au/explore/86725 (accessed 1 March 2018).

Powell B (2015f) *Wadawurrung Language: Bonan Youang.* Open ABC. https://open.abc.net.au/explore/86782 (accessed 1 January 2018).

Powell B (2015g) *Wadawurrung Language: Modewarre.* Open ABC. https://open.abc.net.au/explore/86725 (accessed 1 January 2018).

Powell B (2015h) *Wadawurrung Language: Wurdi Youang. Open ABC.* https://open.abc.net.au/explore/86719 (accessed 1 January 2018).

Powell B (2015i) *Wadawurrung Language: Warrenyeep.* Open ABC. https://open.abc.net.au/explore/86723 (accessed 1 January 2018).

Powell B (2019) *Coolenth Jumbunna*: blackfella talking. In DS Jones and D Low Choy (Eds) *Yurlendj-nganjin: Everyone's Knowledge/Our Intelligence.* Cambridge Scholars Publishing, Newcastle Upon Tyne.

Powell G, Jones DS (2019) 'Kim-barne Wadawurrung Tabayl': you are in Wadawurrung Country. *KERB: Journal of Landscape Architecture* **24**.

Presland G (2008) *The Place for a Village: How Nature has Shaped the City of Melbourne.* Museum Victoria Publishing, Melbourne.

Presland G (2008) *First People: The Eastern Kulin of Melbourne, Port Phillip & Central Victoria.* Museum Victoria Publishing, Melbourne.

Referendum Council (2017) *Uluru Statement from the Heart.* https://www.referendumcouncil.org.au/sites/default/files/2017-05/Uluru_Statement_From_The_Heart_0.PDF (accessed 1 March 2018).

Rhoads JW (1986) *Bellarine Peninsula: Archaeological Site Assessment and Management Study.* Department of Anthropology, University of Sydney, Sydney.

Roberts P (1993) The cultural heritage of the Barwon River. Unpublished paper.

Roberts P (2017) Revisiting the Mount William greenstone quarry: employment specialisation and a market economy in an early contact hunter-gatherer society. *Australian Aboriginal Studies* **2**, 14–26.

Roös PB (2015) Indigenous knowledge and climate change: settlement patterns of the past to adaptation of the future. *International Journal of Climate Change: Impacts and Responses* **7**(1), 13–31.

Roös PB (2017) *Regenerative-adaptive Design for Coastal Settlements: A Pattern Language Approach to Future Resilience.* PhD thesis. School of Architecture and Built Environment, Deakin University, Geelong. www.dro.deakin.edu.au/eserv/DU:30103436/roos-regenerativeadaptive-2017.pdf (accessed 1 March 2018).

Rose DB (1996) *Nourishing Terrains: Australian Aboriginal View of Landscape and Wilderness.* Australian Heritage Commission, Canberra. http://www.environment.gov.au/heritage/ahc/publications/nourishing-terrains (accessed 1 January 2017).

Rowe D, Huddle L (2000) *Greater Geelong Outer Areas Heritage Study Stage 2.* City of Greater Geelong, Geelong.

Rowe D, Jacobs W (2013) *Lara Heritage Review – Draft Document: Thematic Environmental History, Volume 2.* City of Greater Geelong Council, Geelong. https://geelongaustralia.com.au/common/Public/Documents/8d028fb1ee6dc78-DRAFT%20VOL%201%20Lara%20Heritage%20Review%20Report%20-%2026%20Feb%2013.pdf (accessed 1 March 2018).

Ryan S (2016) Creative places in south-west Victoria. Unpublished SRL733 paper. School of Architecture and Built Environment, Deakin University, Geelong.

Ryan S (2017) Conserving our Indigenous landscapes to acknowledge our country's history. Unpublished SRR711 thesis. School of Architecture and Built Environment, Deakin University, Geelong.

Sayers CE (Ed.) (1969) *Letters from Victorian Pioneers: Being a Series of Papers on the Early Occupation of the Colony, the Aborigines, etc. addressed by Victorian Pioneers to His Excellency Charles Joseph La Trobe, Lieutenant Governor of the Colony of Victoria.* Heinemann, Melbourne.

Shaw AGL (1996) *A History of the Port Phillip District: Victoria before Separation.* Miegunyah Press, Melbourne.

Shillingshaw JJ (Ed.) (1972) *Historical Records of Port Phillip: The First Annals of the Colony of Victoria.* Heinemann, Melbourne.

Smyth RB (1878) *The Aborigines of Victoria: with Notes relating to the Habits of the Natives of other Parts of Australia and Tasmania compiled from Various Sources for the Government of Victoria*, Vol. 1. https://en.wikisource.org/wiki/The_Aborigines_of_Victoria/Volume_1/Chapter_20 (accessed 1 March 2018).

Smyth RB (1972 [1878]) *The Aborigines of Victoria: with Notes relating to the Habits of the Natives of other Parts of Australia and Tasmania compiled from Various Sources for the Government of Victoria*, Vol. 1. (facsimile edn). John Currey O'Neil, Melbourne.

Stapylton GWC (1971) The journal of Granville Wilson Chetwynd Stapylton – kept while accompanying Major Sir Thomas Livingstone Mitchell on his Australia Felix expedition, 1836. In MH Douglas and L O'Brien (Eds) *The Natural History of Western Victoria: Proceedings of the Symposium*. Australian Institute of Agricultural Science, Horsham, Victoria.

Su RJ (2014) Narana creations. Unpublished SRL733 paper. School of Architecture and Built Environment, Deakin University, Geelong.

Surf Coast Shire (2014) *Acknowledging and Celebrating Aboriginal Heritage in the Surf Coast Shire*. Surf Coast Shire, Torquay, Victoria.

Thom BG, Roy PS (1985) Relative sea levels and coastal sedimentation in southeast Australia in the Holocene. *Journal of Sedimentary Petrology* **55**, 257–264.

Thom BG, Hespa P, Bryant E (1994) Last glacial 'coastal' dunes in eastern Australia and implications for landscape stability during the Last Glacial Maximum. *Palaeogeography, Palaeoclimatology, Palaeoecology* **111**(3–4), 229–248. doi:10.1016/0031-0182(94)90065-5

Thomas W (2014) *The Journal of William Thomas, Assistant Protector of the Aborigines of Port Phillip and Guardian of the Aborigines of Victoria 1839–1867; Vol. 1, 1839 to 1843*. Victorian Aboriginal Corporation for Languages, Melbourne.

Thurstan R, Brittain Z, Jones DS, Bellgrove A, Dearnaley J (2018) Archival evidence of Indigenous uses of seaweeds in temperate Australia. *Journal of Applied Phycology* **30**(3), 1821–1832. doi:10.1007/s10811-017-1384-z

Todd W (1835) *Journal of William Todd, June to November 1835*. State Library of Victoria, Melbourne.

Tournier D (2010) *Wathaurong*: the Three Sisters. In *Indigenous Creation Stories of the Kulin Nation* (pp. 12–15). Arts Victoria, Melbourne.

Tournier D (2011a) *Bundjil* and the making of the land. *Wathaurong Creation Stories* [DVD]. Wathaurong Aboriginal Co-operative, Geelong.

Tournier D (2011b) The Three Sisters. *Wathaurong Creation Stories* [DVD]. Wathaurong Aboriginal Co-operative, Geelong.

Tournier D (2011c) *Wathaurong Creation Stories* [DVD]. Wathaurong Aboriginal Co-operative, Geelong.

Tournier D (2013) Welcome to Country. Unpublished paper. Wathaurong Aboriginal Co-operative, Geelong.

Tournier D (2014a) *Reading the Country: A Story from the You Yangs*. https://www.youtube.com/watch?v=5LbEcbxhJq0 (accessed 1 March 2018).

Tournier D (2014b) The Three Sisters. *Nyernila: Listen Continuously – Aboriginal Creation Stories of Victoria* (pp. 50–51). Arts Victoria, Melbourne.

Tournier D (2015a) *Wadda Wurrung Ngarrwa Kat-Kat* (to know *Wathaurong* foods). Unpublished paper. Wathaurong Aboriginal Co-operative, Geelong.

Tournier D (2015b), *Ngunhuk Lotjba – Tandop David Tournier*. https://vimeo.com/117989092 (accessed 1 March 2018).

Tournier D with Jones DS (2015) Proposed *Wathaurong* medicinal and food plant list for the REACH Link. Unpublished paper. Deakin University, Geelong.

Townrow K (1997) *An Archaeological Survey of Sealing and Whaling Sites in Victoria*. Victorian Heritage Council, Melbourne.

Tuckey JH (1805) *An Account of a Voyage to establish a Colony at Port Phillip in Bass's Strait, on the South Coast of New South Wales, in His Majesty's Ship Calcutta in the Years 1802–4*. Longman Hurst Rees & Orme, London.

Tudehope CM (1962a) William Buckley. *Victorian Naturalist* **32**, 216–236.

Tudehope CM (1962b) The You-Yangs rock wells. *Victorian Naturalist* **78**(11), 330–332.

Turnbull J (2009) *Armstrong Creek West Precinct: Desktop Cultural Assessment – A Report to Villawood Properties P/L*. Villawood Properties P/L and Ochre Imprints, Melbourne. https://www.geelongaustralia.com.au/common/public/documents/amendments/8ce70eeea3d8008-ACW%20Cultural%20Heritage%20Report.pdf (accessed 1 March 2018).

Waarden NV (1986) *Archaeological Survey of the You Yangs, Victoria*. Occasional Report. Aboriginal Affairs Division, Department of Health and Community Services, Melbourne.

WAC (2018a) *Wurdi Youang Business Plan: Summary*. Wathaurong Aboriginal Co-operative, Geelong. http://www.ccma.vic.gov.au/admin/file/content2/c7/Wurdi%20Youang(Summary)_Spreads-compressed.pdf (accessed 1 June 2018.

WAC (2018b) *Wadawurrung Country of the Victorian Volcanic Plains*. Wathaurung Aboriginal Co-operative, Ballarat.

Walsh N (2016) Murnong: a name for Murnong (*Microseris*: Asteraceae: Cichorioideae). *Muelleria* **34**, 63–67.

Wardle M (2008a) *King Billy: Last of the Tribal Wathaurung – Part 1*. https://www.youtube.com/watch?v=vHPYNd_eCkI (accessed 1 March 2018).

Wardle M (2008b) *King Billy: Last of the Tribal Wathaurung – Part 2*. https://www.youtube.com/watch?v=c9q0Y5gqCbk (accessed 1 March 2018).

Wedge JH (1835) *Narrative of an Excursion amongst the Natives of Port Phillip and the South Coast of New Holland*. Wedge Papers, State Library of Victoria, Melbourne.

Wedge JH (1835–36) Field Book 1835–36, *Port Phillip Papers Digitising Project* (126 leaves: ill.; 9.5 x 19.4 cm. State Library of Victoria, 1835–36) at slide 61.

Whitehead J with Tournier D (2014) *The Waddawurrung People: Did You Know*. Queenscliffe Point Lonsdale Community Enterprise, Point Lonsdale, Victoria.

Wynd I (1971) *Geelong the Pivot: A Short History of Geelong and District*. Cypress Press, Melbourne.

Wynd I (1992) *Barrabool: Land of the Magpie*. Barrabool Shire, Torquay, Victoria.

Zola N, Gott B (1992) *Koorie Plants, Koorie People: Traditional Aboriginal Food, Fibre and Healing Plants of Victoria*. Melbourne Koorie Heritage Trust, Melbourne.

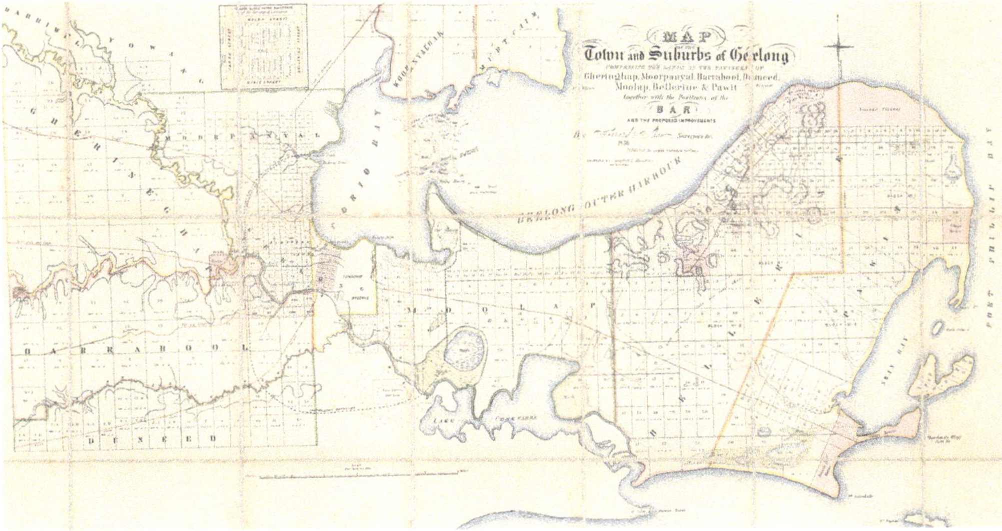

Plate 1.1. Garrard and Shaw, 1838. 'Map of the Town and Suburbs of Geelong comprising the Lands in the Parishes of Gheringhap, Moorpanyal, Barrabool, Duneed, Moolap, Bellerine & Pawit: Together with the Positions of the Bar and the Proposed Improvements / by Garrard and Shaw, Surveyors &c.; engraved by Campbell & Macartney'. http://nla.gov.au/nla.obj-231354483/view, National Library of Australia MAP RM 963. Reproduced with the permission of the National Library of Australia.

Plate 1.2. Jan Senbergs, 2010. 'Geelong Capriccio (if Geelong were settled instead of Melbourne)'. Reproduced with the permission of Jan Senbergs, Niagara Galleries Melbourne, and assistance from the Deakin University Art Collection and Galleries Unit.

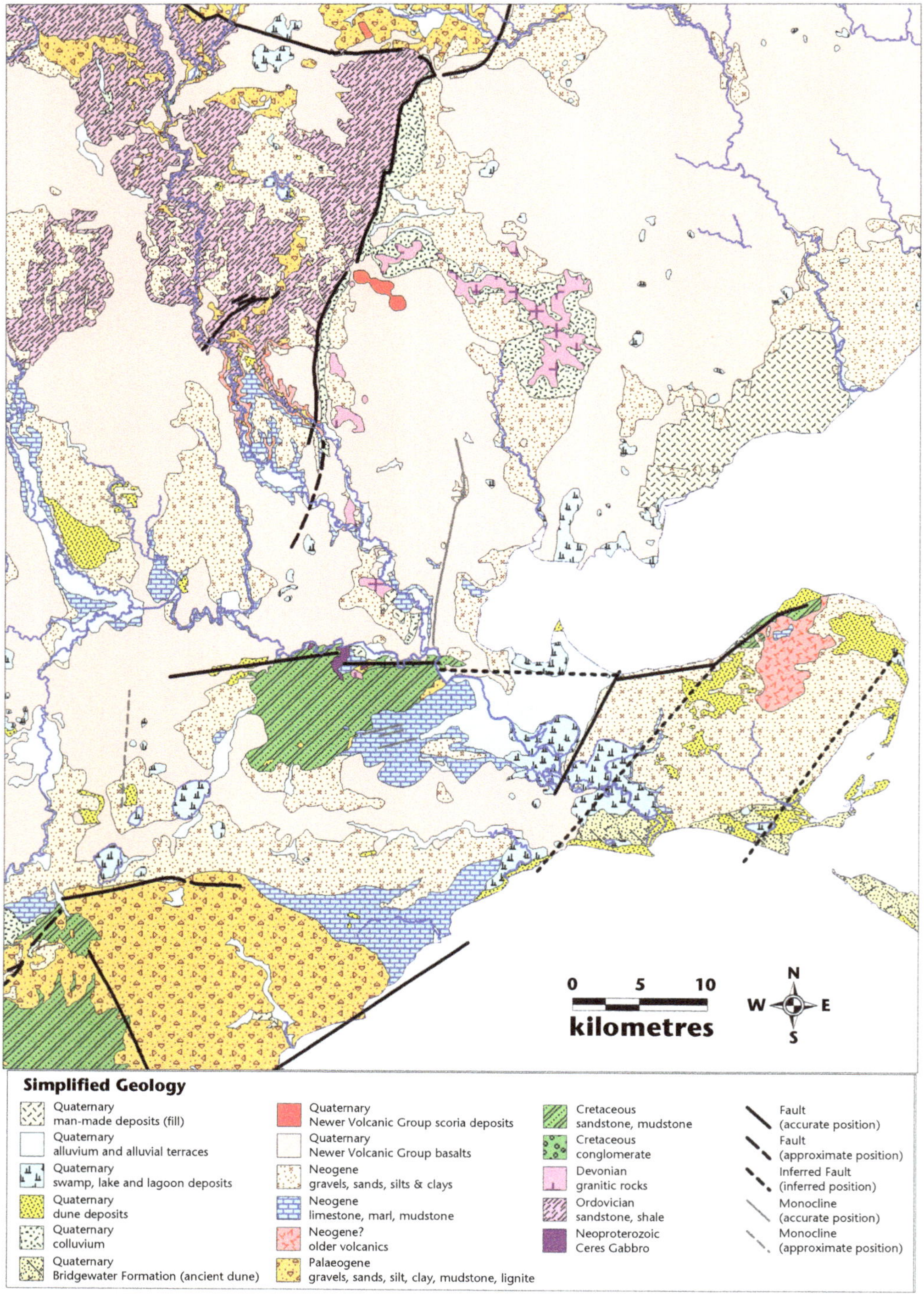

Plate 2.1. Simplified geology of the Geelong region. Modified from DPI (2010).

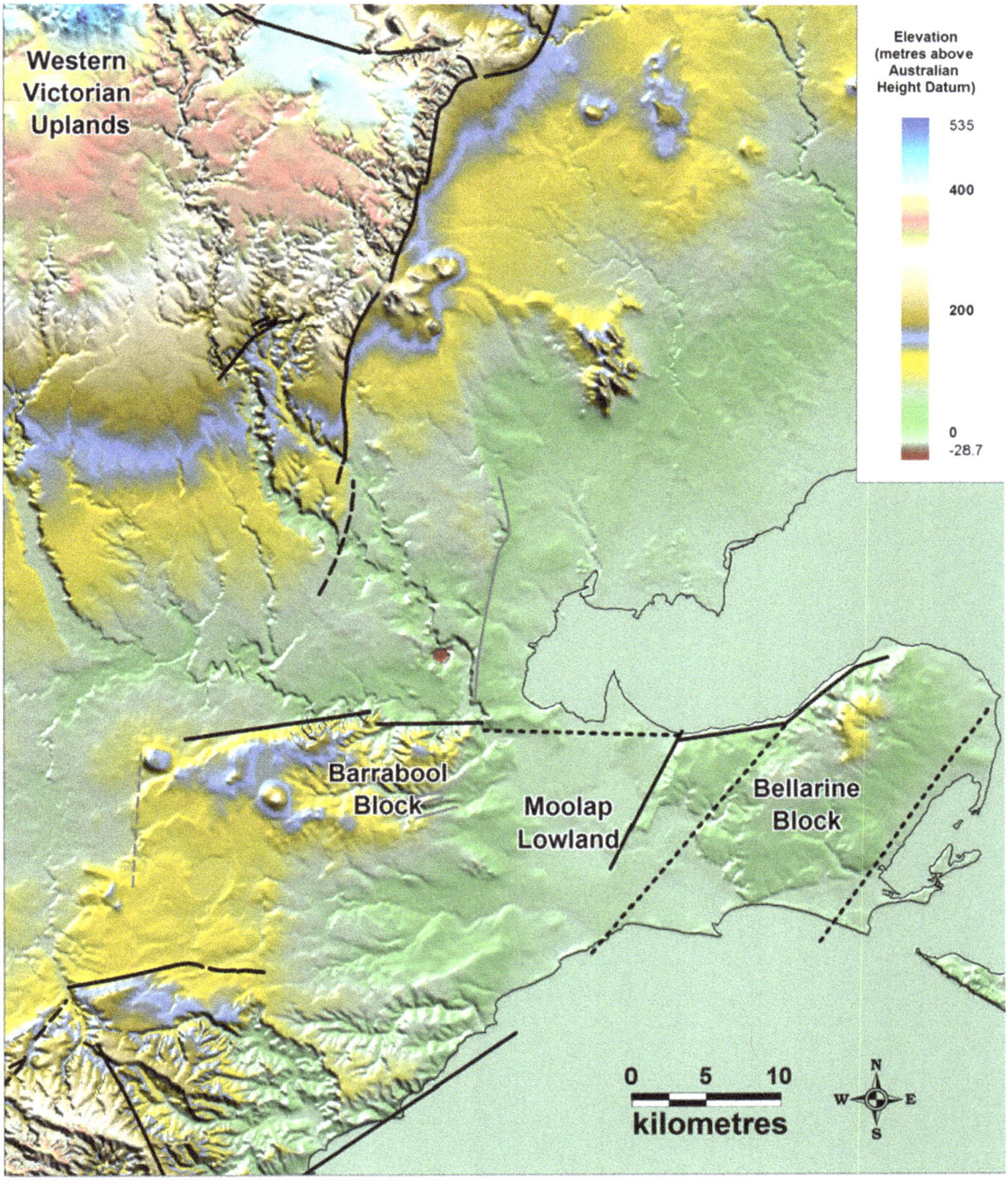

Plate 2.2. Structural elements of the Geelong region.

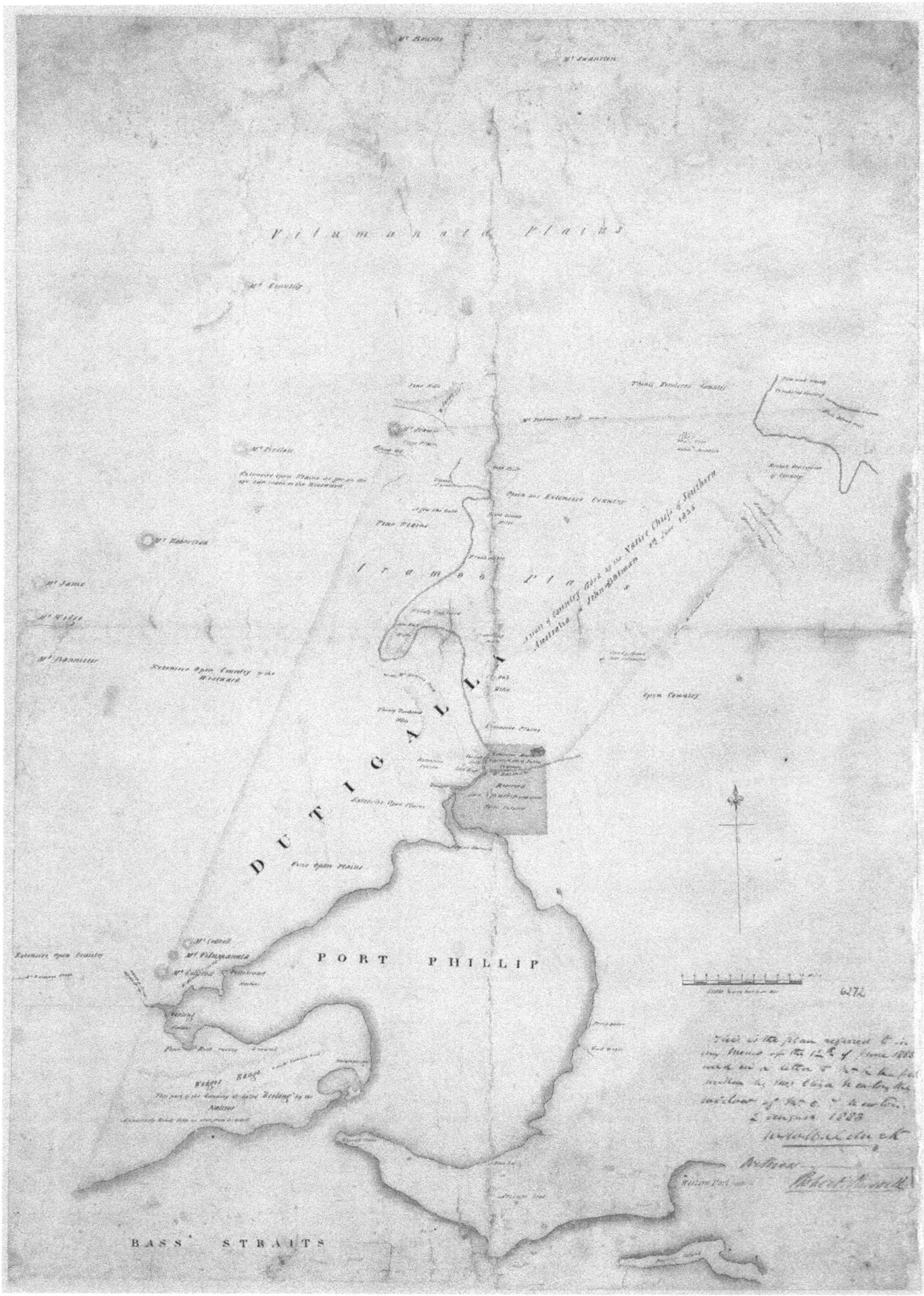

Plate 5.1. J.H. Wedge, c.1835. 'Map of Port Phillip from the surveyor of Mr. Wedge and others'. Source: State Library of Victoria.

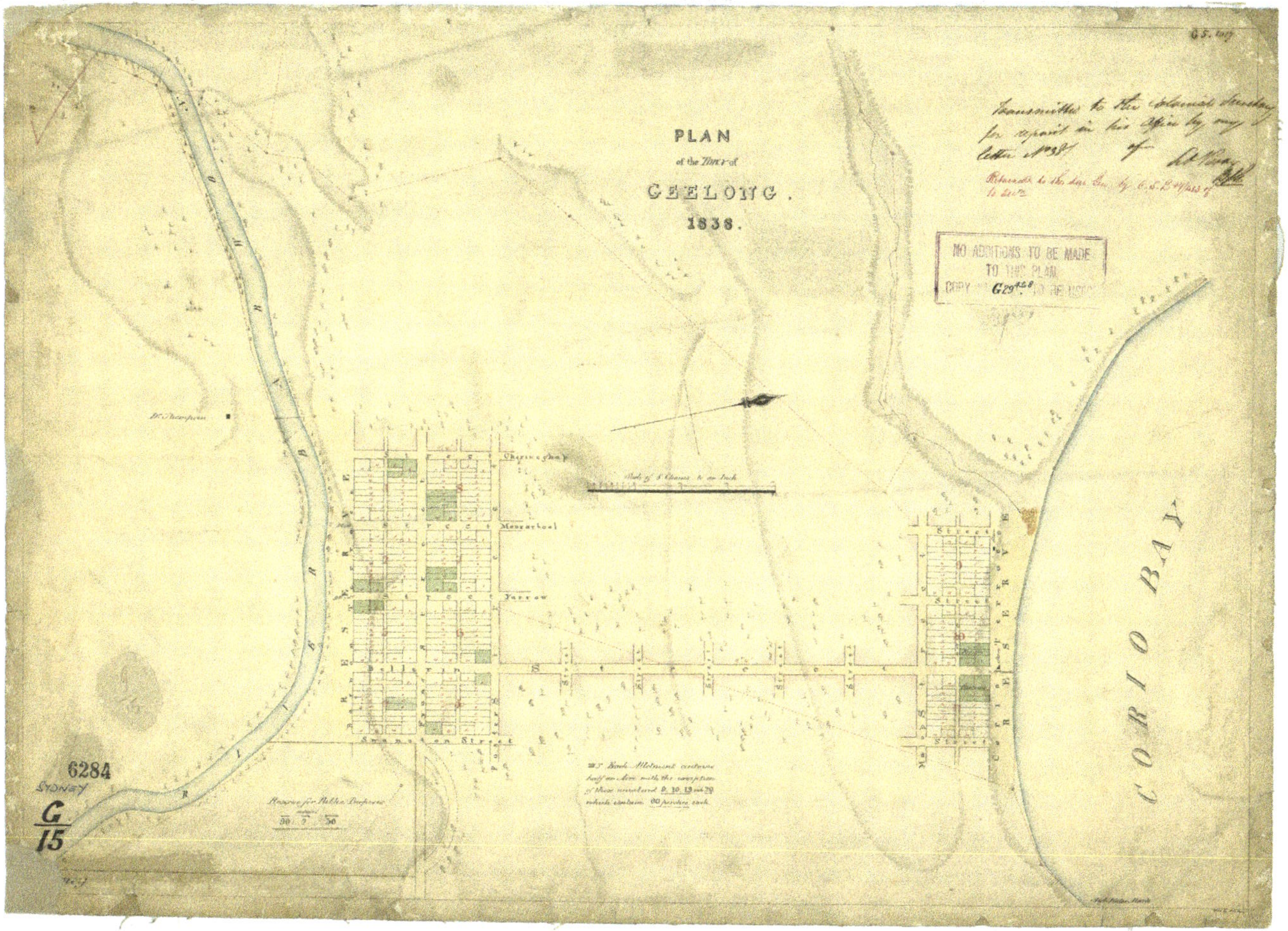

Plate 5.2. 'Plan of the Town of Geelong', 1838. Source: Historic Plan Collection, SydneyG15; Geelong, VPRS 8168/P2 Unit 6127. Public Record Office Victoria, Melbourne.

Plate 6.1. Sue Anderson, 'Horses, grasses', 2009. Lithograph, 60 × 30 cm.

Plate 6.2. Sue Anderson, 'Avalon', 2015. Oil on linen, 137 × 162 cm.

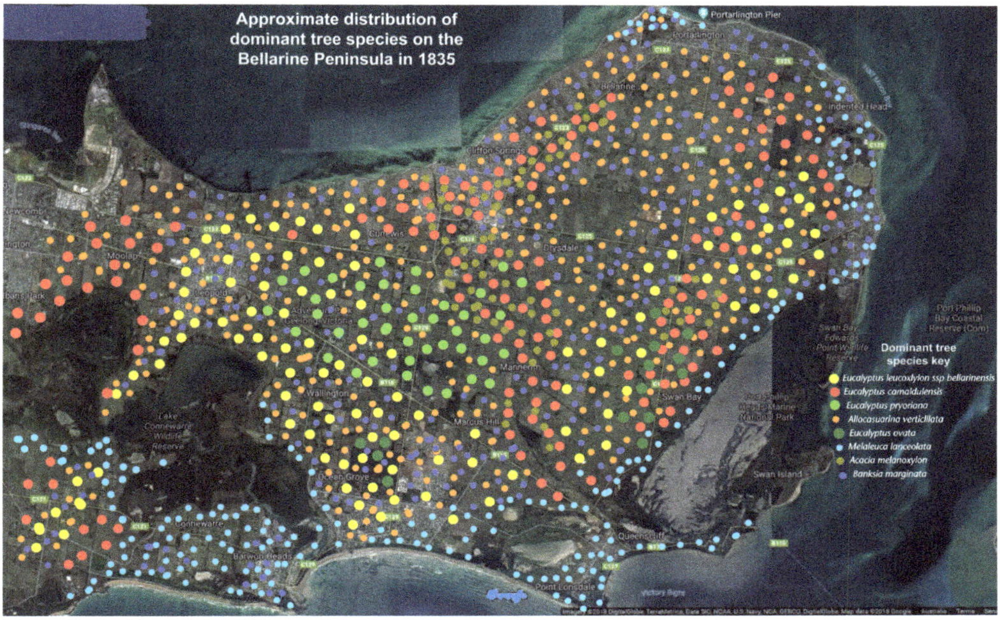

Plate 7.1. The approximate distribution of dominant tree species on the Bellarine Peninsula in 1835.

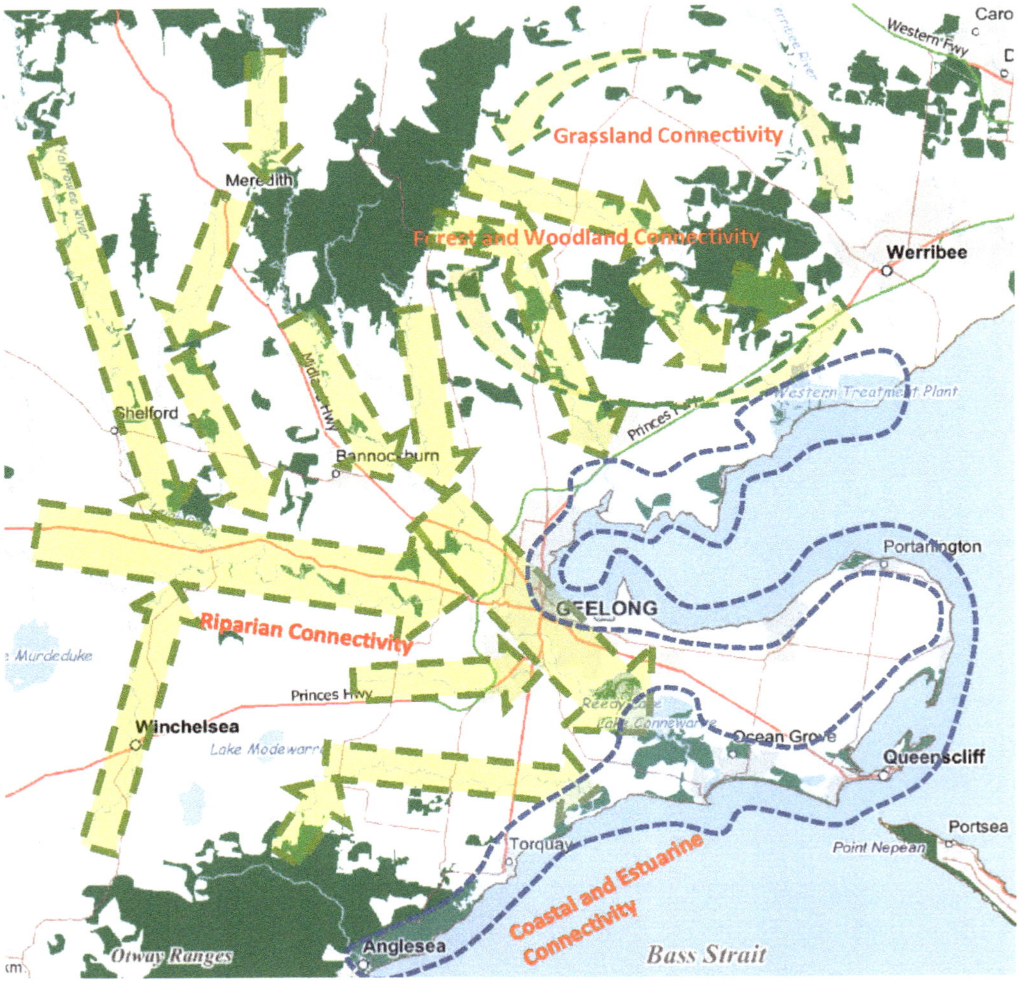

Plate 10.1. KEPs integration in a first-tier green infrastructure plan for the Geelong region. Source: Williamson.

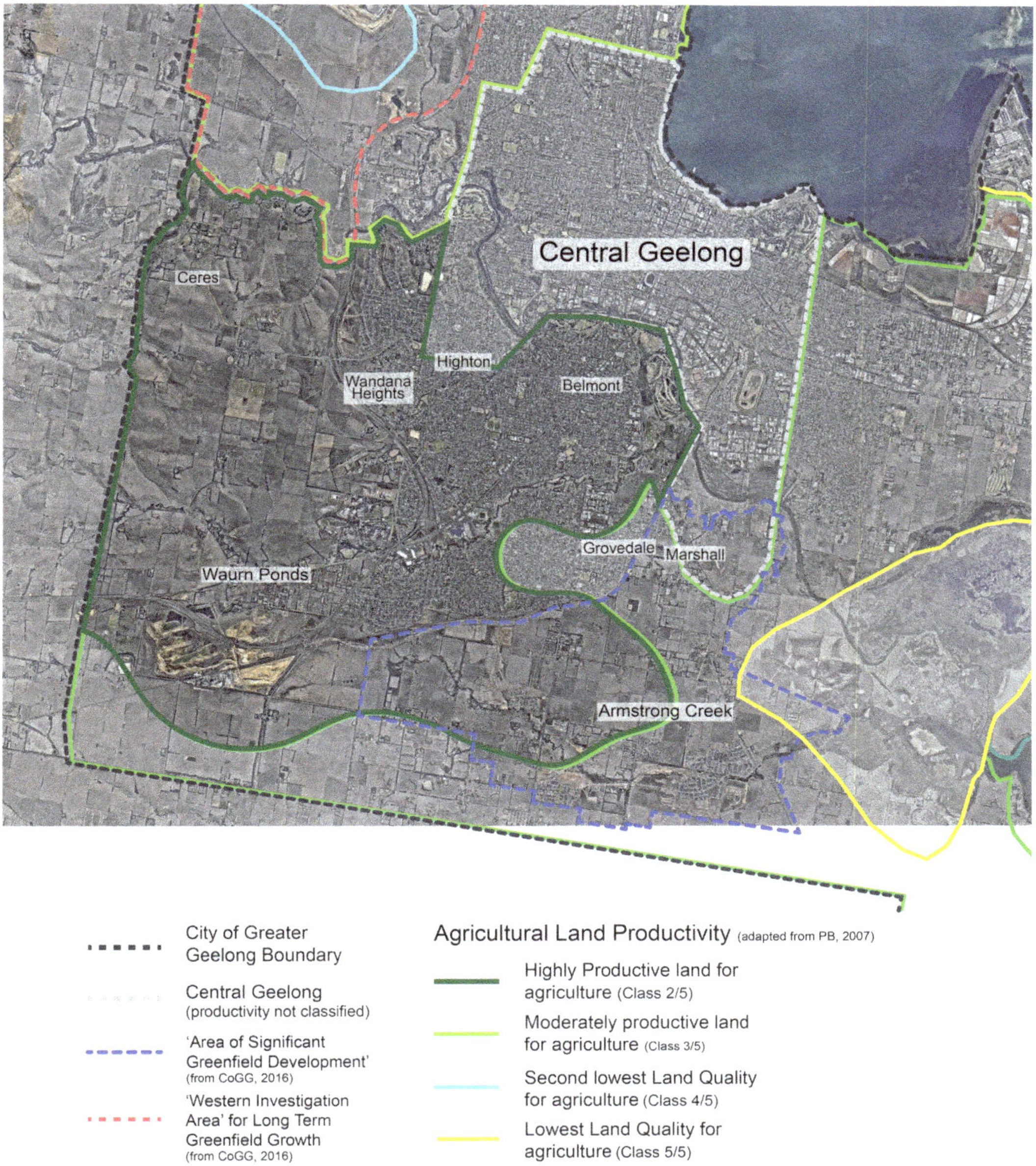

Plate 13.1. CoGG's historic, current and future land use planning.

Plate 16.1. Stephanie Ho, 'Point Henry'. Dream Line, 2014. © Stephanie Ho.

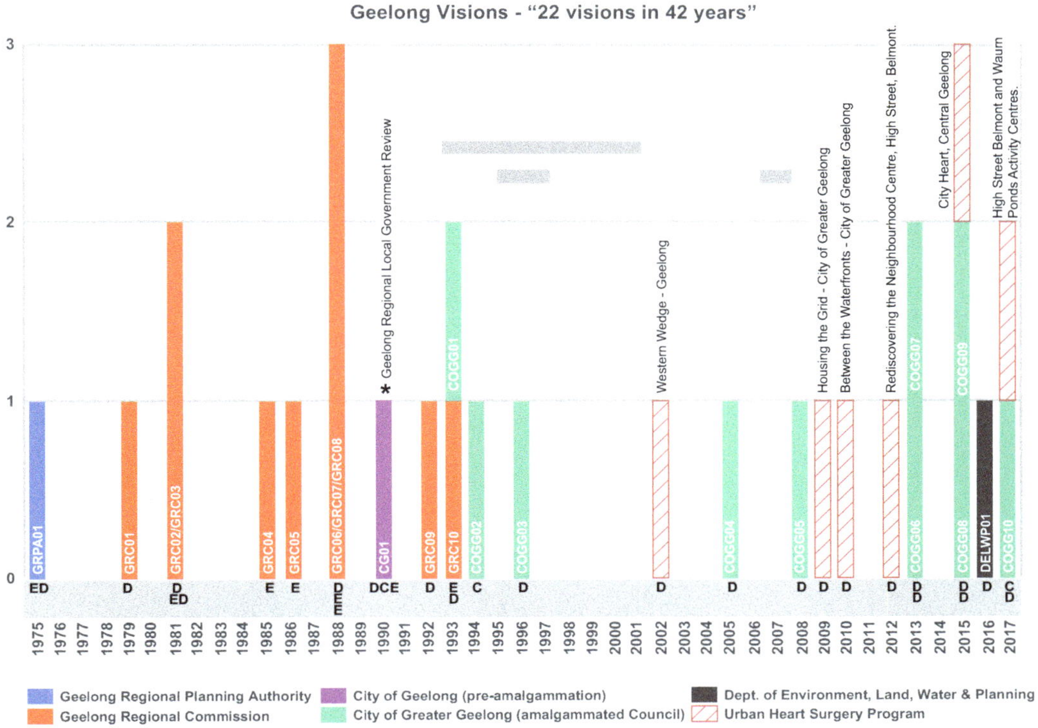

GRPA01	Geelong City by the Bay – Geelong Central Business District Study	1975
GRC01	Geelong Central Area Study	1979
GRC02	Geelong Central Area Strategy	1981
GRC03	Geelong City by the Bay	1981
GRC04	Geelong Regional Development Strategic Plan	1985
GRC05	Geelong Region: Framework for the Future – Development Strategy	1986
GRC06	City by the Bay Foreshore Precinct Project	1988
GRC07	Directions – The Geelong Regional Development Strategy	1988
GRC08	Geelong City by the Bay – The Bay Link	1988
GRC09	Directions – The Geelong Regional Development Strategy	1992
GRC10	Determinants of Growth in the Geelong Region	1993
CG01	Geelong and Foreshore Draft Structure Plan	1990
COGG01	Central Geelong Foreshore and Environs Framework	1993
COGG02	Geelong 2010 A preferred Vision	1994
COGG03	Waterfront Geelong Design and Development Code	1996
COGG04	Geelong Western Wedge Framework	2005
COGG05	Geelong Transit City Station Precinct Urban Design Framework	2008
COGG06	Central Geelong Urban Design Guidelines	2013
COGG07	Central Geelong Vision – Design Studio Report Draft	2013
COGG08	Central Geelong Action Plan	2015
COGG09	Geelong City Centre "City in the Park"	2015
DELWP01	Revitalising Central Geelong Action Plan	2016
COGG10	A Clever and Creative Future	2017

Plate 17.1. 22 visions in 42 years. Source: Figure created by John Rollo and Yolanda Esteban, 2018.

Plate 17.2. Geelong Western Wedge, 2002. Source: Plan drawing by Davin Smith. Urban heart student design team Deakin University, Davin Smith, Shem Kelder, Lee-Anne Manski and Jeremy Schulter (NB: All plan images are not presented at original scale).

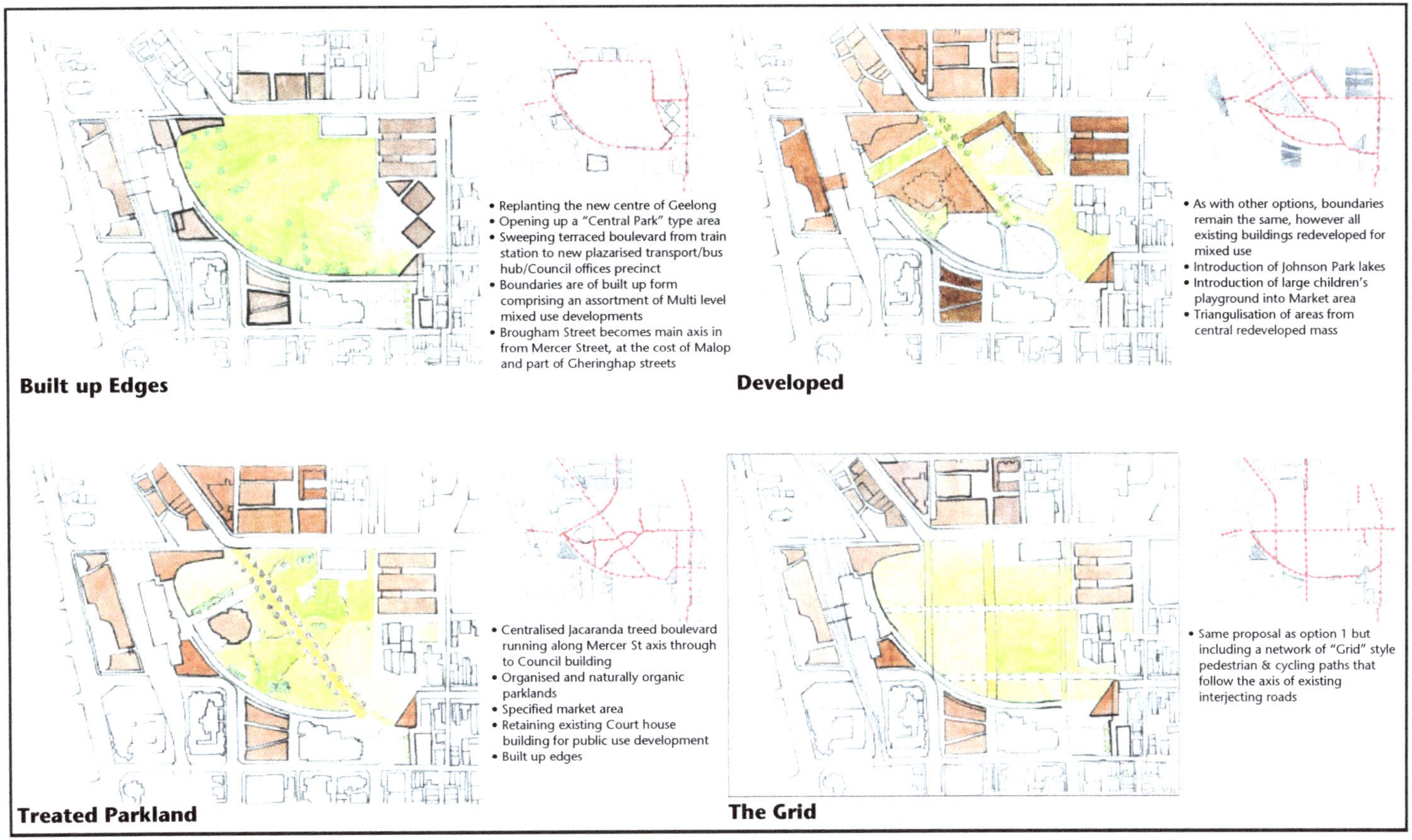

Plate 17.3. COGG, Geelong, Housing the Grid, 2010. Source: Urban heart student design team Deakin University, Jandi Vagg, Tim McErvale, Timothy Fullwood (NB: All plan images are not presented at original scale).

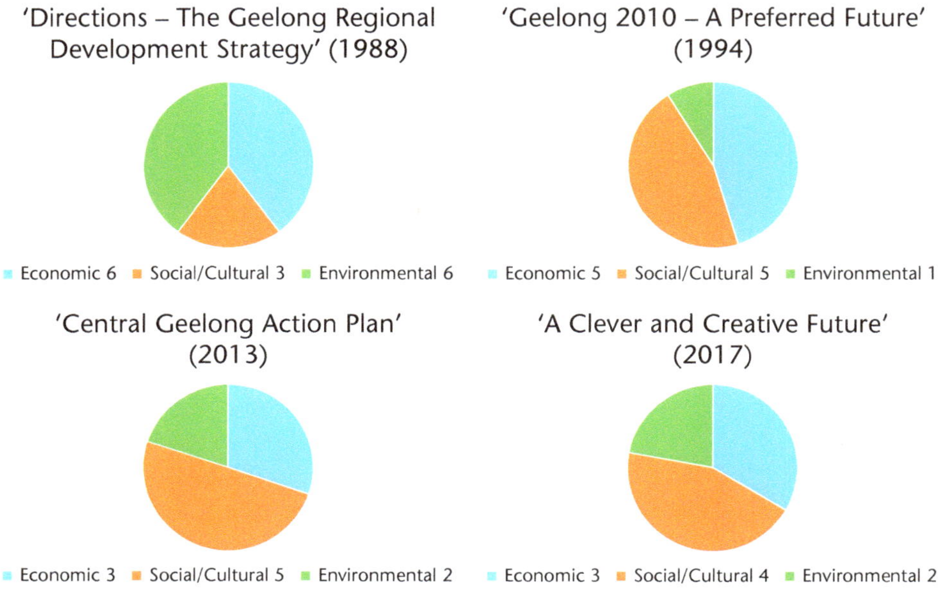

Plate 17.4. Economic (blue), social/cultural (red), and environmental (green) aspirations from four different reports. Source: Rollo & Esteban.

Indigenous plants of the Geelong Region

REGIONAL MAP

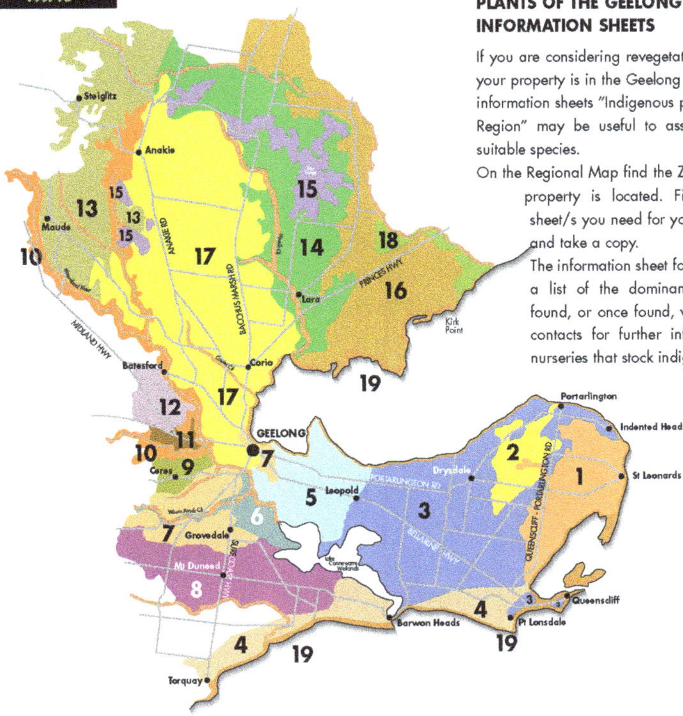

HOW TO USE THE 'INDIGENOUS PLANTS OF THE GEELONG REGION' INFORMATION SHEETS

If you are considering revegetating or planting, and your property is in the Geelong Region, the series of information sheets "Indigenous plants of the Geelong Region" may be useful to assist you in selecting suitable species.

On the Regional Map find the Zone/s in which your property is located. Find the information sheet/s you need for your particular Zone/s and take a copy.

The information sheet for each Zone includes a list of the dominant indigenous plants found, or once found, within that Zone and contacts for further information and local nurseries that stock indigenous plants.

WHY SHOULD I PLANT INDIGENOUS PLANTS?

'Indigenous' means naturally belonging to an area. Indigenous plants are plants which occur naturally, or historically occurred naturally, in an area.

Choosing indigenous plant species and retaining natural vegetation wherever possible helps to maintain the overall ecological balance of an area.

By selecting indigenous plants you can be sure that they are well adapted to the soils, topography and climate of the local area. They will need little watering or fertilising after establishment, although in most cases watering and feeding will increase growth.

Best results are achieved by using indigenous plants that grow naturally in your area (local provenance).

THE ADVANTAGES

Indigenous plants:

- maintain the area's natural beauty and biodiversity.
- encourage local wildlife by providing food and habitat.
- are well adapted to the local soil and climatic conditions, so require little water or fertiliser once established.
- are more able to resist attack by pests and diseases.
- will often regenerate naturally.
- do not become pest plants/environmental weeds.

WHERE CAN INDIGENOUS PLANTS BE OBTAINED?

A number of local nurseries in the Geelong region specialise in the supply of local indigenous plants. These are listed on the information sheet for each Zone.

Alternatively you can propagate and plant out indigenous plants from seed collected locally, or direct seeding can be used to reduce revegetation costs.

It is important to note that under the Flora and Fauna Guarantee Act it is illegal to remove indigenous plants or collect seed from public land without a permit.

Appendix Plate 1. Indigenous plants of the Geelong region. Reproduced with the permission of the City of Greater Geelong Council.

Indigenous plants of the Geelong Region

THE ZONES

The Geelong region has been divided into 19 zones that reflect the different physical features (eg soil type, drainage pattern, rainfall etc) which determine the distribution of indigenous vegetation.
An information sheet on each of the zones provides a description, some history of the flora for the zone, a list of the dominant indigenous plants found or once found within that zone, local indigenous nurseries and contacts for further information.

ZONE 1 **EASTERN BELLARINE PLAINS:** *Indented Head, St Leonards & Murradoc*

ZONE 2 **MT BELLARINE VOLCANIC SOILS:** *Mt Bellarine / Scotchmans Hill*

ZONE 3 **CENTRAL BELLARINE HILLS:** *Portarlington, Clifton Springs, Drysdale, Mannerim, Marcus Hill, Wallington, Leopold, Curlewis, Pt Lonsdale & Queenscliff*

ZONE 4 **LOWER REAR DUNE COMMUNITIES:** *Torquay to Point Lonsdale*

ZONE 5 **MOOLAP PLAINS:** *Moolap, Point Henry, Whittington, Newcomb & Limeburners Point*

ZONE 6 **MARSHALL PLAINS & WAURN PONDS FLATS**

ZONE 7 **GEELONG, GROVEDALE, WAURN PONDS LIMESTONE PLAINS & HILLS**

ZONE 8 **MOUNT DUNEED & SURROUNDING BASALT FLOWS**

ZONE 9 **BARRABOOL HILLS**

ZONE 10 **BARWON & MOORABOOL VALLEYS / RIVERS & MAJOR CREEKS**

ZONE 11 **STONEHAVEN, FYANSFORD & WESTERN BASALT PLAINS**

ZONE 12 **BATESFORD, GRANITIC & TERTIARY SANDS**

ZONE 13 **MEREDITH-STEIGLITZ AREA**

ZONE 14 **COLLUVIAL PLAINS EAST OF THE BRISBANE RANGES & YOU YANGS PLAINS**

ZONE 15 **YOU YANGS GRANITE**

ZONE 16 **WERRIBEE PLAINS**

ZONE 17 **ANAKIE EAST / LOVELY BANKS / CORIO BASALT PLAINS**

ZONE 18 **WERRIBEE PLAINS - RIVER & WATERCOURSES**

ZONE 19 **COASTAL ZONES OF THE GEELONG REGION**

City of GREATER Geelong

FOR MORE INFORMATION

Additional information regarding suitable species can be obtained from the City of Greater Geelong Environment Unit phone 5227 0270, the Department of Natural Resources and Environment phone 5226 4667 or the agencies and nurseries listed on the information sheets.

ACKNOWLEDGMENTS

The City of Greater Geelong acknowledges the Department of Natural Resources and Environment, in particular David Perry and also Mark Trengove, Geelong Indigenous Nursery.
Illustrations: George Stolfo and Brian Bainbridge, courtesy of Hyland House.

Version 1

REGIONAL MAP

Appendix Plate 2. Indigenous plants of the Geelong region. Reproduced with the permission of the City of Greater Geelong Council.

5

Djilang, Corayo and beyond: the Geelong region landscape and its European transformation

David Rowe

> *The town of Geelong stands between the miniature Bay of Corio, with its picturesque green cliffs and comparatively sheltered water, and the river Barwon, which flows from the north to south (at the distance of a long mile from the landing place), towards its outlet in Lake Connewarre, which again joins the sea in Bass' Straits, by a narrow, and, we believe, un-navigable entrance. The scenery of its suburbs is agreeable, its soil as fertile, and its climate as healthful as that of Melbourne.*
>
> – *Sydney Chronicle*, 30 September 1847.

The Geelong environment has been explored, colonised, exploited and enjoyed by Europeans since the beginning of the 19th century, and the transformative effects have been far-reaching. Originally (and erroneously) considered by the British as 'waste lands of the Crown', European explorations of the Country of the *Wadawurrung* Aboriginal peoples from 1802 recognised the region for its grazing potential, leading to the arrival of squatters in 1836. Urbanisation came in 1838, following the setting out of the Geelong township, the landscape influencing the choice of site. Urban and suburban expansion has continued to the present day. Informal and statutory town planning processes in the 20th century also influenced the development of the region.

Exploitation of the landscape began with livestock grazing and progressed to the introduction of exotic flora and fauna, and extractive, manufacturing and other industries. Natural and contrived bayside and coastal foreshore landscapes have been a key part of the Geelong environment, especially for recreational purposes. Other landscapes were created for commemorative reasons.

Exploration of the landscape

Wadawurrung Country and European perceptions

The first European impressions of the Geelong region landscape were of *Wadawurrung* Country, the land of an Aboriginal tribe of the *Kulin* Nation. Extending from the Werribee River to Aireys Inlet on the coastline of Bass Strait, and inland towards Ballarat and

Ararat, the diverse landscape was home to 25 clans (Clark 1990; WAC nd a).[1] At this time, the Geelong landscape was a product of traditional Aboriginal land practices, including the open mosaic-patterned grasslands (caused by fire management) that produced grains and tubers (Pascoe 2014).

Following British occupation from 1788, the Geelong region was officially considered 'waste lands', being beyond the limits of European occupation. Through European eyes, the flora of *Wadawurrung* Country – which had existed for thousands of years – was seen as useless and without form (Halsey 2006).

Such unqualified perceptions of the landscape changed with European explorations, although the Indigenous cultural value of country was hardly recognised. In 1802, Lieutenant John Murray charted part of Indented Head (Clark 1995). In the same year, Matthew Flinders discovered the Geelong area. He was impressed by the vast extent of the harbour and, from the elevated peaks of the You Yangs, he observed the low, grassy plains lightly covered with wood (Brownhill and Wynd 1990). In 1803, Charles Grimes surveyed the Port Phillip region, mapping the coastline and giving a record of the environment (Grimes 1803).

In 1824, Hamilton Hume and William Hovell reached Corio Bay as part of their overland travels from New South Wales. They elicited from the *Wada wurrung balug* clan that the bay was *Djilang* and the land was *Corayo* (VACL 2015; Rowe 2002a).

In 1835, John Batman from Tasmania led a party that landed at Indented Head. They established an encampment at St Leonards. The intention was to establish a permanent European colony there (months later, the settlement relocated to Melbourne) (Rowe and Jacobs 2013). John Helder Wedge, a British surveyor resident in Tasmania, visited Batman's encampment in August 1835 (Wynd 2005). He mapped the Port Phillip area (Plate 5.1) and gave a detailed description:

> *The peninsula of Indented Head first attracted my attention. Its extent is about 100,000 acres* [40 468 ha]. *It is bounded on the west by the Barwon, a river discovered by myself, which empties into Bass strait a few miles to the westward of Indented Head, and in its course passes within about three miles* [4.8 km] *of the western extremity of Port Phillip. The eastern part of this peninsula, for about four or five miles* [6.4–8 km] *from the margin of the Port, is a low and flat surface, the soil being light and sandy, and well covered with grass, thinly wooded with the honeysuckle, sheoak, mimosa, and eucalyptus. The land then swells into low tiers, and alternates with beautiful hill and dale. On these hills the soil is of finer quality, and the grass more luxuriant than the plains. The hills gradually trend to the westward in gentle undulations and terminate at the Barwon, in some places in steep banks, varying in height from 30ft to 60ft* [9–18 m]. *It is a great drawback to the availability of the peninsula that the River Barwon is subject to the tides, and is, consequently, salt up to where it is joined by another river, about three miles from the western extremity of the Port, otherwise it would be one of the finest situations for sheep farming I have ever met with.*
>
> *On the peninsula there are many small waterholes, which afford the natives a supply of water … At the junction of the rivers above alluded to, the one coming from the north-west is called the Yaloak by the natives, the other coming from the westward I have named the Byron, into which, about 10 or 12 miles* [16–19 km] *up, another stream falls, which I have named the Leigh. These rivers pass through very extensive open plains much further than the eye can reach … About 15 miles* [24 km], *in a*

south-west direction, from the junction of the Byron with the Yaloak, is a lake, called by the natives Modewarrie, the intermediate country being grassy hills (called by the native Barrabool), of moderate elevation, thinly covered with she-oak trees; and around the lake an undulating grassy country, thinly timbered, extends to the westward.

... The country between the river coming from the northward and the western extremity of the Port, and from 20 to 40 miles [32–64 km] *inland, is open, and partakes more of the nature of downs. The whole is thickly covered with light growth of grass, the soil being, in general, stiff and shallow. About midway there is a river falling into the Port, which comes from the north west ... About Station Mount* [Station Peak, You Yangs] *(called by the natives Villamanta), the country is wooded, with this exception, and here and there along the shore of the Port, and along the course of the river just mentioned, the plains are quite open, as much so as the heaths of Cambridgeshire (Hobart Mercury, 22 January 1935, p. 8).*

Numerous early accounts also provided an appreciation of Geelong's native wildlife, which was a source of food and aquaculture for the *Wadawurrung*. Kangaroos roamed 'in vast herds of thousands in a flock' with emu (*Dromaius novaehollandiae*) in lesser numbers (*Geelong Advertiser* (*G.A.*), 27 November 1866, p. 2). The lakes boasted swan (*Cygnus atratus*), Cape Barren geese (*Cereopsis novaehollandiae*), wild duck, teal, widgeon and wild turkey, with quail and snipe in lesser numbers. Saltwater fish included schnapper, pike, whiting, garfish, mullet, barracouta, rock cod, flounders and 'a viviparous fish called flat-head' (*G.A.* 27 November 1866, p. 2). Oyster beds were lower in numbers but crayfish were plentiful. In the rivers and creeks were blackfish, species of grayling and eels (*G.A.* 27 November 1866, p. 2).

Dispossession of Country following European explorations

Underscoring European exploration was the British demand for wool, and the productive landscape of the *Wadawurrung* was 're-visioned' for 'alternative' grazing purposes. Legalised European occupation for pastoral purposes began from 1833 and increased with the *Crown Lands Occupation Act* in 1836 (Billis and Kenyon 1930). In June 1835, Batman executed a so-called treaty with Indigenous peoples for the purchase for 600 000 acres (242 811 ha) of territory (Brownhill and Wynd 1990).[2] The deed that included the Geelong area was a fabrication: it named eight clan heads of the *Bun wurrung* and *Wurundjeri* tribes, even though the Geelong area was the territory of the *Wadawurrung* (Clark 1995).[3]

From 1835, *Wadawurrung* Country was occupied by members of the Port Phillip Association led by John Batman. The land was subdivided into several large sheep grazing allotments (Fig. 5.1) (Campbell 1987; Wynd 1981). This had a devastating impact on the lifestyle, culture and agricultural practices of the *Wadawurrung*. 'Collisions'[4] between the *Wadawurrung* peoples and European graziers began from 1836. Although traumatic for all involved, Aboriginal people in particular were killed or wounded, usually as a reprisal for stealing stock for food (Clark 1995).

In 1836, John Helder Wedge recognised the need 'for an establishment for the civilisation of the Aborigines' (Clark 1995, p. 281). In early 1839 a temporary mission was created on the Barwon River at Charlemont (earlier known as Marshall)[5] before the more permanent Buntingdale Mission at Birregurra was established later that year (Clark 1990; Le Griffon 2006; Brewer 1904). The Port Phillip Aboriginal Protectorate, led by George Augustus Robinson, was also established 'to protect Aborigines from any encroachment on their

Fig. 5.1. F. Kruger, 'You Yangs with enclosed sheep paddock in foreground', *c.*1880–89. Source: State Library of Victoria, accession no. H488.

territory, and from acts of cruelty, oppression and injustice' (Clark 1995, p. 3). The Protectorate system failed and it was abolished in 1849 (Clark 1995). In Geelong, from the 1850s, *Wadawurrung* culture was devalued to an existence of confinement through the provision of reserves. At Mount Duneed beside Armstrong Creek, a reserve known as *Dooliebeal* or Stewart Reserve was created in *c.*1856. In 1859, a 640 acre (259 ha) reserve on the north bank of the Little River at Beremboke was set aside and in 1861 the Duneed Aboriginal Reserve at Waurn Ponds was gazetted and occupied by seven surviving *Wadawurrung* members, including *Willem Baa Nip*, also known as 'Willem Gore' and 'King Billy', last member of the *Wada wurrung balug* (Boland 1992; Wynd 2000; *Victoria Government Gazette*, 26 June 1860; Rowe and Huddle 2000). Ultimate control of these reservations resided with Europeans. Following the creation of the 'Central Board Appointed to Watch Over the Interests of Aborigines' in 1860, local Honorary Correspondents were charged with providing rations and clothing, and reporting to the Board on the numbers, condition and location of the *Wadawurrung* (Sumner 1861).[6]

Colonisation of the landscape

Geelong and its inner suburbs

In October 1837, Captain Foster Fyans, first Police Magistrate for the Geelong district, selected the location of the Geelong township (Edmonds 2005). He chose a site where Corio Bay and the Barwon River were close together, despite the river being brackish when it was not flowing strongly (Edmonds 2005). On 21 August 1838, H.W.H. Smyth, Assistant

GEELONG IN 1840.

Fig. 5.2. S. Mossman, 'Geelong in 1840'. Hugh George for Wilson and Mackinnon, Melbourne, 10 June 1876. Source: State Library of Victoria, accession no. A/S10/06/76/45; Mossman (1879).

Surveyor, prepared a plan of Geelong following a grid pattern based on instructions from Robert Hoddle, Government Surveyor. A *Map of Geelong and its Suburbs* was published in 1855 (Proeschel 1855) (Plate 5.2).

'North Geelong' near the bay and 'South Geelong' near the river were the initial locations for development. At 'North Geelong', the earliest buildings lined Corio Street – they were not visible to sailing craft given the high foreshore cliffs. Access from a jetty on the foreshore was through a cutting to Yarra Street (Fig. 5.2). Beyond the formalised town reserve was a well-worn track to Point Henry, the place of disembarkation for arrivals. A sandbar prevented large ships entering Corio Bay and this slowed Geelong's progress until after 1893 when dredging works commenced and the Hopetoun Channel was created (Wild 1950; Thompson Berrill Landscape Design and Mark Trengove Ecological Services 2006). Also partially thwarting development at the height of the Victorian goldrush in 1854 was a false map published by Melbourne merchants, inaccurately showing Ballarat closer to Melbourne than to Geelong (Ball 1983).

Suburban settlement was established adjacent to the township reserve from 1839, on Newtown Hill and at Geelong West (Seaton 1983; Rowe and Jacobs 2010). In 1854, an eastern extension of the original township reserve included an inventive layout with diagonal streets at Sydney Parade and Sydney Avenue (then known as Sydney Place North, South, East and West), and curved streets for Pevensey Crescent and Bourke Crescent (Plate 5.2). These streetscapes alluded to the Baroque geometric planning of Bath, England (Butler and Associates 1991). Years later, in 1890 and south of the Barwon River, Dr Alexander Thomson's 'Kardinia' property at Belmont was subdivided into 250 allotments. Some streets were laid out in terraces, following the contours of the land (Rowe and Jacobs 2007).

Residential subdivisions and tree planting in the early 20th century

Suburban expansion continued in the early 20th century, with some residential subdivisions modelled on Garden City principles.[7] In 1912 at Herne Hill, 300 lots were created in a radiating layout as part of the Manifold Heights Estate (*G.A.* 5 October 1912, p. 3). The focal point was a small park, and trees were to line the streets and be planted on each property (*G.A.* 23 February 1912, p. 2).

An equally remarkable but unrealised scheme was the 'Plan of Greater Geelong' created in 1912 by German engineer Laurell Messy-Rhine (Messy-Rhine 1912). It was to cater for the considerable growth of the Geelong region with a concentric railway network, a ferry line connecting Point Henry to Avalon, and specified community settlements (*G.A.* 22 February 1912, p. 2). The locations of these settlements are now being developed as Geelong's newest suburbs.

Street beautification schemes enhanced the urban landscape. In 1917, the Geelong City Council began a tree-planting program commencing with an avenue of palm trees in Normanby Street, East Geelong (*G.A.* 7 August 1917, p. 2). In 1924, the newly created Geelong Town Planning Association began tree-planting schemes including the planting of 'ornamental' trees beside the Melbourne Road (near Bacchus Marsh Road), and on the bank of the Barwon River (*G.A.* June–July 1912).

Key figures in the City's tree-planting schemes were Dr Frederick Moreton (1861–1948), medical practitioner and the Association's inaugural president, and Cr Robert Purnell (1855–1937), shipping agent and life member of the Geelong Horticultural Society. Their ideas were influential in the formation of the Geelong Town Planning Association (*G.A.* 29 March 1924, p. 5). Purnell's architect-brother, Frederick (1868–1953), and particularly George R. King (1872–1950), Director of the Gordon Technical College and instructor in architecture, and A.J. Laird and T.J. Buchan (of the architectural firm, Laird and Buchan), were noteworthy figures that helped shape Geelong's town planning throughout the 20th century.[8]

At Lara, the Forestry Commission commenced a reafforestation program at the You Yangs which was then considered 'Geelong's National Park' (*G.A.* 8 February 1924, p. 3).[9]

Soldier settlement

In the early and mid 20th centuries, some Crown land and large pastoral estates were compulsorily acquired for Discharged Soldier Settler Schemes to provide returned soldiers from World Wars I and II with agricultural land. Farmland at Barrabool, Bellarine, Belmont, Curlewis, Drysdale, Gnarwarre, Highton, Inverleigh, Lara, Lethbridge, Lovely Banks, Marshall, Meredith, Modewarre, Moolap, Point Wilson, Wallington and Winchelsea was set aside (Battle to Farm 2015; Lake 1987; Smallwood 1992). Yet, the dream of living on the land did not materialise for many. At 'Moranghurk' Homestead, Lethbridge, the land was flat and stony and required clearing by hand (Rowe and Jacobs 2005). The small-sized allotments did not provide sufficient landholdings for sustainable farming (Rowe 2014–16).

Formalised planning

The work of the Geelong Town Planning Association, especially that of some of its leading members from 1924, laid the foundations for the establishment of the Geelong Planning Scheme in 1959 to regulate the development of all land (McLean 2005). Created in 1977, the Geelong Regional Commission functions included 'the planning, management, integration and coordination of development in the region, including the protection of areas of natural beauty of special significance' (McLean 2005, p. 120). In consultation with the City of Geelong, the Geelong Central Area Strategy was developed. It included the City by the Bay plan released in 1981 (McLean 2005; GRC 1981). In 1985, the Commission transformed the central foreshore area between the Moorabool and Yarra Street piers into Steampacket Gardens (McLean 2005). This was a precursor to the development of the waterfront from 1995 by the Victorian state government and City of Greater Geelong (CoGG 2008). It included the installation of over 100 bollards from Limeburners Point to Rippleside Park. Sculpted by Jan Mitchell, the bollards told the story of Geelong's history, from the Indigenous inhabitants, through the introduction of rabbits, to more contemporary characters (CoGG nd).

21st-century developments

In 2013, the City of Greater Geelong released a Central Geelong Action Plan to provide a 15-year strategy for the revitalisation of the central city area. A catalyst for the Plan was the Vision 2 project led by Deakin University. It included 'a distinctive and enduring network of green spines and spaces that support living, working and walking, commencing with the greening of Malop Street' (CoGG 2013, p. 11).

At this time, one of the largest residential developments in Geelong was established at Armstrong Creek, near Mount Duneed. It is to provide housing for 55 000–65 000 people (DELWP 2016). Warralily at Armstrong Creek was Geelong's first and largest master-planned community for which a holistic approach to existing, public and private landscapes was a key design element (Newland Developers 2015; Warralily News, May 2013). Creek and wetland path loops were linked with boardwalks, public art pieces and a park (Fig. 5.3). The public art installations were created by Jeremy Minter, Glenn Romanis and

Fig. 5.3. Warralily Park, 2018. Source: David Rowe.

Mark Trinham as part of a landscape and public art master plan for the area. Initially based on knowledge from Aunty Marlene Gilson, *Wadawurrung* Elder, the narrative for the work was the portrayal of significant untold stories of the past, present and future (*Warralily News*, May 2013). In 2016, their latest installation, *Murrup dja*, was unveiled. It was named by *Wadawurrung* Elder, Uncle Bryon Powell, with *murrup* meaning spirit and *dja* meaning land/earth, which together form the spirit of the land (Millard 2016).

Other residential estates were established in the Armstrong Creek and Mount Duneed area, including Armstrong and Ashbury.

In 2015, a large housing estate commenced at Fyansford following the 'Gen Fyansford' project presented to the City of Greater Geelong in 2001 (Murphy 2015). The previously scrubby land is undergoing transformation and a public parkland is proposed along the banks of the Moorabool River near the old Swan Inn (built 1842) and former pleasure grounds (*G.A.* 21 November 2015).

Exploitation of the landscape: farming and agriculture

From 1836, European farming transformed the outlying areas of the region. Especially transformative were the sheep grazing properties established by the pioneer wool growers. Tangible legacies included those established by Alexander Thompson ('Kardinia', Belmont), Thomas Austin ('Barwon Park', Winchelsea), John Hopkins ('Wormbete', Winchelsea: VHD nd j), Manifold Brothers ('Purrumbete' near Camperdown), Dennis family ('Tarndwarncoort', Warncoort), George Russell ('Golf Hill', Shelford: VHD nd e) and Thomas Armytage ('Ingleby', Winchelsea: VHD nd g) (Brownhill and Wynd 1990; Rowe *et al.* 2001).

Vineyards proliferated near the Barwon and Moorabool rivers, Sutherlands and Cowies creeks, and the Barrabool Hills and Waurn Ponds until 1877, when they were devastated by *Phylloxera* disease (Berry *et al.* 1878; Bonney 1879).

Farming and other establishments brought many changes to the landscape. There was the planting of exotic tree and hedge borders as windbreaks. Drystone walls were prevalent at Sutherlands Creek, Lara, Bannockburn and Lethbridge, being signs of tenure and security, and barriers for the multiplying rabbit (Rowe and Huddle 2000; CAC1995).

Ecological imperialism[10]

From 1837, Europeans re-visioned the land by clearing the native vegetation and introducing exotic plants and animals (LCC 1996). In 1844, in the Western District, the first of the degrading effects caused by imported hoofed livestock came with the disappearance of herbaceous plants, and landslips beside creeks (Robertson 1852).

Exotic plants were introduced for commercial reasons. In 1859, Charles Wyatt established Frogmore Nursery at Fyansford (VHD nd d). It soon boasted thousands of fruit and ornamental trees and shrubs, hedges and flowers, the whole being similar to an English nursery (*The Australasian*, 9 May 1868, p. 25). Other nurseries were established at Batesford, Fyansford and Highton (*The Age*, 17 December 1862, p. 3; *G.A.* 22 October 1866, p. 3, Pescott 1985).

Local squatters and, from 1861, the Victorian Acclimatisation Society also influenced ecological change (Gillbank 1986). Particularly notable was Thomas Austin of 'Barwon Park', Winchelsea (VHD nd a). He introduced blackbirds (*Turdus merula*), pheasant, thrush and wild rabbit (*Oryctolagus cuniculus*) to his property in 1859 (*G.A.* 27 November 1866, p. 2). His rabbits multiplied and spread across Victoria, causing havoc to the environment.

Exotic plants were introduced to remediate coastal erosion. From 1899, marram grass (*Ammophila* sp.) was planted at Thirteenth Beach, Barwon Heads and Ocean Grove to stabilise the fragile sand dunes (Rowe 2002b).

Use of waterways

The Barwon River has been fundamental to the progress and sustainability of the Geelong region. Securing a water supply was critical to this progress. In 1840, Foster Fyans devised a breakwater comprising two large stone walls to provide a natural water supply and prevent salt water from entering the dam. Situated at Breakwater south-east of the Geelong township, it was one of the first major public works in the Port Phillip District and considered a 'work of magnitude and merit' (Edmonds 2005). The river was also exploited by the establishment of tanneries and woollen, flour and paper mills along its banks. It was a water supply and conduit for waste. Noteworthy was the construction of the Barwon Paper Mill at Fyansford in 1877–78 (Brownhill and Wynd 1990). A weir and water race were built above Buckley's Falls to power the mill.

Wool trade, extractive, manufacturing and other industries

Given the large number of sheep grazing stations established in and near the region from 1836, Geelong became the capital of the wool market, where wool was stored and shipped to London (Rowe and Jacobs 2017). From 1840 until the 1960s, the urban landscape of the north-western portion of central Geelong was transformed by the construction of brick and stone woolstores. Early woolbrokers included F.J. Strachan and Co., Swanston Willis and Co., C.J. Dennys, and Dalgety Ibbotson and Co. (Rowe and Jacobs 2017). A Classical Revival Wool Exchange was built in Corio Street in 1928. Today, the surviving adapted woolstores – including the Waterfront Campus of Deakin University on Western Beach Road, the National Wool Museum and Steampacket House on Moorabool Street and the market building on Brougham Street – comprise a unique urban setting (Rowe and Jacobs 2017).

One of the earliest extractive industries in Geelong was limeburning. In 1838, Patrick Melville worked a quarry at Limeburners Point, adjacent to Eastern Park, with James Boucher as limeburner (Harrington 2000). More ubiquitous were the limeburning enterprises at Lara from 1853, their quarries gouging the land (Rowe 2016).

From 1849, Nicholas McCann operated sandstone quarries at Ceres, as did Ezra Firth and Benjamin Holdsworth soon after (Holdsworth also opened another quarry at Waurn Ponds in 1869) (Rowe and Huddle 2000; Wynd 1992; VHD nd f).

In 1853, gold was discovered at Steiglitz (Brownhill and Wynd 1990). The heavily treed hills were denuded and replaced with tracks, tents, buildings, mine shafts, poppet heads and mullock heaps (Washbourne *c*.1870–88). To meet the needs of the mining town, the nearby Brisbane Ranges were rapidly devegetated (Boland 1992). After the demise of gold-mining, the Steiglitz surrounds largely reverted to the former treed environment.[11] At Teesdale, surviving evidence of gold-mining in the 1850s is Chinaman's Lagoon, established as a local water reserve (Heritage Matters 2009).

In 1872, Michael Dongahy relocated his prospering Fairview Ropeworks from Chilwell to a 2.8 ha site in Pakington Street, Geelong West. The complex included a brick factory (now demolished) and ropewalk. In the late 19th and early 20th centuries, this factory was the biggest single source of employment in Geelong West (Rowe and Jacobs 2010).

In 1888 in the swamp at Moolap, Richard Cheetham established a saltern known as Cheetham Saltworks, modelled on those in the south of France (VHD nd b) (Fig. 5.4). The 320 acre (130 ha) site was enclosed and subdivided into paddocks by dykes (Campbell 1893).

At Batesford, from 1867, Peter McCann quarried limestone after opening a quarry on the western side of the Moorabool River north of Fyansford (*G.A.* 13 August 1867, p. 2). The deposits became highly valuable for the manufacture of cement. In 1890, McCann founded the Australian Portland Cement company at Fyansford (McNeill 1990). Shale was secured through the

Fig. 5.4. C. Pratt, Aerial view looking east of Cheetham Saltworks, Moolap, *c.*1927. Source: State Library of Victoria, accession no. H2000.222/9.

alluvial deposits of silicate of alumina on the Moorabool River flats (McCann 1943). Kilns were erected in the escarpment below Herne Hill. Extracted limestone from the Batesford Quarry was initially transported to Fyansford by horse-drawn drays (Rowe and Huddle 2000). This was superseded about 1911 by an aerial ropeway, and from 1921 steam shovels were used on a light-gauge railway system that involved a 1 km tunnel (Rowe and Huddle 2000). In 1966, the railway was replaced by an elevated belt conveyor track on concrete pylons. Remnants of this survive today. The cement product was ubiquitous in the construction of buildings and other structures throughout Victoria until the closure of the operations in 2001 (Batesford Quarry nd). The surviving silos still dominate the skyline today, and the Batesford Quarry is monumental in scale (Rowe and Huddle 2000).

Equally dramatic in the landscape was the transformation of the western shores of Corio Bay with manufacturing and other industries. With more superior port facilities established at Cowies Creek (the port being named Corio Quay) (Wynd 2000), the industries included the Federal Woollen Mill, North Geelong (1913), Ford Motor Co., North Geelong (1924), Corio Distillery, Corio (1928), wheat terminal, North Geelong (1928) fertiliser production at Corio and North Shore (1923 and 1925), farm machinery production by International Harvester, Corio (1939) and the Shell Oil Refinery, Corio (1953) (Wynd 2000).

In 1958, brown coal was found in abundance at Anglesea. A power station was established, burning the coal to produce power for the Alcoa Aluminium smelter at Point Henry (which opened in 1961), greatly transforming this marine landscape (Kellaway and Rowe 2009; Wynd 2005). Other heavy industries established at Point Henry included Primal Chemicals (now Dow Industries), and the Winchester Western Australia ammunition factory in 1967 (Wynd 2005).

Enjoyment of the landscape: recreation and tourism

Public parks in Geelong

Noteworthy in central Geelong from the 19th century was the creation of several public parks. The earliest and most central was the Market Square in 1845 (bounded by Malop, Little Malop, Moorabool and Yarra streets). In 1860, a portion of this swamp land was beautified with the planting of elms and pines (Brownhill and Wynd 1990). Johnstone Park was established in 1848 in Gheringhap Street (Brownhill and Wynd 1990). The park's natural depression formed a water basin known as La Trobe's Dam (Brownhill and Wynd 1990). Criticised as a receptacle of filth in 1865, it was transformed with a program of filling and planting (Brownhill and Wynd 1990). Further transformation occurred after 1916 following a redevelopment scheme by Percy Everett with Laird and Buchan architects (VHD nd h) (Fig. 5.5).

Fig. 5.5. C. Pratt, Aerial view looking south-east over Geelong, 1938. In the upper left foreground are Transvaal Square and Harding and Austin Parks. To the right in the upper middle ground is Johnstone Park. To the upper right is part of Kardinia Park. Source: State Library of Victoria, accession number H91.160/627.

The largest public park was Eastern Park and Geelong Botanic Gardens, reserved in 1851 (Jones 1984). An intricate pattern of pathways and a lake were created by the first curator, Daniel Bunce. He initiated the planting of conifers as the dominant species in the park and formally laid out the parallel paths and beds in the botanic gardens (Chris Dance Land Design *et al.* 1995).

Kardinia Park at South Geelong was used as a cricket ground before 1854 (Taylor 1854), and was set aside as a public park in 1872 (Brownhill and Wynd 1990). Several improvements soon followed, including fencing and exotic tree- and shrub-planting, damming small gullies into a lake, and the laying out of drives. A miniature zoo was established in 1902 (*G.A.* 16 September 1902, p. 2). During the interwar years a sports ground was re-established and in 1947 the oval became the home ground of the Geelong Football Club (Brownhill and Wynd 1990). During the early 21st century, the football stadium underwent a multi-million dollar redevelopment.

Other early parks included Pevensey Gardens, Pevensey Crescent (in two parts, 1878), Harding and Austin Parks in Bellarine Street (both gazetted 1885), and Richmond Place (1854) (Butler and Associates 1991). All except Richmond Place were beautified with exotic trees.[12] In 1910, the Victorian Premier questioned the number of parks in Geelong, given a proposal to excise part of Market Square for commercial development. He claimed that 'Geelong was one of the most over-lunged places he knew' (*The Argus*, 8 March 1910, p. 10). This opinion was rebuked by the public and the editor of the *Geelong Advertiser* for the next five years (*G.A.* 21 March 1910, p. 2; 22 November 1915, p. 2).

Parks and reserves in outer suburban Geelong

Other parks and reserves were established in outer suburban areas. In 1849, Tea Gardens were created at Point Henry (Brownhill and Wynd 1990). They were closed and reopened several times, and were revived for the last time as the Bellarine Tea Gardens in 1892. Hundreds of trees had been planted in the grounds (Brownhill and Wynd 1990).

Although no longer serving as Geelong's principal port, a tramway and truck were erected on the long eastern pier in 1892 for unloading goods to and from steamers. In 1939, a Signal Station was built to control shipping traffic (Huddle *et al.* 1996). From the 1960s, the Point Henry landscape was completely changed with industrial progress. In 2004, the cultural heritage significance of the area was recognised by inclusion in the Victorian Heritage Inventory as the Point Henry Maritime Heritage Precinct (VHD nd i).

In 1857, Queen's Gardens (later known as Queens Park) opened beside the Barwon River at Highton (Brownhill and Wynd 1990) (Fig. 5.6). The 186 acre (75 ha) site was laid out with assistance from the District Surveyor, A.J. Skene, the celebrated botanist Ferdinand von Mueller, and Daniel Bunce. Approximately 2000 ornamental trees and shrubs were planted, a carriage drive formed and seats installed (*G.A.* 16 August 1879, p. 3).

In 1861, a common beside the Barwon River at Belmont was proclaimed for grazing cattle and horses. In the 1920s it was used for aviation and in 1950 it became the site of a migrants' village. From the 1960s, the common was transformed into sports fields, a recreation reserve, a golf course and wetlands (Rowe and Huddle 2000).

Geelong West was bereft of a public park until 1911 when part of the Pineville property at the corner of Pakington and Autumn streets became the Geelong West Park (Rowe and

Fig. 5.6. F. Kruger, 'View on Barwon River, Queen's Park, Geelong', 1880. Source: State Library of Victoria, accession no. H2006.123/31.

Jacobs 2010). In 1936 Sparrow Park was created at the corner of Hope and Elizabeth streets (Rowe and Jacobs 2010). At North Geelong, the grounds of Osborne House were initially the location for polo matches in 1896 (Wynd 1981). In 1919, Osborne Park was created for public purposes (Lovell Chen 2009). Nearby St Helen's Park opened in 1927 (Wynd 1981).

At Torquay, Taylor Park was established in 1882 (Rowe and Jacobs 2009). Many exotic and native trees were planted and pathways laid out (Kellaway and Rowe 2009).

Foreshore reserves

From the 1840s, segregated and fenced-off sea baths were erected in the bayside areas of Corio and Port Phillip Bays, and coastal areas of the Bellarine Peninsula and the Surf Coast. The earliest was in 1844 at Eastern Beach (Rowe 2015). It was followed by others at Western Beach, Geelong (Brownhill and Wynd 1990). Further afield, baths were built at Queenscliff (1862), St Leonards (1865), Portarlington (1868) and Ocean Grove (*c.*1880s) (Rowe and Jacobs 2013). Supplementing the baths and beaches were boat sheds, including those at Ocean Grove, Barwon Heads, Indented Head, Anglesea and Torquay (Rowe and Jacobs 2013) (Fig. 5.7). Victorian government legislation resulted in the removal of boat sheds from the ocean coastal areas in 1936, although bayside boat sheds were permitted to remain (Rowe 2002b; Rowe and Jacobs 2013).

Fig. 5.7. T.T. Holmes, Barwon River foreshore showing boat sheds, Barwon Heads Bridge and the Bluff in the background, n.d. [*c.*1930]. Source: Holmes photograph album, Authentic Heritage Services Pty Ltd.

In 1876, the Ocean Park Reserve from Point Lonsdale to Point Roadknight was permanently reserved from sale (Rowe 2002b). Committees of Management were later established to care for, manage and regulate the use of the coastal reserve (Rowe 2002b; Rowe and Jacobs 2013). A key initiative was the planting of exotic trees and shrubs and the creation of gravelled paths, altering the natural environment to partially conform to European ideals (Rowe 2002b).

Mineral springs

In 1875, the Mineral Springs and Spa Complex was opened on the Clifton Springs foreshore (VHD nd c). It exploited the popular belief in the recuperative and invigorating powers of 'taking the waters'. Two jetties, a bath house, bottling house, manager's cottage, kiosk and other infrastructure were established. Mineral springs at Eastern Beach were also a tourist drawcard (Dearnley 1997).

Eastern Beach

The Eastern Beach Public Park on the Geelong township foreshore was designed in 1925 by the architect Harry Hare (Rowe 2015). Formed with a natural amphitheatre, the design included Art Deco staircases, a kiosk and dressing shed, landscaping and sea walls. In 1938, a curved semi-circular bathing enclosure and circular children's pool were constructed, the whole becoming a remarkable public marine resort (Rowe 2015) (Fig. 5.8).

Fig. 5.8. C. Pratt, Aerial view looking south-east showing Eastern Beach in foreground, with a partial to Eastern Park to the upper left and part of Pevensey Gardens to the upper right, 1939. Source: State Library of Victoria, accession no. H91.160/709.

Brisbane Ranges National Park

In 1973, 3023 ha of bushland between Anakie and Bacchus Marsh was reserved as the Brisbane Ranges National Park. There was a further extension of the reservation in 1979. The Park was created to preserve the rare diversity of flora and to 'primarily provide public enjoyment, education and inspiration in natural environments' (Boland 1992, pp. 61–62).

Wurdi Youang Indigenous Protected Area

In 2006, following investment by the Indigenous Land Council, *Wurdi Youang* at Little River near the foot of the You Yangs was obtained by the Wathaurong Aboriginal Co-operative (Abrahams nd; WAC nd b). The acquisition generated employment and training in grassland restoration and rehabilitation. The area includes the *Wurdi Youang* Stone Arrangement which:

> *... demonstrates the Wathaurong knowledge of the setting of sun solstices and equinoxes, with the east–west axis of the stone arrangement aligning with the solstices. In addition, the most prominent stones at the western end of the arrangement (also the point of the equinoxes) physically match the profile of the hills on the western horizon and form the base of a further arrangement of stones that also align with the setting sun equinoxes and solstices. The stone arrangement represents a significant engineer and intellectual feat (WAC nd b).*

Today, the descendants of the *Wadawurrung* are returning this part of country to its original habitat, and contributing to employment opportunities and cultural experiences.

War memorials

Community landscapes have also been created as war memorials. In 1900, Transvaal Square near the Geelong foreshore frontage was established to commemorate those who served in the South African (Boer) War. The initial design was based on Trafalgar Square, London, but with a less ambitious scheme of grass and exotic tree planting commencing from 1903 (*G.A.* 1900–03; *The Leader*, 17 November 1900).

More common were World War I and II avenues of honour. The earliest planted was on Arbor Day 1916, at the Recreation Reserve, Ceres (*G.A.* 22 November 1916). At Modewarre, an avenue of elm (*Ulmus* sp.) trees was planted in 1918 (replaced with cypress trees (*Cupressus* sp.) in later years) (Rowe and Jacobs 2008; Sheehan and Associates 2003). At Lara in 1918, sugar gums (*Eucalyptus cladocalyx*) and pines (*Pinus* sp.) were planted in the recreation reserve. They were followed by an avenue of cypress trees in Walkers Road in 1948 (Rowe and Jacobs 2016). An avenue of Moreton Bay figs (*Ficus macrophylla*), pines (*Pinus* sp.) and eucalypts (*Eucalyptus* sp.) was planted beside the Colac Road at Waurn Ponds in 1919 (Rowe and Jacobs 2016). At St Helen's Park, an avenue of Ash trees (*Fraxinus* sp.) was planted in 1946 (Rowe and Jacobs 2016).

Other avenues of honour followed. In 1995, a Vietnam Veterans avenue of honour of London Plane trees (*Platanus* x *hispanica*) was planted along the Melbourne Road, North Geelong.[13] At Curlewis, an avenue comprising four separate rows was planted in 2007 to commemorate the soldiers of Ocean Grove who served in all 20th-century wars (Rowe and Jacobs 2016).

Conclusion

European culture has had a profound effect on the Geelong landscape. The Country of the *Wadawurrung* was initially viewed as 'waste lands' and re-visioned for its grazing potential. In 1836, Point Henry was the gateway for squatters who transformed the Western District. In 1838, the Geelong township was established, its location between Corio Bay and the Barwon River being influenced by the landscape. The transformation went beyond urbanisation. It involved the re-landscaping of rural land for pastoral, commercial and cultural pursuits. Extractive industries, including gold-mining at Steiglitz and quarrying at Geelong, Lara and Fyansford, a saltern at Moolap and manufacturing industries at North Geelong, North Shore and Corio, brought major change to the landscape.

From the 1850s, Europeans introduced new plant and animal species. Some exotic species devastated indigenous ecosystems, such as the rabbits brought to Winchelsea. Others, such as the propagation of ornamental trees and shrubs, made an important contribution to the beautification of the newly established landscapes in the Geelong region.

Urban progress in the early 20th century included some notable housing estates and streetscape beautification schemes based on Garden City ideals. Formalised planning, including the creation of statutory bodies such as the Geelong Regional Commission, led to the protection of important landscapes. In the early 21st century, some rural areas were transformed into residential suburbs, including Armstrong, Warralily and Fyansford, where some natural features and settings were retained.

The management of bayside settings in Corio and Port Phillip Bays and foreshore reserves along the Bellarine Peninsula and Surf Coasts for public enjoyment has ensured the retention and enhancement of these environments that are unique to the Geelong region.

Endnotes

1 The *Wadawurrung* clans in the Geelong region were *Bengalat balug* (Indented Head), *Gerarlture* (west of Lake Modewarre), *Neerer balug* (between Geelong and the You Yangs), *Wada warrung balug* (Barrabool Hills) and *Yaawangi* (You Yang Hills) (Clark 1990).

2 Batman apparently paid in blankets, knives, looking glasses, tomahawks, handkerchiefs, shirts, scissors, clothing and flour. A yearly tribute or rent was also agreed.

3 In *John Batman and the Aborigines*, John Campbell (1987, pp. 99–106), gave further details on the dubious nature of the deeds. He concluded: 'There can be no doubt that the deeds were fraudulent and that Batman was untrustworthy. His documentation was highly irregular and his intentions almost certainly devious. These are aspects of Batman's story not previously told by historians, and certainly not part of the popular view of Batman as heroic adventurer, founder of Melbourne.'

4 This was the official term used to describe conflict between the European settlers and the Aborigines by Crown Commissioner Frederick Powlett to Governor Charles La Trobe in 1846 (Powlett 1846).

5 In a newspaper article, G.N. Brewer (1904, p. 4) claimed that a Presbyterian [sic. Wesleyan] Mission Station was located on property then owned by G. Gardiner of Marshalltown. On his death in 1911, George Gardiner's Statement of Assets included part of Crown allotments 4 and 5 in the Parish of Connewarre 'below the Breakwater', and Crown Allotment 15 of Section 1 in the Parish of Connewarre. It would appear to have been on one of these allotments that the temporary Aboriginal mission was located. See G. Gardiner, Probate Administration files, 1911, VPRS 238/P3 Unit 209, Public Record Office Victoria.

6 Mordaunt Maclean, a squatter at Bacchus Marsh and Justice of the Peace, was appointed the first Honorary Correspondent at Bacchus Marsh. Along with the local bench of magistrates, he had supported the creation of the Aboriginal reservation at Beremboke (which became known as the Steiglitz reserve). Maclean had previous experience interacting with Indigenous peoples in New South Wales. His view was that written agreements should be made with the Aboriginal people, with wages recovered before the bench of magistrates for farm work performed. His perception of confinement therefore extended beyond the reservation area, to total control of the Indigenous way of life. The system was considered to protect European squatters 'from being defrauded, and [the Indigenous peoples] were at the same time liable to punishment for absconding or misconduct' (Maclean in Sumner 1861, Appendices 2 and 5; OCLS 1859–60). The other original Honorary Correspondents in the Geelong area were Henry Andrews, J.M. Garratt and Charles Read.

7 Garden City principles were a form of urban planning established by Ebenezer Howard in England in 1898. They involved the provision of self-contained communities surrounded by greenbelts. There was an emphasis on correct orientation, road layouts, public recreation spaces and private gardens for healthy living (Howard 1902).

8 See Page (1990), *An Architectural Apex* (pp. 78–81). Page gives an overview of the town planning work of Laird and Buchan in Geelong in the 20th century. In particular, he states that Thomas Buchan described George King as 'one of the very few qualified men in Australia of the Town Planning Institute of Greater Britain' and that Buchan felt that 'Victoria should take the opportunity to show Australia the way in town planning'.

9 In 1992, the You Yangs Regional Park was gazetted to conserve and protect the natural ecosystems (Pescott 1995).

10 This term was first used by Alfred Crosby (2009) in *Ecological Imperialism: The Biological Expansion of Europe, 900–1900*, in which he proffered the theory that European settlement led to major shifts in the ecology of indigenous environments because of the accidental or deliberate introduction of flora and fauna.
11 In 1979, Steiglitz was proclaimed a National Park (Heritage Matters 2009).
12 In 1999, Harding Park was sold to a private consortium under controversial circumstances, with commercial and residential developments being carried out (Auditor-General of Victoria 2001).
13 The London Plane (*Platanus* x *hispanica*) trees were replaced with the existing Ironbark (*Eucalyptus sideroxylon*) trees in 1999–2000 (Krastins 2015–16).

References

Abrahams R (nd) *Wurdi Youang Indigenous Protected Areas: Enhancing the Natural and Cultural Heritage of Wathaurong* Country. Online brochure. Wathaurong Aboriginal Co-operative. http://www.wathaurong.org.au/ (accessed 1 March 2018).

Askew M (1991) *Environmental History: Geelong City Urban Conservation Study*, vol. 2. City of Geelong, Geelong.

Auditor-General of Victoria (2001) *Report on Public Sector Agencies: Results of 30 June 2001 Financial Statement Audits, Parliament of Victoria*. Government Printer, Melbourne.

Ball G (1983) *Shadows on the Wall*. Graham James Ball, Geelong.

Batesford Quarry (nd) *Batesford Quarry History*. http://www.batesfordquarry.com.au/Main.asp?_=Our%20History (accessed 1 March 2018).

Battle to Farm (2015) *Battle to Farm: WW1 Soldier Settlement Records in Victoria*. Public Record Office Victoria, Melbourne. http://soldiersettlement.prov.vic.gov.au/map/ (accessed 1 March 2018).

Berry G, Hopton E, Wallis AR (1878) *Phylloxera vastatrix: Report of the Inspectors of Vineyards in the Geelong District*. Parliamentary Paper no. 17. Parliament of Victoria. John Ferres, Government Printer, Melbourne.

Billis RV, Kenyon AS (1930) *Pastures New: An Account of the Pastoral Occupation of Port Phillip*. Macmillan and Co., London.

Boland S (1992) The history of the Brisbane Ranges National Park. Unpublished BA thesis. Monash University, Melbourne.

Bonney SB (1879) *Plan showing the Vineyards of the Geelong District: also showing the Natural Features of the Locality*. Prepared for the Department of Crown Lands and Survey, Victoria. Steam Lithographers, Government Printing Office, State Library of Victoria, Melbourne.

Brewer GN (1904) Memories of old Geelong: the 'Connewarre Flatties', their home and cruising grounds. *Geelong Advertiser*, 10 September.

Brownhill WR, Wynd I (1990) *The History of Geelong and Corio Bay with Postscript 1955–1990*, postscript edn. *Geelong Advertiser*, Geelong.

Butler G and Associates (1991) *Geelong City Urban Conservation Study*, vol. 3. City of Geelong, Geelong.

CAC (1995) *If These Walls Could Talk: Report of the Corangamite Dry Stone Wall Conservation Project*. Corangamite Arts Council, Terang, Victoria.

Campbell AJ (1893) *Tourists' Guide to Geelong and the Southern Watering Places on the Bay and Coast and Popular Holiday Resorts: including Torquay (late Spring Creek), Barwon Heads, Ocean Grove, North Shore, Point Lonsdale, Queenscliff, Portarlington, Drysdale and Clifton Springs &c., &c., Together with a Chapter on the Geology of the District by T.S.H.; First Settlers and Early Days by Rev. Dr. Campbell; and a Trip Down the Lakes from Geelong to Barwon Heads by J.L.C.* M.L. Hutchinson, Melbourne.

Campbell AH (1987) *John Batman and the Aborigines.* Kibble Books, Malmsbury.
Chris Dance Land Design et al. (1995) *Geelong Botanic Gardens and Eastern Park Conservation Management Plan.* City of Greater Geelong and Friends of the Geelong Botanic Gardens, Geelong.
Clark ID (1990) *Aboriginal Languages and Clans: An Historical Atlas of Western and Central Victoria, 1800–1900.* Monash Publications in Geography No. 37. Department of Geography and Environmental Science, Monash University, Melbourne.
Clark ID (1995) *Scars in the Landscape: A Register of Massacre Sites in Western Victoria, 1803–1859.* Australian Institute of Aboriginal and Torres Strait Islander Studies, Canberra.
CoGG (nd) *Geelong Bollard Trail Walk.* City of Greater Geelong, Geelong. http://www.geelongaustralia.com.au/bollards/article/item/8d250a1f01e3425.aspx (accessed 25 October 2016).
CoGG (2008) *The Waterfront Story.* City of Greater Geelong, Geelong. https://www.geelongaustralia.com.au/armstrongcreek/armstrong/article/item/8cfafd49ea31e3f.aspx (accessed 1 March 2018).
CoGG (2013) *Central Geelong Action Plan.* Central Geelong Task Force, City of Greater Geelong, Geelong. https://www.geelongaustralia.com.au/common/public/documents/amendments/8d2d0901dd9c6bb-CentralGeelongActionPlan.pdf
Crosby AW (2009) *Ecological Imperialism: The Biological Expansion of Europe, 900–1900.* Cambridge University Press, New York.
Dearnley C (1997) *The Development of Eastern Beach, Geelong: A Chronicle, 1891 to 1940.* Colin Dearnley, Geelong.
DELWP (2016) *Revitalising Central Geelong Action Plan.* Department of Environment, Land, Water and Planning, Melbourne. https://www.planning.vic.gov.au/__data/assets/pdf_file/0021/95250/Revitalising-Central-Geelong-Action-Plan.pdf
Edmonds L (2005) *Living by Water: A History of Barwon Water and its Predecessors.* Barwon Region Water Authority, Geelong.
Gardiner G (1911) Statement of Assets and Liabilities, Probate Administration files, VPRS 28/P3 Unit 209, Public Record Office Victoria.
Geelong Advertiser, 'Flesh, fish and fowl', 27 November 1866.
Geelong Advertiser, 22 October 1866; 13 August 1867; 21 June 1900; 16 November 1900; 7 February 1901; 25 February 1902; 5 June 1902; 16 September 1902; 21 March 1910; 22 February 1912; 23 February 1912; 5 October 1912; 7 July 1915; 22 November 1916; 7 August 1917; 19 June 1924; 2 July 1924; 12 July 1924; 21 November 2015.
Gillbank L (1986) The origins of the Acclimatisation Society of Victoria: practical science in the wake of the gold rush. *Historical Records of Australian Science* **6**(3), 359–374. doi:10.1071/HR9860630359
GRC (1981b) *Geelong: City by the Bay.* Geelong Regional Commission, Geelong.
Grimes C (1803) *Map of Port Phillip.* Lithographed at the Department of Lands and Survey by T. Salter, 1879. State Library of Victoria, Melbourne.
Halsey M (2006) *Deleuze and Environmental Damage: Violence of the Text.* Ashgate Publishing, Burlington.
Harrington J (2000) *An Archaeological and Historical Overview of Limeburning in Victoria.* Heritage Council of Victoria, Melbourne.
Heritage Matters (2009) *Golden Plains Shire Heritage Study Phase 2.* Report prepared by Heritage Matters Pty Ltd for Golden Plains Shire, Bannockburn, Victoria.
Hobart Mercury, 22 January 1935.
Howard E (1902) *Garden Cities of Tomorrow.* Faber, London.
Huddle L, Howe R, Lewis R, Francis K (1996) *Bellarine Heritage Study.* City of Greater Geelong, Geelong.
Jones G (1984) *Growing Together: A Gardening History of Geelong – Extending to Colac and Camperdown.* George Samuel Jones, Geelong.
Kellaway C, Rowe D (2009) *Farmland, Forest and Surf: Environmental History.* Surf Coast Shire, Torquay, Victoria.

Krastins K (2015–16) *Vietnam Veterans Memorial and Avenue of Honour and Osborne House Gates Heritage Assessment*. City of Greater Geelong, Geelong.

Lake M (1987) *The Limits of Hope: Soldier Settlement in Victoria 1915–38*. Oxford University Press, Melbourne.

LCC (1996) *Historic Places: Special Investigation: South-western Victoria Descriptive Report*. Land Conservation Council, Melbourne.

Le Griffon H (2006) *Campfires at the Cross: An Account of the Bunting Dale Aboriginal Mission 1839–1851 at Birregurra, near Colac, Victoria: with a Biography of Francis Tuckfield*. Australian Scholarly Publishing, Melbourne.

Lovell Chen (2009) *Osborne House and Stables North Geelong, Victoria: Conservation Management Plan*. Report prepared by Lovell Chen Pty Ltd for City of Greater Geelong, Geelong.

McCann WB (1943) *History of Descendants of Peter McCann who landed in Australia in 1799 and the Establishment of the Cement Industry and its Development in Victoria*. WB McCann, Melbourne.

McLean GA (2005) The history of the Geelong Regional Commission. Unpublished PhD thesis. Deakin University, Geelong.

McNeill J (1990) *A Journey to Destiny: 100 Years of Cement Manufacturing at Fyansford by Australian Cement Limited*. Australian Cement Ltd.

Messy-Rhine L (1912) Plan of Greater Geelong. *News of the Week*, 5 December.

Millard J (2016) Sculpting history. *Weekly Review*, 25 February.

Mossman S (1879) Geelong in 1840. *Geelong Advertiser*, 19 December.

Murphy N (2015) *Geelong Indy*, 7 July.

Newland Developers (2015) *Warralily Design and Siting Guidelines*. Newland Developers Pty Ltd, Melbourne. http://warralily.com.au/getmedia/275a8163-7a64-4354-a737-233666aea0a7/Warralily-Design-Guidelines

OCLS (1859–60) *Correspondence to and from the Office of Crown Lands and Survey*. VPRS 242/P0 Unit 466, file C15690. Public Record Office Victoria, Melbourne.

Page M (1990) *An Architectural Apex*. Buchan Laird, Melbourne.

Pascoe B (2014) *Dark Emu: Black Seeds Agriculture or Accident?* Magabala Books, Broome, WA.

Pescott J (1985) *South Barwon 1857–1985*. Neptune Press with City of South Barwon, Geelong.

Pescott T (1995) *The You Yangs Range*. Yaugher Print, Geelong.

Powlett FA (1846) *Letter to Lt. Govt. C.J. La Trobe*, 11 September. VPRS 19/P1, file 47/1348. Public Record Office Victoria, Melbourne.

Proeschel F (1855) *Map of Geelong and its Suburbs*. Campbell & Fergusson Lithographers, State Library of Victoria, Melbourne.

Robertson JG (1852) *Letter to Lt. Gov. C.J. La Trobe*, 26 September. In TS Bride (Ed.) *Letters from Victorian Pioneers*. Trustees of the Public Library by Robert S. Brain, Government Printer, Melbourne.

Rowe D (2002a) *Hume and Hovell Monument, Lara: Conservation Management Plan*. City of Greater Geelong, Geelong.

Rowe D (2002b) *The Pleasure Grounds of the Barwon Coast: A History*. Barwon Coast Committee of Management, Barwon Heads, Victoria.

Rowe D (2015) Eastern Beach, Geelong's aquatic playground: its physical history, context and conservation. *Historic Environment* **27**(3) 78–90.

Rowe D (2016) *Lara Heritage Review: Thematic Environmental History*, vol. 2. City of Greater Geelong, Geelong.

Rowe D (2014–16) Forest, farmland and gold: thematic environmental history of the western region of the Moorabool Shire, vol. 2. In D Rowe and W Jacobs (Eds) *West Moorabool Shire Heritage Study Stage 2A*. Moorabool Shire Council, Ballan, Victoria.

Rowe D, Huddle L (2000) *Greater Geelong Outer Areas Heritage Study Stage 2*. City of Greater Geelong, Geelong.

Rowe D, Jacobs W (2005) *Moranghurk Homestead Conservation Management Plan*. For the owners, Lethbridge.

Rowe D, Jacobs W (2007) *Belmont Heritage Review*. City of Greater Geelong, Geelong.

Rowe D, Jacobs W (2008) *Surf Coast Shire Heritage Study: Stage 2B*. Surf Coast Council, Torquay, Victoria.

Rowe D, Jacobs W (2009) *Surf Coast Shire Heritage Study*. Surf Coast Council, Torquay, Victoria.

Rowe D, Jacobs W (2010) *Ashby Heritage Review*, vol. 1. City of Greater Geelong, Geelong.

Rowe D, Jacobs W (2013) *Batman Park Heritage Review*. City of Greater Geelong, Geelong.

Rowe D, Jacobs W (2016) *Lara Heritage Review*. City of Greater Geelong, Geelong.

Rowe D, Jacobs W (2017) *Former Dennys Lascelles Woolstore Heritage Assessment, 20 Brougham Street, Geelong*. City of Greater Geelong, Geelong.

Rowe D, Huddle L, Jacobs W, Jellie P (2001) *Purrumbete Homestead Conservation Management Plan*. Privately commissioned.

Seaton G (1983) *Model Borough: An Introduction to old Newtown and Chilwell*. City of Newtown, Newtown, Victoria.

Sheehan and Associates (2003) *Colac Otway Shire Heritage Study Thematic History*, vol. 1. Shire of Colac Otway, Colac, Victoria.

Smallwood R (1992) *Hard to go Bung: World War Two Soldier Settlement in Victoria 1945–1962*. Hyland House, Melbourne.

Sumner TJ (1861) *First Report of the Central Board Appointed to Watch Over the Interests of the Aborigines in the Colony of Victoria. Presented to both Houses of Parliament*. John Ferres, Melbourne.

Sydney Chronicle, 30 September 1847.

Taylor JH (1854) *Survey of Geelong, reduced and drawn at the Survey General's Office by J Kearney, Draughtsman*. Victorian Surveyor General's Department, Melbourne.

The Age, 17 December 1862.

The Argus, 27 June 1901; 5 May 1903; 29 August 1904; 8 March 1910.

The Australasian, 9 May 1868.

The Leader, 17 November 1900.

Thompson Berrill Landscape Design, Mark Trengove Ecological Services (2006) *Point Henry Foreshore Management Plan*. Prepared for the City of Greater Geelong, Geelong.

VACL (2015) Victorian Aboriginal Corporation for Languages. http://www.vaclang.org.au/Search.html?searchphrase=all&searchword=djilang (accessed 14 July 2015).

VHD (nd a) *'Barwon Park', Winchelsea*. Victorian Heritage Database. http://vhd.heritagecouncil.vic.gov.au/places/1234 (accessed 1 March 2018).

VHD (nd b) *'Cheetham Saltworks', Moolap, H1157*. Victorian Heritage Database. http://vhd.heritagecouncil.vic.gov.au/places/4820 (accessed 21 October 2016).

VHD (nd c) *'Former Mineral Springs', Clifton Springs, H2088*. Victorian Heritage Database. http://vhd.heritagecouncil.vic.gov.au/places/13999 (accessed 21 October 2016).

VHD (nd d) *'Frogmore', Fyansford, H696*. Victorian Heritage Database. http://vhd.heritagecouncil.vic.gov.au/places/512 (accessed 21 October 2016).

VHD (nd e) *'Golf Hill Homestead and Outbuildings, Shelford'*. Victorian Heritage Database. http://vhd.heritagecouncil.vic.gov.au/places/23706

VHD (nd f) *'Holdsworth Quarry', Ceres, H7721-0060*. Victorian Heritage Database. http://vhd.heritagecouncil.vic.gov.au/places/6409 (accessed 21 October 2016).

VHD (nd g) *'Ingleby Homestead and Outbuildings'*. Victorian Heritage Database. http://vhd.heritagecouncil.vic.gov.au/places/1233 (accessed 21 October 2016).

VHD (nd h) *'Johnstone Park', Geelong*. H1114. Victorian Heritage Database. http://vhd.heritagecouncil.vic.gov.au/places/1992 (accessed 21 October 2016).

VHD (nd i) *'Point Henry Maritime Heritage Precinct'*. Victorian Heritage Database. http://vhd.heritagecouncil.vic.gov.au/places/111689

VHD (nd j) *'Wormbete Homestead', Winchelsea*. Victorian Heritage Database. http://vhd.heritagecouncil.vic.gov.au/places/69773 (accessed 1 March 2018).

Victoria Government Gazette, 26 June 1860. Government Printer, Melbourne.

WAC (nd a) *History*. Wathaurong Aboriginal Co-operative, Geelong. http://www.wathaurong.org.au/history-profile/ (accessed 1 March 2018).

WAC (nd b) *Indigenous Protected Area*. Wathaurong Aboriginal Co-operative, Geelong. http://www.wathaurong.org.au/indigenous-protected-area/ (accessed 1 March 2018).

Warralily News, May 2013.

Washbourne TJ (*c*.1870–88) Photograph of Steiglitz. Accession no. H38303. State Library of Victoria, Melbourne.

Wild D (1950) *The Tale of a City: Geelong 1850–1950*. F.W. Cheshire, Melbourne.

Wynd I (1981) *So Fine a Country: A History of the Shire of Corio*. Shire of Corio, Geelong.

Wynd I (1992) *Barrabool: Land of the Magpie*. Barrabool Shire, Torquay, Victoria.

Wynd I (2000) *Greater Geelong Outer Areas Heritage Study Environmental History*, vol. 2. City of Greater Geelong, Geelong.

Wynd I (2005) *Bella-Wein: A History of the Shire of Bellarine*, 2nd edn. Bellarine Online for Bellarine Historical Society, Drysdale.

PART 2

ECOLOGY OF THE GEELONG REGION

6

Vegetation changes since European arrival

Mark Trengove

> *... It is to be feared and regretted that many species of herbs and annuals which at present adorn the hillsides and valleys of locality, may in the course of a few years disappear owing to pasturage of stock. Even now some of these and the rare shrubs and other small plants must be looked for within the numerous wire fence enclosures or paddocks where traffic is limited ...*
>
> – William Guilfoyle (1894).

The story of the changes in the native plant communities of the Geelong locality is, for the most part, the story of the loss and degradation of native vegetation. The salient point is that decline is ongoing (Carr *et al.* 2001).

Overview of impacts

Europeans noted changes in the vegetation of the Geelong region relatively early. Direct impacts, such as clearing of trees for firewood and construction and to improve grazing land, were noted by 1842 (Wynd 1986).

> *In the first 10 years the axe was used ruthlessly to slay trees and to clear the land for settlement and pasture. The destruction went on unchecked* (Brownhill 1955).

Indirect impacts, such as the loss of diversity of native plant species and the introduction of weeds, were also noted relatively early. By 1858 there were complaints concerning thistles on the Bellarine Peninsula (Wynd 1988).

Over time, impacts have continued and often increased in intensity as the European population has grown and sought higher living standards. Regional resources have also been exploited, particularly the native grasslands which were heavily grazed for wool production.

> *The early settlers brought 1000s of sheep and cattle. The rich pastoral lands in the neighbourhood of Geelong were invaded and used profitably* (Brownhill 1955).

As a result of European land practices, it is estimated that only 5% of the City of Greater Geelong now supports remnant indigenous vegetation, of which the majority is degraded (Carr *et al.* 2001).

The dominant causes of native vegetation loss and degradation include:

- agricultural impacts, including the loss of native vegetation due to grazing by introduced hard-hoofed animals, clearing, the introduction of exotic plant and animal species, in particular the European Rabbit (*Oryctolagus cuniculis*) (DEWHA 2008), and the application of fertilisers and pesticides
- mining impacts, including the mining of limestone, sandstone, granite and basalt for building, road construction and so on (Brownhill 1955), the mining of limestone for construction (lime) (Harrington 1996) and for agriculture, the mining of shell grit and sand deposits (Brownhill 1955), and mining for gold (Sumner 1982)
- urban and industrial impacts (the direct and indirect impacts of development)
- the modification of water flows, including the loss of wetlands, as a result of the construction of dams and diversions (Edmonds 2005)
- changes in land management practices, such as loss of native herbivores and changes due to altered fire regimes. The *Wadawurrung* people who, at the time of European arrival, had lived in this region for tens of thousands of years, regularly used fire as a tool to manage the landscape (see Chapter 4).

Perhaps the major consequence of these changes has been the fragmentation of native vegetation. Ecological theory predicts that, in general, fragmentation leads to decline in the ecological condition of vegetation (Debinski and Holt 2000). Consequences of fragmentation for native plant populations include:

- altered landscape processes such as increased erosion and increased salinity
- reduction of plant populations, for example the Large-flowered Crane's-Bill (*Geranium* sp.), which is now reduced to one small population on the Geelong–Ballarat Railway line at Bell Post Hill
- loss of plant taxa. The Lara Grevillea (*Grevillia* 'Lara dwarf'), is now extinct in the wild but is kept in horticulture at the Geelong Botanic Gardens (Pescott 1989; Carr 1989)
- altered environmental conditions and 'edge effects'
- increased invasive species, both plant and animal
- changes in disturbance regimes, such as herbivore grazing and fire patterns
- reduced pollination and seed set, as well as altered seed predation and dispersal patterns due to impacts of diminished habitat size and isolation (Hobbs and Yates 2003).

Additionally, there are significant impacts from fragmentation for native faunal populations.

Grasslands

Loss and degradation due to agricultural exploitation has occurred more extensively in areas with more productive volcanic soils that carried native grassland and grassy woodland vegetation. By the end of the 1830s the grasslands of western Victoria, including those of the Geelong region, were being exploited by squatters for sheep grazing (Nicholas *et al.* 2015). This grazing altered the composition of native grasslands and led to the loss of more palatable species such as the Murnong (*Microseris scapigera*). However, it was the introduction of superphosphate, the sowing of 'improved' pasture species and the intensified use of heavy farm machinery after World War II that caused the rapid loss of much of the native grasslands (Nicholas *et al.* 2015). The current practice of removing basalt rock and laser-levelling the land for sowing crops, facilitated by the availability of larger machinery, continues to impact upon native grasslands.

It is estimated that more than 99.5% of the grasslands of western Victoria have been lost (Kirkpatrick *et al.* 1995).

Changes in the grasslands are documented in several ways. Local artist Sue Anderson has painted these landscapes for 30 years (Plates 6.1 and 6.2; Anderson 2018). Her work records changes in the composition of plant and animal species over time, including variations in abundance of plant species such as native Windmill Grass (*Chloris truncata*) and the proliferation of weed species such as Patterson's Curse (*Echium plantagineum*). These indicators can be traced to seasonal weather variation, disturbance to and loss of native habitat, and the drying of the land due to climate change, 'the Corangamite region's climate is changing' (CCMA 2016).

Across this agricultural landscape, small areas such as cemeteries and rail and roadside corridors have occasionally avoided significant disturbance. These sites are often the most diverse, and hence most significant, remnants of native grassland vegetation. In the region, significant grassland remnants occur along the Geelong–Ballarat and Geelong–Melbourne railway lines, the Old Melbourne Road, the Murtcaim Flora and Fauna Reserve, the Corio Grassland Reserve (Carr *et al.* 2001) and the Bannockburn Cemetery (Trengove 2002).

Woodlands

Much of the Bellarine Peninsula and the Otway plain between Geelong and Torquay consisted of a mosaic of heathy woodlands dominated by Manna Gum (*Eucalyptus viminalis*) and grassy woodlands dominated by the Melbourne Yellow Gum (*Eucalyptus leucoxylon* subsp. *connata*) and the Bellarine Yellow Gum (*Eucalyptus leucoxylon* subsp. *bellarinensis*) as well as River Red Gum (*Eucalyptus camaldulensis*) (see Chapters 3 and 7; DNRE 2000; DELWP 2018a).

Grassy woodlands

Grassy woodlands dominated by the endemic Bellarine Yellow Gum (*Eucalyptus leucoxylon* subsp. *bellarinensis*) are naturally limited to the Torquay and the Ocean Grove/Wallington areas (Trengove 2001a, 2013a). Due in large part to the location of this endemic vegetation, the pressures associated with development continue to impact on this community, although the removal of stock grazing has benefited individual trees in a few instances.

River Red Gum-dominated grassy woodlands typically occur on relatively fertile low-lying and riparian areas. These woodlands have been subjected to extensive impacts from agriculture and firewood collection. Large tracts of River Red Gum woodlands on the flats of the eastern Bellarine Peninsula were harvested to supply firewood to Melbourne. For ~80 years until the 1940s, large quantities of wattles (*Acacia* spp.) were harvested for the leather tanning industry (Edwards 1952).

Heathy woodlands and heathlands

In contrast to the grassy ecosystems, the majority of the native vegetation that remains in the region occurs in areas of relatively poor soils, such as the Brisbane Ranges, the Eastern Otways and areas of the Bellarine Peninsula, that were not considered worthy of clearing for agriculture. The vegetation that occurs in these areas typically consists of heathland

and heathy woodland or forest, usually dominated by Messmate Stringybark (*Eucalyptus obliqua*). Although the vegetation of the Brisbane Ranges has been extensively modified, initially during the gold rush (Sumner 1982) and subsequently for firewood collection (Parks Victoria 1997), it is still essentially intact and now constitutes the largest area of native vegetation within the City of Greater Geelong (Carr *et al.* 2001; Weste and Taylor 1971). Due to its rich floristic diversity, the Brisbane Ranges were proclaimed a National Park in 1973 and substantially enlarged in 1979 (Parks Victoria 1997).

Impacts of weeds

Invasion of indigenous vegetation by environmental weeds is arguably the most serious conservation problem in Victoria (Frood and Calder 1987). From the first days of European settlement in Victoria, deliberately and accidentally introduced exotic plants have naturalised and become weeds (Carr 1993; Parsons and Cuthbertson 1992). In the Geelong region, the invasion of weeds into native vegetation and agricultural land continues to be a major concern. Within the City of Greater Geelong there are 1119 exotic naturalised plant species (62% of the total flora) (Carr *et al.* 2001). In inland Geelong at Buckley Falls, weeds constitute 56% of the total flora, while in the coastal Borough of Queenscliffe weeds constitute 53% of the total flora (Trengove 1992).

Gorse (*Ulex europaeus*) and Sweet Briar (*Rosa rubiginosa*) are woody weeds that were deliberately introduced into Australia in the early 1800s (Parsons and Cuthbertson 1992). Other deliberately introduced weeds such as Sweet Pittosporum (*Pittosporum undulatum*), a native plant that is not indigenous to this region, Hawthorn (*Crataegus monogyna*) and, nearer to the coast, Mirror Bush (*Coprosma repens*), Boneseed (*Chysantemoides monilifera*) and Italian Buckthorn (*Rhamnus alaternus*) are all shrubs that produce berries. By eating the berries and voiding the seeds, exotic animals such as European Red Fox (*Vulpes vulpes*) and Common Blackbird (*Turdus merula*) are vectors for spreading the weeds (Carr 1993).

At the Edwards Point Fauna Reserve, an area of relatively isolated and undisturbed native vegetation, a total of 87 weed species are recorded (49% of the total flora). Thirty-one of those species are horticultural escapes, with 28 of the 31 identified as 'serious weeds' that require control (Trengove 1999).

Recently, several serious environmental weed species appear to have become more widespread in the region. Weeds like African Thistle (*Berkheya rigida*), Blanket Weed (*Galenia pubescens*) and species of *Nassella* (Serrated Tussock, Chilean Needle-grass, Texas Needle-grass, etc.) are now more widespread. These changes are most likely due to altered land management practices, including more intensive and diverse settlement and management of the peri-urban areas, and to changing climate.

Changes in vegetation due to grazing regimes

At Batesford, a significant area of Drooping Sheoke (*Allocasuarina verticillata*) and Yellow Gum/Manna Gum woodland remains. The Dog Rocks Flora and Fauna Sanctuary has been managed primarily for conservation since it was purchased by the Belcher family in 1856 (Trengove 2013b). The adjacent area to the south has been grazed by sheep but not subject to pasture improvement or significant disturbance (Bill Honey, landholder, pers. comm. October 2016). Both sites still support significant native vegetation, but are now quite different in terms of plant species composition and vegetation structure. The grazed area is now open and dominated by large old Eucalypts with an understorey dominated by native grasses, while the vegetation of the sanctuary is dominated by a dense shrub layer. Plant species that are palatable to sheep, such as Common Everlasting (*Chrysocephalum apiculatum*), are absent from the open grazed area, while other plant species that are sensitive to overcrowding, such as Kangaroo Grass (*Themeda triandra*), are largely absent from the sanctuary (Trengove 2013b). Although they are adjacent, the two areas have different weed floras. They also provide differing habitat values for fauna.

Similar vegetation patterns are found locally where town commons were established in agricultural areas such as at Teesdale, Bannockburn and Inverleigh, which are now gazetted nature reserves. The town commons allowed practices such as firewood collection but were only occasionally grazed, thus retaining relatively intact native grassy woodland understorey vegetation. The surrounding privately owned areas were heavily grazed and have relatively degraded understorey vegetation, but often retain large old trees (LCC 1987).

Changes in coastal vegetation due to altered management practices

At Point Lonsdale, two centrally located adjacent properties, 'Ballara' and 'Arilpa', have strong historic associations with former Prime Minister Alfred Deakin. Alfred and his wife Pattie had strong environmental beliefs and were keen to preserve what they recognised as significant native vegetation at 'Ballara' (Granddaughter Judith Harley, pers. comm. 2016). At the time, most of the native vegetation at 'Ballara' and 'Arilpa' was heathland dominated by Austral Grass-tree (*Xanthorrhoea australis*). In 1882, George Crouch described the land, 'the scrub was low and spread with wildflowers' (Dunn 1949). In the same publication, Nancy Dunn observed, 'The brushwood that was once scarcely knee high then, now shuts out any view of the rip'. This change in vegetation structure is what we now acknowledge to be invasion of the native shrub Coast Tea-tree (*Leptospermum laevigatum*) beyond its natural distribution. Early records such as those by Byrne (1930) state that 'At that time the beach tea-tree was confined to a narrow fringe in the sand just above the high water mark'.

The underlying mechanisms of woody plant encroachment are varied, but changes in key drivers such as climate, disturbance regimes and land use practices (Bennett 1994) may favour some woody species that are normally held in check by competitive interactions (Britton *et al.* 2000). Regardless of the mechanism, the invasion or encroachment of woody plants into herbaceous-dominated ecosystems generally results in the impoverishment of the original community (Price and Morgan 2008).

Coast Tea-tree is a species prone to becoming ecologically out of balance, meaning that where it occurs naturally it has a tendency to outcompete the other naturally occurring plant species, leading to the degradation of that vegetation. It has been shown that heathland vegetation requires relatively regular burning to maintain ecological health (Gill 1996). Further, Coast Tea-tree is also known to be an environmental weed, meaning that it has become naturalised beyond its former strictly coastal range (Arnold 1981; Carr 1993).

The current owners of 'Ballara' and 'Arilpa', descendants of the Deakin and Brookes families, have realised that Coast Tea-tree invasion, along with a host of environmental weed species, has had significant impact on the values of the native vegetation, including senescence of Austral Grass-tree. Both families (independently, but in consultation) commenced works to address the decline (Sharp 1996; Trengove 1991, 1995, 2006). At 'Ballara' an ecological burn was undertaken on Guy Fawkes Day 1992, a time of year selected to inhibit regeneration of Coast Tea-tree. The selected area was dominated by thick Coast Tea-tree vegetation with minimal native understorey species. Prior to the fire, 13 indigenous plant species were recorded (Trengove 1991). Post fire, 61 indigenous plant species were recorded, including 15 species not otherwise present at 'Ballara' or within the Borough of Queenscliffe. These 'new' species had been present as soil-stored seeds waiting for the right conditions to germinate. Newly germinated plants of Austral Grass-tree were also recorded (Trengove 1995). In the remaining areas of native vegetation at both properties, restoration works have focused on the removal of environmental weeds and the physical removal of Coast Tea-tree (without burning) (Anon 1993). At 'Ballara' smaller areas of Coast Tea-tree have been retained, while at 'Arilpa' Coast Tea-tree has been totally removed. The works at 'Arilpa' have led to a more gradual but impressive increase in species diversity.

Fragmentation of box woodlands: Box Forest Road

After 1835, the factors changing the vegetation changed abruptly. While the fundamental ecological processes related to climate, soils and competition continued; new and overwhelmingly powerful forces changed the vegetation markedly, to produce the largely cleared landscape we see today. The most obvious of these changes to affect tree distribution was the cutting of timber. The removal of discrete 'chunks' of grey box forest, often following the edges of paddocks or properties, is an obvious result. Less obvious is the cutting of Allocasuarina species from the plains beyond the forests (Sinclair and Atchison 2012).

Box Forest Road, north of Anakie, now traverses treeless cropping land. A *c.*1950 photograph of the Box Forest Road landscape shows relatively dense Grey Box-dominated vegetation on either side of the road as far as the eye can see. In the early 1960s this entire woodland, an area of ~100 ha, the last extensive stand of Grey Box in the region, was cleared. Although referred to historically (Wedge 1835) and by locals as a 'forest', current ecological thinking would class this vegetation as a woodland Ecological Vegetation Class (DELWP 2018b). The best timber was used to make fence posts and the rest was sold as firewood (Peter Bufton, local farmer, pers. comm. October 2016). Previously the area had been lightly grazed by sheep and was relatively intact. Subsequent to the clearing, the site was used for more intensive sheep grazing. More recently the land has been 'improved' and cropped. Bufton said that this vegetation, dominated almost exclusively by Box Eucalypts with occasional shrubs and 'rough native grasses', gave way to open natural grasslands to the east.

This distribution corresponds with the soil types of the area, the Box occurring on colluvial deposits from the Brisbane Ranges that give way to underlying basaltic clays further away from the ranges (GSV 1963). Currently the only Grey Box trees in this vicinity occur along the roadside reserves of Box Forest Road and Buftons Road. Similar patterns can be seen along much of the eastern flank of the Brisbane Ranges. South of Anakie the landscape was formerly dominated by Melbourne Yellow Gum (*Eucalyptus leucoxylon* subsp. *connata*); the remnants of this species are now mostly confined to the roadside verges. Within the region, box woodlands are now much restricted, with some of the best examples occurring on the northern and north-western flanks of the You Yangs, at the Ford Proving Ground and at the Mount Rothwell Biodiversity Interpretation Centre (Carr *et al.* 2001).

A significant loss of grassy woodland vegetation in the region occurred in 1997 at Stephens Road, Bannockburn. Barwon Water removed 8.5 ha of old growth Yellow Gum woodland, listed as of state significance, to build sewerage ponds (Ecology Australia 1996).

Drooping Sheoke and the Barrabool Hills

There is a degree of confusion concerning the natural vegetation of the Barrabool Hills. Hammond stated, 'The soil on these hills is exceedingly rich … unlike most hills and mountains in this primitive region, they are clear of trees' (Hammond 1898). However, referring to the same land, he continued, 'A blacksmith was kept on the station to repair just the ploughs alone, the ground being full of stumps'. The Port Phillip Association's map of Port Phillip 'from the Survey of Mr Wedge and others' describes the hills as 'Lightly timbered with oaks' (Clint 1955).

Early paintings by artists William Duke ('Geelong from Mr Hiatts, Barrabool Hills', 1851), Eugene von Guérard ('View of Geelong', 1856 and 'Mr Levien's Hut on the Barwon', 1860) (Geelong Art Gallery collection) as well as 'View of Fyans Ford on the Marabool – The Barabool Hills in the distance' (artist unknown, Brown 1941) all show the Barrabool Hills to be clothed in what is apparently Drooping Sheoke (*Allocasuarina verticillata*). It is instructive to note that the earliest painting has the greatest cover of trees; in contrast, the latest painting shows less cover. When viewed chronologically, these works of art depict the attrition of Drooping Sheoak woodlands over time.

Although remnants of Drooping Sheoke are now uncommon on the Barrabool Hills they can still be seen on sites such as Merrawarp Road and on Georges Hill (Trengove, pers. obs.). A report in the *Geelong Field Naturalist* (Anon 1911) describes the vegetation of the Barrabool Hills as 'on this continually denuded slope grows … some very old Murray River Pines, Callitris, a few Casuarinas and Boobyalas or Myoporum'. The *Myoporum* would have been Common Boobialla (*Myporum insulare*) or Sticky Boobialla (*M. viscosum*), both of which still occur along the Barwon River. The *Callitris* was probably *C. gracillis*, which still occurs at Dog Rocks, and the Casuarina would have been *Allocasuarina verticillata*.

One explanation for this confusion might be that the original parish of Barrabool included the adjacent Mount Pollock, a volcanic hill that likely would have been a native grassland (Brownhill 1955).

The Barrabool Hills also support smaller areas of Eucalypt woodland, dominated by Swamp Gum (*Eucalyptus ovata*) in gullies, Manna Gum in areas of deeper sands, and River Red Gum in riparian zones (DNRE 2000; Trengove, pers. obs.; Wynd 1992).

The removal of Drooping Sheoke woodland from the Barrabool Hills, and probably also from Waurn Ponds, Geelong and the Bellarine Hills, coincides with the sudden increase in the demand for lime to build Melbourne, Launceston (Harrington 1996) and the newly wealthy gold rush towns. From the 1830s onwards, coastal lime kilns were established to exploit the ready supply of Sheoke, Banksia and possibly Acacia. At one time, 17 lime kilns were in operation at the heads of Port Phillip Bay (Harrington 1996).

Drooping Sheoke was locally common and a convenient size for cutting to feed the lime kilns. The traffic of the drays bringing timber to the kilns was constant enough to warrant complaints from residents, 'The Sheoke has gone before the woodman's axe' (Brownhill 1955). Today, the largest Drooping Sheoke woodland remaining in the City of Greater

Geelong is located in a single paddock at Mannerim, where it is grazed by stock and in decline (Carr *et al.* 2001).

Coastal Moonah woodlands

Coastal Moonah (*Melaleuca lanceolata*) woodlands, like Bellarine Yellow Gum woodlands, have a distribution naturally limited to coastal and near-coastal areas, although they are not endemic like the Bellarine Yellow Gum woodlands (DSE 2010). Consequently, they have been subject to depletion and fragmentation to the extent that 'If coastal development continues at the current rate, all but a few remnants of this community will be surrounded by the urban landscape' (Moxham and Turner 2011). In the Geelong region, areas of relatively intact Moonah woodland vegetation still occur on Swan Island and to a lesser extent at Torquay, Barwon Heads and Point Lonsdale (Trengove 2001b).

Reversing the trends

Actions such as the proclamation of the Brisbane Ranges National Park and the creation of the Dog Rocks Flora and Fauna Sanctuary recognise the significance of native biota. Works undertaken by authorities and the community in relatively recent times are addressing environmental concerns.

- Bellarine Landcare has revegetated most of the previously cleared creeks draining into Swan Bay.
- Brisbane Ranges Landcare has revegetated significant stretches of Little River.
- The Corangamite Catchment Management Authority, City of Greater Geelong and Friends of Buckley Falls have revegetated and restored significant stretches of the Barwon River.
- At the Dog Rocks Flora and Fauna Sanctuary, volunteers have undertaken weed control and revegetation works.
- Barwon Coast Committee of Management undertakes vegetation restoration works and monitoring programs in the extensive sand dunes between Collendina and Black Rock.
- A coordinated control program for the weed Sea Spurge (*Euphorbia paralia*) has been implemented by management agencies and community groups along the Surf Coast.

Conclusion

Changes in vegetation since European arrival have been profound. The nature and extent of European land use has dictated the degree of impact on native vegetation. Fertile land has been heavily exploited for agriculture; consequently, the naturally occurring grassy ecosystems are now grossly depleted and fragmented. Additionally, areas near the coast and rivers, most in demand for settlement, have been subject to ongoing urban and industrial sprawl. Conversely, relatively large areas of native heathy vegetation remain on less fertile land that was in lower demand for agriculture. However, these areas have also been impacted, typically as a result of mining and firewood collection.

The main consequences are fragmentation of native vegetation and loss of species diversity. Efforts are being made to reverse degradation; however, to a large extent, regional ecosystems are still in decline.

References

Anderson S (2018) *S Anderson*. http://www.sueanderson.com.au/ (accessed 1 March 2018).

Anon (1911) Barrabool Hills. Geelong Field Naturalist, np.

Anon (1993) Burning to restore our Coastal Heathlands. *Victorian Landcare Magazine.*

Arnold AH (1981) Exotic plant invasions in coastal ecosystems. In *Proceedings of National Workshop in Coastal Management*. Institute of Parks and Recreation, Canberra.

Bennett LT (1994) The expansion of *Leptospermum laevigatum* on Yankie Isthmus, Wilsons Promontory, Victoria. *Proceedings of the Royal Society of Victoria* **114**, 43–58.

Britton AJ, Carey PD, Pakeman RJ, Marrs RH (2000) A comparison of regeneration dynamics following gap creation at two geographically contrasting heathland sites. *Journal of Applied Ecology* **37**, 832–844. doi:10.1046/j.1365-2664.2000.00533.x

Brown PL (Ed.) (1941) *Clyde Company Papers*. Oxford University Press, Melbourne.

Brownhill WR (1955) *The History of Geelong. Geelong Advertiser*, Geelong.

Byrne G (1930) Early days of the Mornington Peninsula. In *Coastal Vegetation of Port Phillip Bay*. Port Phillip Authority, Melbourne.

Carr GW (1989) Notes on Grevillea 'Lara Dwarf'. *Indigenotes*. Indigenous Flora and Fauna Association, Melbourne.

Carr GW (1993) Exotic flora of Victoria and its impact on indigenous biota. In P Foreman and NG Walsh (Eds) *Flora of Victoria*. Inkata Press, Melbourne.

Carr GW, McMillan SE, Trengove MH, White MD, McMahon ARG, Silveira CE (2001) *City of Greater Geelong Biodiversity Management Plan*. Ecology Australia, Melbourne.

CCMA (2016) *Corangamite Natural Resource Management Plan for Climate Change*. Corangamite Catchment Management Authority, Colac, Victoria.

Clint R (1955) Map of Port Phillip from the survey of Mr Wedge and others. In WR Brownhill (Ed.) *The History of Geelong. Geelong Advertiser*, Geelong.

Debinski DM, Holt RD (2000) A survey and overview of habitat fragmentation experiments. *Conservation Biology* **14**, 342–355. doi:10.1046/j.1523-1739.2000.98081.x

DELWP (2018a) *Biodiversity Interactive Map 3.2*. Department of Environment, Land, Water and Planning, Melbourne. http://mapshare2.dse.vic.gov.au/MapShare2EXT/imf.jsp?site=bim (accessed 1 March 2018).

DELWP (2018b) *Biodiversity Assessment Guidelines*. Department of Environment, Land, Water and Planning, Melbourne. http://www.depi.vic.gov.au/__data/assets/pdf_file/0011/198758/Permitted-clearing-of-native-vegation-Biodiversity-assessment-guidelines.pdf (accessed 1 March 2018).

DEWHA (2008) *Threat Abatement Plan for Competition and Land Degradation by Rabbits*. Department of Environment, Water, Heritage and the Arts, Canberra.

DNRE (2000) Plant species lists by M Trengove. In *Indigenous Trees and Shrubs of the West Port Phillip Region*. Department of Natural Resources and Environment, Melbourne.

DSE (2010) *A Field Guide to the Coastal Moonah Woodland in Victoria*. Arthur Rylah Institute, Department of Sustainability and Environment, Melbourne.

Dunn NA (1949) *History of Point Lonsdale*. Publisher unknown.

Ecology Australia (1996). *Stephens Road Bannockburn Vegetation Assessment*: Report prepared for Barwon Water. Ecology Australia, Melbourne.

Edmonds L (2005) *Living by Water: A History of Barwon Water and its Predecessors*. Barwon Water, Geelong.

Edwards RG (1952) *The History of Ocean Grove*. Edina Press, Melbourne.

Frood D, Calder M (1987) *Nature Conservation in Victoria*. Victorian National Parks Association, Melbourne.

Gill AM (1996) *How Fires affect Biodiversity*. Australian National Botanic Gardens. http://www.anbg.gov.au/fire_ecology/fire-and-biodiversity.html (accessed 1 March 2018).

GSV (1963) *Geelong Geology Map Sheet 1 mile to 1 inch*. Geological Society of Victoria. Government Printer, Melbourne.

Guilfoyle WR (1894) A contribution to the botany of the Bellarine District. From Ocean Grove to Clifton Springs. *Geelong Naturalist* 4.

Hammond W (1898) Early history of the Barrabool Hills. *Geelong Advertiser*, 24 December 1898.

Harrington J (1996) The lime-burning industry in Victoria: an occupance approach. *Australian Historical Archaeology* **14**, 19–24.

Hobbs RJ, Yates CJ (2003) Impacts of ecosystem fragmentation on plant populations: generalising the idiosyncratic. *Australian Journal of Botany* **51**(5), 471–488. doi:10.1071/BT03037

Kirkpatrick J, McDougall K, Hyde M (1995) *Australia's Most Threatened Ecosystem: The South-eastern Lowland Native Grasslands*. Surrey Beatty & Sons, Sydney.

LCC (1987) *Melbourne Area District 1 Review: Final Recommendations*. Land Conservation Council, Melbourne.

Moxham C, Turner V (2011) The effect of fragmentation on the threatened plant community coastal Moonah woodland in Victoria. *Urban Ecosystems* **14**, 569–583. doi:10.1007/s11252-011-0171-x

Nicholas SGW, Marshall A, Morgan JW (2015) *Land of Sweeping Plains*. CSIRO Publishing, Melbourne.

Parks Victoria (1997) *Brisbane Ranges National Park Management Plan*. Parks Victoria, Melbourne.

Parsons WT, Cuthbertson EG (1992) *Noxious Weeds of Australia*. Inkata Press, Melbourne.

Pescott T (1989) Believed extinct in its natural environment: the Lara Grevillea. *Geelong Advertiser*, 4 May.

Price JN, Morgan JW (2008) Woody plant encroachment reduces species richness of herb-rich woodlands in southern Australia, *Austral Ecology* **33**(3), 278–289.

Sharp R (1996) Ballara: the restoration of an ecosystem. *Coastline Edition 8*. Department of Natural Resources and Environment, Melbourne.

Sinclair SJ, Atchison K (2012) The pre-colonial distribution of grasslands, woodlands and forests on the Werribee Plains, Victoria. *Cunninghamia* **12**, 213–227. doi:10.7751/cunninghamia.2012.12.017

Sumner R (1982) *Steiglitz Memories of Gold*. National Parks Service, Victoria.

Trengove MH (1991) An assessment of the remnant vegetation of Ballara, Point Lonsdale, Victoria. Unpublished report prepared for the Sharp family.

Trengove MH (1992) The vegetation and management of the coastal remnants of the Borough of Queenscliffe. Unpublished report prepared for the Borough of Queenscliffe.

Trengove MH (1995) An assessment of the remnant vegetation of Ballara, Point Lonsdale, Victoria. Unpublished report prepared for the Sharp family.

Trengove MH (1999) Edwards Point Fauna Reserve vegetation management guidelines. Unpublished report prepared for the Friends of Edwards Point.

Trengove MH (2001a) Bellarine Yellow Gums in the Surf Coast Shire. Unpublished report prepared for the Surf Coast Shire.

Trengove MH (2001b) The Narrows Queenscliffe vegetation management guidelines. Unpublished report prepared for the Queenscliff Community Association.

Trengove MH (2002) Plant species list for the Bannockburn Cemetery. Unpublished report.

Trengove MH (2006) Arilpa Point Lonsdale vegetation management plan. Unpublished report prepared for the Trust for Nature.

Trengove MH (2013a) City of Greater Geelong Bellarine Yellow Gum Mapping Project. Unpublished report prepared for the City of Greater Geelong.

Trengove MH (2013b) Dog Rocks Flora and Fauna Sanctuary vegetation management plan. Unpublished report prepared for the Batesford Fyansford Stonehaven Landcare Group.

Wedge JH (1835) *Field Book*. MS10768. State Library of Victoria, Melbourne.
Weste G, Taylor P (1971) The invasion of native forest by *Phytophthora cinnamomi* in the Brisbane Ranges, Victoria. *Australian Journal of Botany* **19**, 281–294. doi:10.1071/BT9710281
Wynd I (1986) *Geelong: The Pivot*. Geelong Historical Society, Geelong.
Wynd I (1988) *Balla-wein: A History of the Shire of Bellarine*. Shire of Bellarine, Drysdale, Victoria.
Wynd I (1992) *Barrabool: Land of the Magpie*. Barrabool Shire, Torquay, Victoria.

7

The ecological history of the Bellarine Peninsula: native plant associations before 1835

Stephen Murphy

> *The Bellarine Hills contain about twenty thousand acres [8094 ha] of land of the finest description. They consist of hill and dale and although we did not see any water in the valleys, I am satisfied water could be easily obtained. The land is thinly timbered; the soil appeared very rich and fit for any purpose. The Kangaroo Grass was up to my middle and with a thick bottom. It is as fine a tract of land as any I have yet passed over.*
>
> – Joseph Tice Gellibrand, 1 February 1836 (cited in Sayers 1983).

The early settlers' diaries often described the Bellarine Peninsula as a land that was ideal for grazing and cropping (Wynd 2005). They were delighted with its open woodlands, its extensive grasslands and its fertile soils. John Batman, one of the founders of Melbourne, described it as 'softly undulating hills and plains with the richest grass and verdure so delightful to the eyes of a sheep farmer' (Batman 1835). Batman and others also described it as being like a gentleman's park, suggesting that it was an open landscape with established well-spaced trees and uniform grassy plains. What they did not appreciate was that a unique geomorphology and over 2000 generations of human occupation had shaped this living landscape.

The soil was varied, as was the vegetation it supported. On the highest ground the black 'older volcanic' clay loam was shaded by Drooping Sheoaks (*Allocasuarina verticillata*) and other small trees (Fleming 1803). To the west and south, fertile marine sandy loams, elevated by movement along the Bellarine and Leopold faults, provided the perfect conditions for eucalypt grassy woodlands to thrive (Carr *et al.* 2001) (Plate 7.1). Open eucalypt woodlands also developed on the east side of the Bellarine Peninsula on the gentler slopes bordering Swan Bay (Flinders 1814). On the most recent alluvial soils in low-lying areas at Lake Connewarre wetlands complex and the numerous salt lakes on the south side of the Peninsula, there were unique and hardy plant communities. Examples of these are the Moonah (*Melaleuca lanceolata*) woodlands and White Mangroves (*Avecinnia marina*). Remarkably, they also support 12 types of salt marsh vegetation, the glassworts and seablite (Yugovic 1985) – important foods for the critically endangered Orange-bellied Parrot (*Neophema chrysogaster*) that has migrated in winter from Tasmania to the Bellarine Peninsula for millennia (BCN *c.*2016a, b).

The *Wadawurrung* people are the Traditional Owners and stewards of the lands and waters that extend from the Geelong–Lorne coastline inland to Ballarat and Werribee. They shaped its vegetation with sophisticated management practices for more than 40 000 years (Stockton 1983). The people who lived on the Bellarine Peninsula were the *Bengalat balug* clan. They hunted kangaroo (*Macropus giganteus*), emu (*Dromaius novaehollandiae*), wetland birds and fish. They gathered freshwater (*Hyridella* sp.) and marine mussels (*Mytilus edulis galloprovinicialis*), harvested wild fruit and cultivated a diverse variety of root vegetables. They used medicinal plants to treat wounds and disease. They fashioned spears, boomerangs and bowls from the woody plants and made baskets from sedges, grasses and lilies. They lived well and sustainably on the diverse resources that the vegetation of the Bellarine Peninsula provided.

A diverse landscape

Events of geological and geomorphological history are fundamentally responsible for the overall land topography, which in turn influences climate and vegetation (Costermans 1986).

The vegetation that grew on the Bellarine Peninsula before 1835 had evolved and adapted to its unique coastal location, topography and soils. These soils and topography mostly developed from sedimentary rocks, though the ancient volcanic rocks played an important part in shaping the raised landscape (Rosengren 2009).

An open grassy woodland

An open grassy woodland is what the early explorers described on their exploratory visits to the Bellarine Peninsula in the early 1800s. James Fleming, the gardener on the Grimes survey team, explored the basalt hills behind Portarlington in late summer of 1803 (Fleming 1803). He wrote in his diary that 'the land is good and fine pasture from ten to eighteen inches deep, fine black earth with white sandy clay or gravel bottom. The timber small and the same as before mentioned'. Fleming, frustratingly, does not clearly list the species on Mount Bellarine, though in his previous diary entry near Werribee he recorded 'gum, banksia, oak and mimosa of sorts'. He appears to be saying that the vegetation at the two locations was similar. The 'oak and mimosa of sorts' would have been Drooping Sheoak (*Allocasuarina verticillata*) and various wattles. It is generally agreed that the banksia was Silver Banksia (*Banksia marginata*) (Trengove 2018, pers. comm.). An interesting question is whether there was a mystery gum.

In 1803 Matthew Flinders described the vegetation on the northern side of the Bellarine Peninsula, including casuarina and eucalyptus (note: Flinders had named the Bellarine Peninsula 'Indented Head'):

> *Indented Head, at the northern part of the western peninsula, had an appearance particularly agreeable; the grass had been burned not long before and had sprung up green and tender. The wood was so thinly scattered that one might see to a considerable distance and the hills rose one over the other to a moderate elevation but so gently that*

a plough might everywhere be used. The vegetable soil is a little mixed with sand, but good, though probably not deep, as I judged by the small size of the trees ... the most common kinds of wood are the casuarina and eucalyptus.

The native grasses Flinders described are deep-rooted perennials that thrive on fertile clay soils. Kangaroo Grass (*Themeda triandra*) was a significant grass on the Peninsula and is well adapted to basalt clay soils, as can be seen on the 'newer volcanic' plains west of Geelong.

There would have been a rich diversity of grasses, herbs, lilies and orchids mixed with the Kangaroo Grass. Batman (1835, p. 25) described these in his diary entry on 29 May 1835, travelling west from St Leonards towards Mount Bellarine:

We passed into the country and at a distance of four miles [6.4 km], *commenced travelling over land a little sandy in places, but of the finest description for grazing purposes; nearly all parts of its surface covered with Kangaroo and other grasses of the most nutritive character, intermixed with herbs of various kinds.*

A few days later Batman anchored near Point Henry, walked 20 miles [~32 km] and climbed the Bellarine Hills. His description focused on the grazing potential of the land but it provides some startling insights into the vegetation:

... very rich light black soil covered in Kangaroo Grass two feet high [0.6 m] *and as thick as it could stand, good hay could be made in any quantity. The trees were not more than six to the acre and those small sheoak and wattle. I walked for a considerable extent and all of the same description, this land forms an isthmus which is about 20 miles* [32 km] *long and 10* [16 km] *across it. I could see five or six miles* [8–9.6 km] *in every direction and most of the high hill was covered with grass to the summit and not a tree* (Batman 1835).

Batman was describing a very open and grassy environment with a small tree every 25–30 m. He makes no mention of eucalypts, which suggests that they were either absent or in small numbers. Thirty-two years before James Fleming supported the assessment that 'the timber was small'. Batman wrote that he was 'never so astonished in my life' when he saw the open woodlands of the Bellarine Hills with their bald summits (Batman 1835). What he and the other explorers didn't realise was they were seeing a designed and managed landscape.

A managed landscape

It was Major Thomas Mitchell who first understood why the Australian landscapes were so open:

The native applies fire to the grass at certain seasons, in order that a young green crop may subsequently spring up and so attract and enable him to kill or take the kangaroo with nets. In summer, the burning of the long grass also discloses vermin, birds' nests, etc., on which the females and the children, who chiefly burn the grass, feed. But for this simple process, the Australian woods might have been as thick a jungle, like those of New Zealand or America, instead of open forests (Mitchell 1839).

Anthropologist Rhys Jones changed the way the world thought about Indigenous Australians when he published 'Fire stick farming' in 1969 (Jones 1969). He began his article about Aboriginal use of fire with a question. 'We imagine that the country seen by the first colonists, before they ring-barked their first tree, was natural. But was it?' (Jones 2012, p. 3). Jones went on to describe an Australian landscape that had been changed by burning to produce a land full of wild vegetables and rich hunting grounds. He proposed that fire had many uses for the First Australians, including the important effect of altering the vegetation.

Bill Gammage argues that the whole of Australia was managed with strategically planned cool fires designed to improve the efficiency and convenience of hunting and gathering food (Gammage 2011).

Archaeologist Louis Lane believed that the *Wadawurrung* people had:

> *... developed a sophisticated and disciplined society that utilised every element of their surroundings while practising conservation strategies in their own clans-land. Their yearly calendar was marked by the flowering of different trees and shrubs and the movement of birds* (Lane 1988).

Security too was important to the *Bengalat balug* clan. William Buckley, who lived with the *Wadawurrung* for 32 years, described how different neighbouring tribes would seek justice or revenge for perceived wrongdoing. This revenge could end in the death or injury of clan members. Keeping the hills clear of vegetation enabled 360° views and adequate warning of pending danger (Morgan 1857).

Trees and shrubs of Mount Bellarine: Portarlington

Mount Bellarine is a relic of an earlier volcanic age called the Oligocene. The basalt lava and emissions of pyroclastic material lit up the night skies over 25 million years ago. The soil that developed from the basalt is a fertile clay loam that shrinks and cracks in the dry seasons.

These basalt hills are 120 m above sea level at their highest point, which exposes trees and shrubs to punishing winds. A tree that survives today and was most widespread and adapted to these local conditions is the Drooping Sheoak, (*Allocasuarina verticillata*).

Another durable small tree that has adapted well to the clay soils is the broadleaf Sweet Bursaria (*Bursaria spinosa* var. *macrophylla*) (Fig. 7.1). Three contorted trees, which are likely to be over 200 years old, grow in the Portarlington cemetery and represent a widespread community of past years (Trengove 1992).

The mimosa noted by early explorers would have been Black Wattle (*Acacia mearnsii*), Golden Wattle (*A. pycnantha*) and Hedge Wattle (*A. paradoxa*). All can still be seen on the elevated dry roadsides and would have been widespread throughout the basalt hills (pers. obs. 2018). Lightwood (*A. implexa*) grows on the newer basalt clay soils west of Geelong (pers. obs. 2018) and is a prominent small tree in the Point Richards Flora and Fauna Reserve at Portarlington (pers. obs. 2018). It would likely have been part of the original grassy woodland plant community on the heavier soils (Kendell 2018, pers. comm.).

The River Red Gum (*Eucalyptus camaldulensis*) and Blackwood (*Acacia melanoxylon*) grow along drainage lines on the volcanic soils between Portarlington and Drysdale. Examples of mature River Red Gums can be seen at Drakes Bushland Reserve east of Drysdale. These majestic trees would have been widespread at the time of settlement.

Fig. 7.1. Ancient Sweet Bursaria (*Bursaria spinosa* var. *macrophylla*) in the Portarlington cemetery.

The mystery 'gum'

Eucalypts appear to have been uncommon on the elevated basalt soils of the Bellarine Hills. The 'mystery gum' mentioned by Fleming and Flinders is unlikely to be Coast Manna Gum (*Eucalyptus viminalis* subsp. *Pryoriana*) because it favours sandy soil elsewhere on the Bellarine Peninsula. River Red Gum and Bellarine Yellow Gum (*E. leucoxylon* subsp. *bellariensis*) grow on clay soils on the Peninsula; however, they mature into large trees and would not have been easily overlooked. The fourth local species, Swamp Gum (*E. ovata*), grows in low swampy areas of the Peninsula and is not suited to the exposed drier soils of the hills.

Archaeologist James Rhoads argued that the fertile Bellarine Hills were an 'open forest of Messmate, *Eucalyptus obliqua* ... with Manna Gum and Southern Blue Gum, *E. globulus* subsp. *bicostata*) comprising the forest canopy' (Rhoads 1986). There are no Messmates or Southern Blue Gums surviving on the Bellarine Peninsula and the current opinion is that these species of eucalypt were never part of the Peninsula ecology (Trengove 2018, pers. comm.).

Marine sandy loams: Drysdale

The elevated hills to the north, south and west of Drysdale are composed of an extensive thin sheet of sands, silts, clays and gravels. They formed around 5 million years ago in the Pliocene from marine sediments left behind as the sea retreated (Rosengren 2009). These deposits became sandy soils and sandy loams that suited a more diverse mix of grassland plants, shrubs and trees. The topography and quantity of clay in the soil determined which plants would thrive, and where.

In January 1803 Surveyor Charles Grimes described the vegetation on hills between Clifton Springs and Point Richards as 'gentle rising hills of good land thinly wooded with low and decayed timber' (Fleming 1803). This observation, at the end of a dry summer, was of vegetation regularly exposed to hot north winds. There are only a few remnant plants surviving along the steep coastal cliffs that hint at what Grimes would have seen. These are Drooping Sheoak, Golden Wattle, Moonah, Common Boobialla (*Myoporum insulare*) and Seaberry Saltbush (*Rhagodia candolleana*) (BCN 2014).

Inland, the protected gullies and ephemeral creeks would have supported River Red Gum. For example, McLeod's Waterhole Reserve in Drysdale is fed by permanent springs. Originally it consisted of bogs and marshes that fed a creek that flowed into Corio Bay. River Red Gum and associated wetland species like Swamp Gum, Blackwood, Black Sheaok (*Allocasuarina littoralis*), Moonah (*Melaleuca lanceolata*), Hop Goodenia (*Goodenia ovata*), Prickly Tea-tree (*Leptospermum continentale*), Shrub Everlasting (*Ozothamnus ferrugineus*) and Golden Spray (*Viminaria juncea*) would have thrived around this wetland environment (Perry 2000). Only 11 River Red Gum specimens remained in 1962 (Lane 1962).

Bengalat balug clan at *Balla-wein*

McLeod's Waterhole Reserve was called *Neerim bolac* (the long lakes) by the *Bengalat balug* clan. The area which is now central Drysdale was called *Balla-wein* (a place where we rest upon our elbow by the fire after a good meal). Drysdale was the clan's sheltered winter home, protected by sand ridges and low thick vegetation (Lane 1988). Another place of significance in Drysdale for the *Bengalat balug* clan was *Yerai:ee:oo* (the place for rejoicing), which is located near the High Street/Murradoc Road roundabout. This was where corroborees were regularly held (Lane 1991).

Louis Lane, a tireless researcher and Geelong archaeologist, described *Balla-wein/Yerai-oo* as 'a paradise for hunter and forager as well as bird and beast' (Lane 1988; Wynd 2005). It was a place with ample fresh water fed by permanent springs of 'clear icy water' that ran into deep ponds. Lane's archaeological studies of McLeod's Waterhole suggest 'both lake and marsh must have been much bigger in the days before settlement, providing a permanently damp environment conducive to the proliferation of flora and fauna'. Her historical research showed that even after a long dry summer there was abundant water for 'Swan, duck, dabchick, water-hens, cranes and pelicans, the surrounding ridges thickly covered with vegetation' (Lane 1991).

The vegetation around *Balla-wein* supplied important foodstuffs for the *Bengalat balug* clan. Bull-rushes (*Typha orientalis*) provided energy from the starchy corms, leaf fibres for making twine and fluff from the flower heads for packing open wounds to stem bleeding. They ate the starchy succulent roots of lilies and orchids. Kangaroo Grass grains were ground and baked as a scone-loaf in the ashes. Bracken Fern fiddleheads (*Pteridium aqulinum*) and Murnong roots (*Microcerus lanceolate*) were staple vegetables in their diet.

Seeds harvested from wattles could be steamed green or collected dry for a protein-rich additive to Kangaroo Grass scones. Acacia gum was eaten as a protein-rich snack. Silver Banksia nectar was washed from the flower spikes or tea-tree blossoms for a sweet nourishing drink for toddlers. Juvenile cones from the Drooping Sheoak were a good source of ascorbate and the branchlets (leaves) were eaten to suppress thirst during walks in the hot weather. The curved wood from sheoaks was used for making boomerangs (Lane 1988).[1]

Curlewis

Two kilometres west of McLeod's Waterhole Reserve near Curlewis is 'Coriyule', the original home of Anne Drysdale and Caroline Newcombe. This was a flat landscape of well-spaced trees and grasslands. The low areas would have been dominated by River Red Gum and the drier areas by Coast Manna Gum. The usual mix of co-dominant species would have included Drooping Sheoak, Black Wattle, Lightwood, Golden Wattle and Silver Banksia.

Anne Drysdale wrote in her journal that 'there was an abundance of wildlife in the area in the form of kangaroos, emus and scrub turkeys'. She also recalled acacias, sheoaks and wildflowers, 'the place is so beautiful, clean and fresh, it's like a nobleman's park' (Richardson 1986).

John Richardson, who wrote *The Lady Squatters* about the lives of Drysdale and Newcombe, said that 'until in the early 1920s the "Coriyule" Estate was thickly covered with gnarled old gums, most with hollow trunks or limbs housing countless possums, owls, jackasses (kookaburras) and parrots in a veritable sanctuary for wildlife' (Richardson 1986). These gnarled old gums would likely have been River Red Gums.

Eucalypt woodlands: Ocean Grove and Wallington

Eucalypts were an important canopy tree on the Pliocene sandy loams. In protected areas they grow into substantial trees. The eucalypts usually had an understorey of Drooping Sheoak, Silver Banksia and four acacias - Golden Wattle, Black Wattle, Lightwood and Blackwood – in the low moist areas. Other tree species that are less common today are Black Sheoak and Cherry Ballart (*Exocarpos cupressiformis*). All of these tree species are still growing on roadsides and in reserves today.

In a plant survey conducted at Hutchinson's Bush Block, Wallington by the Geelong Field Naturalists Club in August 1975, 22 species of orchid and 34 species of herb, lily and sedge were identified. Also recorded were Austral Grass-tree (*Xanthorrhoea australis*) and four shrubs, Hedge Wattle, Coast Tea-tree (*Leptospermum laevigatum*),[2] Prickly Tea-tree (*L. continentale* and Snowy Mint-bush (*Prostanthera nivea*).

The Bellarine Yellow Gum (*Eucalyptus leucoxylon bellarinensis*) was widespread. It thrived on the undulating landscapes of Ocean Grove where the soil is sandy loam with a clay subsoil. There is still a significant stand north of Thacker Street, that has been recognised by Ecology Australia as a site of National Conservation Significance (BCN 2017; Trengove 1999). The Yellow Gum has been recorded on the higher north slopes of Lake Connewarre (Billows 2003).

The Bellarine Yellow Gum is an important food source for the endangered Swift Parrot (*Lathamus discolor*). These colourful birds breed in Tasmania and have migrated to the mainland in winter for millennia to feed on the nectar of this flowering eucalypt species.

Usually associated with the Yellow Gums is the Coast Manna Gum in well-drained areas, and Swamp Gums where the drainage is poor. The latter association can still be seen in the Ocean Grove Nature Reserve where the Swamp Gum is the dominant eucalypt species. Other species that are part of the ecological mix in the Nature Reserve are Black Sheoak, Austral Grasstree, Hedge Wattle, Golden Wattle and Coast Tea-tree (BCN 2017).

Areas of the Nature Reserve have not been burnt for over 90 years, which has resulted in the Drooping Sheoaks becoming dominant. The deep Sheoak leaf litter has suppressed germination of the eucalypts, other understorey and shrub species. This site would have been regularly burnt before 1835 by the *Bengalat balug* clan, creating the more open grassy landscape described by early explorers.

> *Koories regularly fired the bush to promote the growth of new shoots. This practice was random enough to maintain a wide variety of plant species and to keep the woodlands of the Bellarine Peninsula open and grassy* (Ferrier 1989).

It is likely that the Yellow Gum was widespread throughout the elevated Pliocene soils with a clay base. This would give it a potential distribution from the raised areas of Leopold, round the higher northern and eastern slopes of Lake Connewarre and then east to a line between Marcus Hill and Curlewis. It is east of this line that the sandy soils become deeper and the Manna Gum becomes the dominant eucalypt species.

The Coastal Manna Gum usually has an understorey of Drooping Sheoak, Black Sheoak, Black Wattle and Austral Grasstree. This ecological mix of species can be seen at Basin Reserve east of Drysdale. This is a 4 ha undulating remnant of indigenous vegetation where Cherry Ballart, Kangaroo Apple (*Solanum aviculare*), Silky Tea-tree (*Leptospermum myrsenoides*) and Prickly Tea-tree also occur.

Mannerim sheoaks

Towards Mannerim, the Drooping Sheoak once again becomes dominant as the Pliocene soil becomes hard and gravelly (Kendell 2018, pers. comm.). The large trunks on some of the Drooping Sheoaks in a private reserve suggest that they are very old trees (Fig. 7.2).

This open woodland of Drooping Sheoak may be a reflection of its original condition as it provided the ideal habitat for the locally extinct Bush Stone-curlew (*Burhinus grallarius*). This shy ground-dwelling bird was a resident of the Mannerim Sheoak until the 1950s (Pescott 2017). Known for its eerie, high-pitched wailing call at night, it inhabits open woodlands with short or scattered grass. The Bush Stone-curlew nests each year in the same location and pairs of curlews occupy 10–20 ha when breeding. Loose flocks may range over 100 km^2 when not breeding (Blakers *et al.* 1985). These specific habitat requirements suggest that a large area of Drooping Sheoak open woodlands would have been essential to support a stable curlew population.

In *Bengalat balug* language, Mannerim means ‘long hand’. This perfectly describes the shape of the land formation curled around Swan Bay – the incurved fingers are Edwards Point and the curved thumb is Point Lonsdale and Swan Island. This curved protective hand suggestive the importance of Swan Bay to the *Bengalat balug* clan as a place for hunting water birds and foraging for fresh eggs.

Fig. 7.2. Remnant Drooping Sheoaks (*Allocasuarina verticillata*) on private property at Mannerim.

Fig. 7.3. Grasstree (*Xanthorrhoea australis*) in a Coast Manna Gum woodland (*Eucalyptus pryoriana*) at the Basin Reserve near Drysdale.

Grasstrees were more widespread

John Batman recorded in his diary on 29 May 1835 a description of the landscape and trees west of St Leonards:

> *... as a relief to the landscape, the rising eminences were adorned with wattle, native honeysuckle* [banksia] *and the sheoak whose short straight stumpy butts and round heads resembled a number of pins sticking in a lady's pincushion* (Batman 1835).

Batman's colourful description of Sheoak sounds very unlike the usual form of the Drooping Sheoak. He could have been describing Grass Trees, which are still found on the sandy soils near Drysdale at the Basin Reserve, Ocean Grove, north-east of Wallington and on the heaths and heathy woodlands of the Borough of Queenscliffe (Trengove 1992) (Fig. 7.3). As Batman was standing 8 km inland on Mount Bellarine overlooking the land towards St Leonards and Swan Bay, it is likely that Grass Trees (*Xanthorrhoea* sp.) were far more widespread on the sandy soils than they are today.

Younger alluvial sands: a gentleman's park – Swan Bay, St Leonards, Indented Head and Point Richards

More recent marine alluvial sand deposits and coastal calcium rich dunes were laid down during a time of rapid and repeated climate change 10 000–2.6 million years ago in the Pleistocene. The softer alluvial sands formed flat plains that can be seen at St Leonards, west of Swan Bay and at Portarlington's Point Richards Flora and Fauna Reserve.

Calcium-rich deposits also formed in the Pleistocene. These have hardened to a natural cement that resists erosion, forming the higher ground west of Indented Head, at the Bluff at Point Lonsdale and the dune formations along Thirteenth Beach (Dunn 1949; Rosengren 2009).

James Fleming described the trees of the land near St Leonards in February 1803:

> *The land is light sand from the point of the hill and in some places swampy; the timber something larger consisting of gum, oak, banksia and mimosa and some small pine, one half of it dead by the country being lately burnt* (Fleming 1803).

The low-nutrient loams that developed from these alluvial sand deposits are well-drained. On the flat plains, Bellarine Yellow Gums grew into very imposing trees. Healthy mature examples of these can be seen on private property 100 m east of Manifold Road near Andersons Road. Some very significant trees remain at the southern end of Grassy Point Road (Fig. 7.4). These trees are remnants of a much larger community of Yellow Gums that would have occupied the higher ground across the eastern end of the Bellarine Peninsula (Small 2018, pers. comm.; Kendell 2018, pers. comm.). A grassland shaded by well-spaced eucalypt giants clearly evokes Batman's image of 'a gentleman's park'.

On the flat plains west of Swan Bay, St Leonards and Indented Head, River Red Gum (*Eucalyptus camaldulensis*) occupied the moister drainage lines. Mature examples of these are growing on the creek behind Indented Head. Where the soil is moister, Swamp Gum was co-dominant with the River Red Gum (Small 2018, pers. comm.).

Other significant species that can be seen on the roadsides are Drooping Sheoak, Cherry Ballart (*Exocarpus cupressiformis*), Golden Wattle, Black Wattle, Sweet Bursaria and

Fig. 7.4. Majestic relic of the pasts: Bellarine Yellow Gums (*Eucalyptus leucoxylon* subsp. *bellerinensis*) on Grassy Point Road.

Seaberry Saltbush. These would have formed the open woodland community, supported by Manna Gum, Silver Banksia, Black Sheoak (likely to be the 'small pine' mentioned by Fleming), Common Boobialla (*Myoporum insulare*) and Hop Goodenia (*Goodenia ovata*). Kangaroo Grass, orchids, lilies and herbs would have been prominent among the grassland flora.

Sailing with Matthew Flinders in May 1802 were naturalist Robert Brown and botanical artist Ferdinand Bauer. They explored the land behind St Leonards and recorded 95 plant species. Some of the specimens collected around Edwards Point and St Leonards were Native Holly (*Lomatia ilicifolia*), Wirilda (*A. retinodes*), Moonah, Creeping Brookweed (*Samolus repens*), Cherry Ballart, Sword Sedge (*Lepidospermum congestum*), Bidgee Widgee (*Acaena novae zelandiae*), Native Pea (*Glycine clandestine*), Coast Beard Heath (*Leucopogan parviflorus*), Bower Spinach (*Tetragonia tetragonioides*) and Coast Tea-tree. A collection of Bauer's drawings is held at the Deakin University Waterfront Library in Geelong (Munster 2013).

Captain George Cole was a colonial entrepreneur. He purchased large areas of land around St Leonards in 1857, that was described in the advertising as 'thickly wooded with gum, honeysuckle and wattle' (Munster 2013). One of his many enterprises was selling firewood into Melbourne where local supplies were already exhausted. Clearing started:

> *... and axes bit deep into the trees. Piles of wood stood awaiting transport. St Leonards became noted for its timber trade. At one period it is said there were over 300 men employed in this work. Timber-cutting, as a big industry, lasted until the 1870s* (McLeod 1962).

Herbs, orchids and lilies

The open areas on the sandy dry soils of Indented Head were a source of herbs, lilies and orchid species. These plants have evolved succulent roots to store water and energy to sustain them through drought and extreme heat (Smithyman and Longmore 2006).

These succulent roots became an important food source for the *Bengalat balug* clan (Lane 1988). For example, in a few hours *Bengalat bulag* women could harvest enough Murnong tubers to fill a large basket (Wedge 1835; Lane 1988) (Fig. 7.5):

> *The native women were spread out over the plain as far as the eye could see, collecting Murnong, or in this language pannin … I inspected their bags and baskets on return and each had a load as much as she could carry* (Pascoe 2014, p. 23)

The Derwent Co. started a large sheep run on the Bellarine Peninsula in 1836. The sheep quickly developed a taste for the Murnong roots and by 1840 the *Wadawurrung* women were distraught because 'the ngamadjig (sheep) eat it all'. There was no Murnong left for them to harvest anywhere near Geelong (Clark 1990).

Lieutenant James Tuckey surveyed Port Phillip Bay in October 1803 and walked over 2 km inland from near St Leonards. He described the natural environment as 'beautiful country, thickly wooded, but the soil sandy and without water'. He was interested in trees that could be used to build battleships and wrote in some thought-provoking detail (Tuckey 1803):

> *The Honeysuckle (banksia) is the largest tree found here; it grows near the shore and in most dry and sandy places. The shaft seldom exceeds eight feet* [2.4 m], *the branches large and straggling and if sound of sufficient growth for a line of battleships knees and timbers. The grain of the wood resembles beech. The Red Gum grows to a large*

J.H.W. Native Women getting Tam bourn Roots 27 August 1835.

Fig. 7.5. Sketch of women harvesting Murnong (*Microseris lanceolata*) at Indented Head (Wedge 1835).

size and in its grain has a nearer resemblance to pine than any other timber, the branches are large and straggling but the wood appears porous and shaky, and the largest trees are commonly decayed, probably from the native's fires which are usually kindled near them. The sheoak or Beef Wood is the most common tree and grows all over the country. Few of these trees grow to any considerable size, and when they do are generally unsound in the heart. Boat's timbers might however be procured from them (Tuckey 1803).

Tuckey confirms that the Drooping Sheoak had a wide distribution and that it had an association with River Red Gum and a species of banksia. His description of the banksia sounds like Coast Banksia (*Banksia integrifolia*). If that was shown to grow near St Leonards, it would send ripples of controversy through the community of plant ecologists.

The Banksia debate

John Currey, the editor of Tuckey's journal, suggests that the size and shoreline position of the banksias makes them very likely to be Coast Banksia. This suggestion is at odds with the view of most ecologists, which is that Coast Banksia is not indigenous to the Bellarine Peninsula. Trengove suggests that its distribution was limited to the Borough of Queenscliffe (Trengove 1992).

It is generally agreed that the tree form of Silver Banksia was 'once very abundant on the Bellarine Peninsula' (Carr *et al.* 2001, p. 91) and that the Coast Banksia was restricted to the other side of Port Phillip Bay (Trengove 2018, pers. comm.). However, James Tuckey's description of banksia trees with 2.5 m trunks, robust enough to make braces (knees) and other structural timbers for battleships, does sound more like Coast Banksia. So, did they occur on the Bellarine Peninsula before European settlement?

Adding to the debate, Dr Steve Sinclair, an expert on the occurrence of Silver Banksia in Victoria, has written that 'the almost extinct tree-form' was a very large tree. The Silver Banksia can also grow in the sand on the coast and Sinclair knows a Silver Banksia that is larger than any Coast Banksia he has seen. However, he cautioned that it is still possible James Tuckey could have been describing Coast Banksia considering that Port Phillip Bay was a grassy plain only 12 000 years ago and the seeds could have been carried across from the Mornington Peninsula by cockatoos. It seems that this perplexing Banksia question remains unresolved (Sinclair 2018, pers. comm.).

Point Richards Flora and Fauna Reserve

Silver Banksia was part of the original flora at the Point Richards Flora and Fauna Reserve (Kendell 2018, pers. comm.). This 62 ha reserve was set aside in the 1970s to protect a population of Southern Brown Bandicoot (*Isoodon obesulus*) that was surviving in the sheltered environment created by extensive areas of Spiny-headed Mat-rush (*Lomandra longifolia*) and Bracken Fern.

The ground flora in this reserve would have been much more diverse in the past (Small 2018, pers. comm.). Formerly, tree and shrub species would have reduced the dominance of the Mat-rush and Bracken Fern. Coast Manna Gum, Lightwood, Black Wattle, Golden Wattle, Wirilda, Hedge Wattle and Tree Everlasting (*Ozothamnus ferrugineus*) would have been part of a heathy woodland which would have provided ideal bandicoot habitat. Hop

Goodenia (*Goodenia ovata*) and Blackwood would have sheltered the edges of the wetland areas.

The freshwater ephemeral wetlands in the reserve support the rare and endangered Growling Grass Frog (*Litoria raniformis*) as well as other species of frog. Their deafening roar in November (Small 2018, pers. comm.) would have been a familiar sound throughout the Bellarine Peninsula's freshwater wetlands before they were drained by the first settlers. These shallow wetlands would have been an important ecological feature of the alluvial sands.

Indented Head, Barwon Heads Bluff and Thirteenth Beach

Calcium-rich dune deposits have bonded into very hard landscape features that are higher than the surrounding alluvial plains and jut into Port Phillip Bay. They can be seen at Indented Head, the Bluff at Barwon Heads and as the coastal dunes that border Thirteenth Beach:

> *The coastal dunes are covered with shrubby plants such as the Coastal Tea-tree (Leptospermum laevigatum). These areas represent the areas of least change since settlement* (Wright 1973, p. 6).
>
> *The original vegetation of the surrounding calcarenite dunes, including Barwon Heads golf course, Thirteenth Beach residential estate and Macafee property was Coast Moonah Woodland* (McMahon and Quin 2009, p. 16).

Only the toughest of plants can grow in these exposed dry coastal conditions, and once again Drooping Sheoak and Moonah were the dominant species. They were supported by a rich suite of hardy shrubs, grasses, herbs and groundcovers. Examples of these are Coast Wattle (*Acacia sophorae*), Common Boobialla, Coast Wirilda (*Acacia uncifolia*), Sea Box (*Alyxia buxifolia*), Coast Saltbush (*Atriplex cineria*), White Correa (*Correa alba*), Cushion Bush (*Leucophyta brownie*), Coast Daisy-bush (*Olearia axillaris*), Thyme Rice-flower (*Pimelea serpyllifolia* subsp. *serpyllifolia*), Running Postman (*Kennedia prostrata*) and the remarkable Coast Beard Heath (*Leucopogon parviflorus*) (Carr 2007).

Today the Coast Beard Heath is a small to medium-sized shrub that grows to 1.5 m. tall. Remarkably, there is a very ancient Coast Beard Heath on the Swan Island Army training area that has grown into a small tree over 4 m tall (Kendell 2018, pers.comm.). It is likely that many of these supersized ancient shrubs would have existed on the mainland before 1835.

Alluviums of the Holocene: Lake Connewarre

Less than 10 000 years old, these younger sediments are a combination of freshwater stream deposits and saline marine sediments (Rosengren 2009). The 860 ha Lake Connewarre wetlands complex receives fresh water from the Barwon River to the west and salty tidal flows at Barwon Heads to the south. It is one of the largest estuarine systems on the central Victorian coast, covering 3300 ha (Billows and Gwyther 2007). It consists of:

- the freshwater Reedy Lake, once an ephemeral wetland and part of the Barwon River floodplain (Pescott 2017). Reedy Lake is regarded as an important link for the passage of birds like Brolga (*Grus rubicunda*) and Lesser Yellowleg (*Tringa flavipes*) between local wetlands (Gray 1934)

- the saline Lake Connewarre and lower Barwon River and Hospital Swamp. The latter was 'once a natural wetland … that has been extensively modified to provide improved waterfowl habitat for hunters' (Pescott 2017)
- Salt Swamp on the south side of the main lake, which is a large saline marsh that extends to Lake Murtnaghurt near Thirteenth Beach. The *Wilsonia* herbfield and *Distichlis* grasslands are considered to be among the best examples of their kind in Australia (Pescott 2017) and a likely indication of salt marsh plant communities of the past (Billows and Gwyther 2007)
- Lake Murtnagurt, which supports the best and largest remaining example of coastal dry saltmarsh in Victoria. It provides an important storm shelter refuge for up to 20 'threatened and vulnerable' Hooded Plover (*Thinornis rubricollis*) (McMahon and Quin 2009).

At the time of European settlement, Lake Connewarre was larger and twice as deep as it is today. The silt-laden Salt Swamp to the south was then part of the deeper main lake. An important feature of this larger saline wetland was its widespread seagrass (*Zostera muelleri*) and seaweed cover (Yugovic 1985).

This was an important fish nursery and provided food for thousands of water birds including the Black Swan (*Cygnus atratus*) after which the lake was named by the *Bengalat balug* people (*Connewarren* means swan water). Water birds were so numerous that in the early 1870s commercial hunters using 30 swivel guns shot as many as 5000 ducks annually with each gun. 'Despite the enormous toll, birds remain abundant, with a species list of ~150, including large numbers of shorebirds and waterfowl' (Pescott 2017, p. 21).

Bird species once prominent at Lake Connewarre that did not survive the hunters were the Cape Barren Goose (*Cereopsis novaehollandiae*) and migratory waders like the large flocks of Sharp-tailed Sandpiper (*Calidris acuminata*) and Little Stint (*Calidris minuta*). The Sandpipers and Stints were sold at local markets for threepence a pair until the 1950s. Many of the migratory waders are from the northern hemisphere and migrate annually to the Lake Connewarre wetland complex for the southern spring and summer (Gray 1934).

Orange-bellied Parrot

The critically endangered Orange-bellied Parrot (*Neophema chrysogaster*), of which there are now only 50 pairs in the wild (Zoos Victoria website 2018) and its close relative the Blue-winged Parrot (*Neophema chrysostoma*) spend the winter in the coastal saltmarsh vegetation at Lake Connewarre, Swan Bay, Freshwater Lake, Lake Victoria and several other sites.

Their most important food plants are Beaded Glasswort (*Sclerostegia quinqueflora*), Southern Sea-heath (*Frankenia pauciflora*), Austral Seablite (*Suaeda australis*), Shrubby Glasswort (*Sclerostegia arbuscular*) and Grey Glasswort (*Halosarcia halocnemoides*). The latter provides an important backup food source for the parrots between the seeding periods for Beaded Glasswort and Shrubby Glasswort at a time of the year when there is a general shortage for food (Yugovic 1985).

A study of Lake Connewarre recorded a total of 137 native plant species that included the very salt-tolerant salt marsh species, Chaffy Saw-sedge (*Gahnia filum*) sedgelands, Coast Tussock Grass (*Poa poiformis*) grasslands, Tangled Lignum (*Muehlenbeckia florulenta*) shrubland and Moonah woodlands which is now localised to Barwon Heads at 'The Sheepwash' (Yugovic 1985) (Fig. 7.6). Billows (2003) did record two ancient very large Moonah on the southern shores of Hospital Swamp, suggesting a much wider distribution at the time of European settlement.

Fig. 7.6. Salt marsh sp. and Tangled Lignum (*Muehlenbeckia florulenta*) at Lake Connewarre.

Majestic trees near Lake Connewarre

Subdivision surveys by Smythe in 1841 (PROV, Historic Plan Series, Moolap, Sydney M16) and 1852 (PROV, Historic Plan Series, Moolap, Featr 442) from the Bellarine Highway, 2 km east of Leopold and south to Lake Connewarre and Reedy Lake, recorded large gums, Drooping Sheoak and Silver Banksia. The large gum would likely have been the Bellarine Yellow Gum, later recorded by Billows on the northern and eastern slopes of Lake Connewarre (Billows 2003). This suggests that the Yellow Gum woodlands would have extended at least as far as the Portarlington Road, where the River Red Gum became the dominant eucalypt. Other species recorded by Smythe in 1852 were Box (Yellow Gum?), Blackwood and Black Wattle.

In 1855 Skene (PROV, historic Plan Series Sale 263, Beyerley; Sale 269, Fynmore) surveyed south of Lake Connewarre. He recorded the land was 'moderately to thickly timbered with sheoak and gum … other blocks were considerably encumbered by stumps of sheoak which have been used for firewood, the gum trees still remain'.

In 2017 Trengove recorded one Bellarine Yellow Gum on Winklers Lane, 1.5 km south of Hospital Swamp (Trengove 2017). There is a small remnant patch of River Red Gum on private property south of this lone Yellow Gum (Ashton-Smith 2018, pers. comm.). This may indicate that both the Yellow Gum and River Red Gum shared the canopy on these low flat plains, the Yellow Gum occupying the higher ground with Drooping Sheoak and acacias.

Barwon River estuary

At the coastal end of the Lake Connewarre wetlands complex is the Barwon River estuary. A significant feature of the estuary is the White Mangrove (*Avicennia marina*) that grows on mudflats in the littoral zone between high and low tides. The pneumatophores of the mangroves provide a protected nursery environment for many fish species and crustaceans that feed on the diverse macroalgae (Yugovic 1985). The Barwon Heads mangroves appear to be a healthy stable population that may have been more widespread before settlement.

Learning from the past

When Batman's party arrived at St Leonards on the Bellarine Peninsula on 8 June 1835, he was surprised at the lightly timbered grasslands that he found. Except for inland areas, around St Leonards, Drysdale and north of Lake Connewarre that were more densely wooded and dominated by large eucalypt species, the more elevated areas of the Peninsula were open and parklike. This was no chance occurrence but a landscape created by the *Bengalat balug* clan. Using fire as a tool to manage the vegetation, they had developed a sustainable and productive Country that they had cared for more than 40 000 years.

The Bellarine Peninsula has a rich undulating landscape because of the geomorphology that shaped it. The evolving soils, combined with its unique coastal environments, gave birth to a diverse array of well-adapted plant species. Plants evolved for protected and harsh windy environments, for extreme salinity and crystal-clear freshwater springs, for droughts, for mild winters and for extreme heat. The wildlife that lived and migrated there were just as diverse and remarkable. They adapted and thrived on the extraordinary range of foods and habitats the Peninsula provided.

The European settlers quickly made their mark on this remarkable but fragile landscape. They introduced sheep, cleared native trees, planted crops, drained and filled many wetlands. It took only 18 months for the *Wadawurrung* to be dispossessed from their land within a 25 mile radius of Geelong, four years for agriculture and sheep to remove all the Murnong (Clark 1990), and 40 years for the Peninsula to be cleared of most of its native vegetation (McLeod 1962). These radical landscape changes have put the unique ecologies of the Bellarine Peninsula under significant stress (Carr *et al.* 2001).

It is time to learn from the past and appreciate that this remarkable ecology of wild plants and animals was sustainably managed by the *Bengalat balug* people who had a spiritual connection with this, their Country. Their land care is unmatched and could show us the way forward in our efforts to restore the Bellarine Peninsula to an ecological balance.

Endnotes

1 All the above-mentioned plants were available throughout the Bellarine Peninsula and would have been harvested by the *Bengalat bulag* clan.
2 Coast Tea-tree is not considered to be indigenous to this area (Small 2018, pers. comm.).

Acknowledgements

I wish to acknowledge the advice of Dr Steve Sinclair (Arthur Rylah Institute), Sophie Small (Bellarine Landcare Group), Gary Kendell (Bellarine Bayside Coastal Management), and Elissa Ashton-Smith (Geelong Landcare Network) in the preparation of this chapter.

References

Batman J (1835) *Rebecca Journal*, 10 May–11 June.

BCN (2014) *Flora of Clifton Springs Coastal Reserve: Beacon Point, Boat Harbour, The Dell.* Bellarine Catchment Network, Drysdale, Victoria.

BCN (*c*.2016a) *Inland Plants of the Bellarine Peninsula*. Bellarine Catchment Network, Drysdale, Victoria.

BCN (*c*.2016b) *Coastal Plants of the Bellarine Peninsula*. Bellarine Catchment Network, Drysdale, Victoria.

BCN (2017) *Discover the Real Ocean Grove: Woodlands Wetlands Coast Estuary*. Bellarine Catchment Network, Drysdale, Victoria.

Billows CA (2003) *Initial Contact with Landholders adjoining the Lake Connewarre State Game Reserve: Wetland Site Assessment and Landholder Survey*. Report prepared for the Bellarine Peninsula Ramsar Wetlands Project. Wetland Care Australia, Ballina.

Billows C, Gwyther J (2007) *Ecological Study of Lake Connewarre Wetlands Complex*. School of Life and Environmental Sciences, Deakin University, Geelong.

Blakers M, Davies SJJF, Reilly PN (1985) *The Atlas of Australian Birds*. Melbourne University Press, Melbourne.

Carr G (2007) Proposed shard trail along 13th Beach Road, Barwon Heads. Unpublished report.

Carr GW, McMillan SE, Trengove MH, White MD, McMahon ARG, Silvereira CE (2001) *Biodiversity Management Plan for City of Greater Geelong*: Report by Ecology Australia Pty Ltd. City of Greater Geelong, Geelong. https://www.geelongaustralia.com.au/common/Public/Documents/8cbc8661b5926ee-BiodiversityStrategy.pdf

Clark ID (1990) *Aboriginal Languages and Clans: An Historical Atlas of Western and Central Victoria, 1800–1900*. Monash Publications in Geography, 37.

Costermans L (1986) *Native Trees and Shrubs of South-eastern Australia*. Rigby, Melbourne.

Dunn NA (1949) History of Pt Lonsdale. Unpublished paper.

Ferrier S (1989) *Wathaurong Medicines*. Wathaurong Aboriginal Co-operative, Geelong.

Fleming J (1803) *A Journal of Grime's Survey. The Cumberland in Port Phillip January–February 1803*. John Currey (Ed.). Banks Society Publications (2002), Melbourne.

Flinders M (1814) *A Voyage to Terra Australis*. G & W Nicol, London.

Gammage B (2011) *The Biggest Estate on Earth. How Aborigines made Australia*. Allen & Unwin, Sydney.

Gray JM (1934) Bird life on the Connewarre Lakes. *Victorian Naturalist* **50**(9), 214–216.

Jones R (1969) Fire stick farming. *Australian Natural History* **16**(7), 224–228.

Jones R (1969) Fire stick farming. *Fire Ecology* **8**(3), 3–8.

Lane L (1962) 'McLeods Holes' Reserve and surrounds. Archaeological survey map.

Lane L (1988) The Bengali hearth-groups of the Wathaurung tribe. Unpublished paper.

Lane L (1991) Aboriginal place names: Geelong & district. Unpublished paper.

McLeod E (1962) *Portarlington, Indented Head, St Leonards History and Reminiscences*. Bellarine Historical Society Publication, Portarlington, Victoria.

McMahon A, Quin D (2009) *Murtnagurt Lagoon & Barwon Heads, Flora and Fauna Assessment*. Ecology Australia, Melbourne.

Mitchell T (1839) *Three Expeditions into the Interior of Eastern Australia: With Descriptions of the recently explored Region of Australia Felix, and of the Present Colony of New South Wales*. T & W Boone, London.

Morgan K (1857) *The Life and Adventures of William Buckley: Thirty-two years a Wanderer amongst the Aborigines of the then unexplored Country around Port Phillip, now the Province of Victoria*. Heinemann, London.

Munster P (2013) *Little of Interest? A History of St Leonards, Victoria*. St Leonards Progress Association, St Leonards, Victoria.

Pascoe B (2014) *Dark Emu Black Seeds: Agriculture or Accident?* Magabala Books Aboriginal Corporation, Broome, WA.

Perry D (2000) *Indigenous Trees and Shrubs of the West Port Phillip Region*. Department of Natural Resources and Environment, Melbourne.

Pescott T (2017) *Birds and Botanists: A Field Naturalist's History of Geelong.* T. Pescott, Geelong.

Rhoads J (1986) *Bellarine Peninsula: Archaeological Site Assessment and Management Study.* Department of Anthropology University of Sydney, Sydney.

Richardson J (1986) *The Lady Squatters Miss Anne Drysdale and Miss Caroline Elizabeth Newcomb, 'Boronggoop' and 'Coriyule'.* Shire of Bellarine, Drysdale, Victoria.

Rosengren NJ (2009) *Murtnaghurt Lagoon, Bellarine Peninsula & Related Landforms: Nature, Origin & Geoscience Significance.* La Trobe University & Environmental GeoSurveys Pty Ltd, La Trobe University, Melbourne.

Sayers CE (Ed.) (1983) *Letters from Victorian Pioneers: A Series of Papers on the Early Occupation of the Colony, the Aborigines etc addressed by Victorian Pioneers to His Excellency Charles Joseph Latrobe Esq, Lieutenant Governor of the Colony of Victoria 1853.* Lloyd O'Neil for Currey O'Neil, Melbourne.

Smithyman S, Longmore S (2006) *Inland Plants of the Bellarine Peninsula.* Bellarine Land Care Group, Drysdale, Victoria.

Stockton J (1983) *The Bellarine Peninsula Archaeological Survey, A study funded by the Australian Heritage commission.* Victorian Archaeological Survey, Melbourne.

Trengove MH (1992) *The Vegetation and Management of the Coastal Remnant of the Borough of Queenscliffe on the Bellarine Peninsula, Victoria.* Report prepared for the Borough of Queenscliffe.

Trengove MH (1999) Reilly's Property Thacker Street Ocean Grove. Yellow Gum Reserve. Vegetation management and restoration prescriptions. Unpublished report prepared for Tony Reilly, 'Goandra' Pty Ltd.

Trengove MH (2017) Yellowgum on Bellarine roadsides map. City of Greater Geelong, Geelong.

Tuckey J (1803) *Memoir of a Chart of Port Phillip surveyed in October 1803.* J Currey (Ed.). Colony Press, Melbourne.

Wedge JH (1835) *The Diaries of John Helder Wedge.* J Crawford, WF Ellis and GH Stancombe (Eds). Royal Society of Tasmania, Hobart.

Wright McKG (1973) *The Bellarine Peninsula. An Environmental Study.* Geelong Regional Planning Authority, Geelong.

Wynd I (2005) *Balla-wein. A History of the Shire of Bellarine, to 1993.* Council of the Shire of Bellarine, Drysdale, Victoria.

Yugovic J (1985). *The Vegetation at the Lake Connewarre State Game Reserve.* Technical Report Series 18. Arthur Rylah Institute for Environmental Research, Melbourne.

8

Riverine ecology

Ross Wissing

> *Water is one of the greatest forces of nature. It can be as gentle as a brook or pond or spring, or as powerful as a flooding river or a storm-lashed lake.*
>
> – Leigh Edmonds (2004).

Water is fundamental to healthy ecological function and processes. Waterways – rivers, streams, wetlands and lakes – are the lifeblood of local ecosystems, as well as their barometer. This is because waterways are generally the lowest point of a catchment, as all water from a catchment drains to waterways. Therefore the health of the waterway and the health of the catchment are intimately linked and need to be considered together.

The riparian zone – the interface between rivers, wetlands and their terrestrial catchments (Boulton and Brock 1999) – plays a critical role in these and other ecosystems. Although only occupying a small proportion of the landscape, riparian areas are among the most productive ecosystems on Earth (Lovett and Price 1999). They are extremely important in landscape-scale processes as they allow the movement of energy, materials and living organisms through the landscape (Ede and Hunt 2008). However, because riparian zones are at the interface of land and water, they usually bear the brunt of human activities in both these environments (Boulton *et al.* 2014).

Geelong, or *Djilang* in the language of the Traditional Owners, the *Wadawurrung*, is the second-largest city in Victoria. Geelong is located in the lower reaches of the Barwon River catchment, which is part of the Corangamite Catchment Management Authority (CCMA) region. Despite accounting for just 1.5% of the CCMA land area, around three-quarters of people in the Corangamite CMA live in the urban centres of Ballarat and Greater Geelong (CCMA 2013). Ballarat is located in the headwaters of two of the major tributaries of the Barwon catchment, the Moorabool River and Leigh/Yarrowee River.

The extent of change beginning 200 years ago was so rapid that there was little scientific recording of the waterways, and their plants and animals, present at the time of European settlement. Yet this knowledge is vital if Australian society is to live within its ecological means. As Boulton *et al.* (2014) identify, to manage impacts on running waters effectively, we must understand how they function so that when we try to restore them we are more likely to be reinstating the original, pre-contact ecological process than treating the symptoms.

Such an approach is known as 'historical ecology', which Lunt and Spooner (2005) define as an interdisciplinary paradigm in which ecologists view ecosystems as historically and spatially influenced non-equilibrium systems that are complex and open to human inputs. This allows more realistic goals to be set for future research and restoration activities (Lunt and Spooner 2005). A historical ecological approach will be utilised to provide a description of the key ecosystems found in and around current urban Geelong. Geelong has been chosen because it is in the lower part of the Barwon catchment. Therefore, its ecological condition provides an insight not only into land and water management activities in urban Geelong but also those in the upstream catchment. This will be achieved by a combination of historic accounts and contemporary scientific understanding of these ecosystems.

Geelong's ecosystems at the time of European settlement

Basins, now more commonly called catchments, are the primary ecological boundaries for waterways. However, waterways are also influenced by other ecological units such as bioregions and Ecological Vegetation Classes (EVCs).

Three bioregions are found in the Barwon catchment – the Victoria Volcanic Plains, the Otway Plain and the Central Victorian Uplands. Bioregions are a landscape-scale approach to classifying the environment using a range of attributes such as climate, geomorphology, geology, soils and vegetation.

In Victoria, bioregions are broken down into EVCs. There are three main ecological vegetation communities in urban Geelong around the Barwon River. These are plains grasslands, grassy woodlands and riparian forests. Prior to European settlement, temperate woodlands such as grassy woodlands occupied over 7 million ha in Victoria, or 32% of the state, grasslands 8% of Victoria and riverine grassy woodlands such as riparian forests ~4% (Lunt and Bennett 1999).

Landscape zones are geographic units used for catchment planning purposes, based on major waterway subcatchments, bioregions and existing remnant vegetation (CCMA 2014). The Barwon catchment comprises the Leigh, Murdeduke, Upper Barwon, Mid Barwon and Bellarine landscape zones. Geelong, and the lower Barwon River, is located in the Bellarine landscape zone and will be the main focus of this chapter (CCMA 2014).

Human interactions with local ecosystems

Historical anthropogenic disturbances have a major influence on current ecosystem condition and diversity (Lunt and Spooner 2005). Waterways have been identified as the ecosystem most impacted by European settlement in Melbourne (Presland 2009) and Victoria (OCE 1988). Lindenmayer (2007) believes that land clearing, primarily for agriculture, is perhaps the single most important cause (or driver) of environmental degradation, loss of species and depletion of ecological communities, both in Australia and worldwide. Wilson (1990) states that heavy grazing by stock and feral animals had the most widespread adverse impacts (Lunt and Spooner 2005). What this identifies is that waterway health can be influenced by many factors, which often interact.

Early European settlements in Australia were, for all intents and purposes, transplanted microcosms of 19th-century British society (DWR 1989b). Such rapid environmental change was heavily influenced by the official view of the European colonisers of Australia that it was *terra nullius* (nobody's land), as well as importing a land management approach that had caused substantial pollution in English cities (Luckin 2000), underpinned by a philosophy that people and the environments in which they lived were separate. This was in direct contrast to the Aboriginal conception so eloquently captured in the notion of Country, that people and the environment are one.

The assessment of waterway condition and the integrated management of threats and impacts on land use has been a significant focus in Victoria over the past three decades. It arose through numerous investigations into the state of Victorian waters and culminated in the formation of regional Catchment Management Authorities via the *Catchment and Land Protection Act 1994*. Such an approach is more consistent with traditional Aboriginal management philosophy. Central to this approach is ecological risk assessment, which evolved from the need to develop processes that better deal with the complexity of aquatic ecosystems, that is, the difficulties in assessing multiple stressors for a wide range of species within inherently variable ecosystems (EPA 2004). In Victoria, ecological risk assessment is predominantly delivered through the DPSIR (Drivers–Pressure–State–Impact–Response) model. This chapter focuses on the ecological drivers and threats (pressures) components.

The conditions that Europeans first encountered in Australian rivers were the result of tens of thousands of years of Aboriginal involvement (Gammage 2012; Humphries 2007; Pascoe 2014). The archaeological record indicates that 40 000 years of Aboriginal use of the environment probably had a lesser impact on the continent's natural ecosystems than did the first 200 years of European settlement (OCE 1988). Consequently, aquatic communities were probably more sustainable in pre-European times (Barmuta 2003; Humphries 2007).

Central to the Aboriginal approach of designing and managing the Australian landscape was land responsibility based on particular water resources (OCE 1988) and acceptance that these water sources were highly variable. Waterways on the Western Basalt Plains, in particular wetlands and associated Aboriginal occupation, have come and gone with the changing climate, while Aboriginal settlement in the resource-rich river valleys has been constant around Melbourne for at least 30 000 years (Canning 2009). The same would have occurred around Geelong. The Western Plains had one of the highest population densities at the time of European arrival (OCE 1988). Three Aboriginal nations were responsible for the Barwon catchment at the time of European settlement – the *Gadabanud*, the *Gulidjan* and the *Wadawurrung*. The *Wada wurrung balug* clan, with whom William Buckley, the 'Wild White Man', lived for more than 30 years, were based around what is now Geelong.

Basalt plains grasslands

The vegetation of south-western Victoria was dominated by grasslands. This includes the vegetation of inner and western Geelong, which forms the eastern boundary of the Western Volcanic Plains. Despite having lower rainfall and runoff than the Otways to the south and the Great Australian Divide to the north, waterbodies are a constant and varied feature of these basalt plains grasslands. They include major rivers such as the Barwon that contained water all year, to smaller streams like Waurn Ponds Creek and Hovells Creek that formed chains of ponds during the drier months. Such chains of ponds were generally fed by groundwater and ranged from fresh to brackish. Likewise, fresh to hyper-saline wetlands were common throughout the region, especially around Colac. Close to Geelong, they included Lake Murdeduke and Lake Modewarre.

The grasslands were naturally low in nutrients – up to 12 times lower in extractable nutrients such as nitrogen, phosphorus and potassium than rainforests (Wijesuriya 1999). At the time of European settlement, grasslands contained three times less nutrients than English soils (Jones 2010). They were maintained by burrowing animals (Wong and Morgan 2007) and more than 60 000 years of deliberate human intervention through fire and cultivation. Such relationships probably changed local water regimes and transport materials by altering catchment vegetation and runoff (Boulton *et al.* 2014). They also meant that soil was soft enough to push a finger into and water was able to soak in rather than run off, which meant that less rain sustained more plants (Gammage 2012). This was fundamental in attracting squatters, who ran sheep throughout much of the Barwon catchment by 1837.

Fire frequency, typically every three to five years, has the most profound influence on grassland composition and structure. The absence of fire for more than 12 years results in poor grassland state and regenerative ability (Lunt and Morgan 2002). No bushfire has been recorded in urban Geelong since 1835 (Smail 1998).

The removal of Aboriginal management and use, combined with imported British agricultural techniques and tools, had a devastating impact on grasslands within 15 years of settlement. By this time, half of the organic soil content had been lost (Jones 2010) – the soils of the basalt plains grasslands were six times less fertile than British soils by the time the gold rush started in 1851 (Wissing and Jones 2018).

Riparian ecosystems

On a global scale, Australian ecosystems typically have low primary production, or biomass production, levels. Floodplains are crucial for the ecology of Australian rivers, being significant sources of carbon and nutrients for the river ecosystem (Davis and Koop 2006). Most primary production within rivers is carried downstream (Boulton and Brock 1999), meaning that estuaries have higher nutrient levels and are more productive than ecosystems further upstream. The Barwon River in Geelong was estuarine at the time of European settlement. The rivers and wetlands of what is now urban Geelong were described as 'useless' swamps in the 1860s (The Argus, 20 July 1866, 8 December1866), and were subjected to extensive clearing from this time onwards (The Australasian, 2 July 1921). Such attitudes had evolved in the countries of origin over many generations (Presland 2014) and still persist in some sections of the community today (Dobbie and Green 2013).

Floodplain riparian woodland covered most of the floodplain of what is now urban Geelong. DSE (2004) describes it as:

> *An open eucalypt woodland to 20 m tall over a medium to tall shrub layer with a ground layer consisting of amphibious and aquatic herbs and sedges. They occur along the banks and floodplains of the larger meandering rivers and major creeks, often in conjunction with one or more floodplain wetland communities. Elevation and rainfall are relatively low and soils are fertile alluviums subject to periodic flooding and inundation.*

Flooding is the main regeneration event (Cheal 2010) in riparian forests, with the desired interval between floods varying between five and 10 years (White 2012). GHD (1982) found that generally 75 mm of rain from the Barwon, Leigh and Moorabool catchment over three days on a wet catchment, or 150 mm of rain on a dry catchment, will cause significant flooding in Geelong.

While fire does occur in riparian forests, White (2012) identifies that this ecosystem type is flammable only occasionally, and fires tend to be of low severity and patchy. The minimum frequency for low-severity fires is 10 years and for high- severity fires is 30 years (White 2012). Fire in wetlands is rare (Cheal 2010).

A range of wetland communities were found in the floodplain around Geelong. These included aquatic herbland/plains sedgy wetland mosaic adjacent to Jerringot Living

Station, swampy riparian woodland found at the junction of Waurn Ponds Creek with the Barwon River, and lignum swamp around Leather Street. Further downstream, Reedy Lake, Hospital Swamp and Lake Connewarre were found. Lake Connewarre forms part of the Port Phillip Bay (western shoreline) and Bellarine Peninsula Ramsar site as a wetland of international importance. Much of the wetland area is part of the Lake Connewarre State Game Reserve.

The Barwon River was typical of other large rivers in southern Victoria, characterised by low annual discharges (Boulton *et al.* 2014), high colour and turbidity, high salinity and low nutrients. Baseline flows in Australian rivers and streams are nearly three times more variable than the world average (Boulton and Brock 1999). As Otto (2011) describes, a 'once in 100 years' Australian flood is on average five times, but sometimes 20 or 30 times, the size of an annual flood; whereas in the rest of the world a 100-year flood is usually only two or three times the annual average size'.

Streams are classified as perennial or permanent if they flow for most of the year (DWR 1989a). In Victoria, only 16% of rainfall flows into streams and 1% filters through to recharge groundwater aquifers (Yates 2010), which by global standards is low. Flows in the Barwon and Leigh rivers are strongly seasonal, with long periods of very low flow in the summer and autumn and maximum flows in August or September (CCMA 2005). Edmonds (2004) commented on the Barwon and Leigh rivers drying up into pools in the major drought of 1837–42. This characteristic was recognised by early explorers and settlers of the region in the naming of waterways such as Duck Ponds Creek in Lara (now Hovells Creek) and Waurn Ponds Creek in Grovedale. As summer continued, nutrient levels increased and blue-green algal blooms occurred in the remaining waterholes and wetlands.

The Barwon catchment covers an area of ~8590 km^2 (CCMA 2017) and includes two major river systems – the Barwon River, which drains the northern slopes of the Otway Ranges in the south, and the Leigh River, draining the southern slopes of the Great Dividing Range (OCE 1988). The Moorabool River enters at the western end of urban Geelong.

The Barwon River was recognised as a very important river by both Aboriginal and European people. In February 1836, Gellibrand identified that the Barwon, along with the Plenty River in Melbourne, were the only ones deserving the name of river (Sayers 1983). While Aboriginal people seldom named streams, instead naming the stopping places along the banks where they camped, they did name the Barwon *Barre:Warre:N: Yallock*, meaning 'the Great River flowing from the uplands [Otway Ranges] to the ocean [Bass Strait]' (Lane 1991). The Barwon was a principal river-highway, a border separating clans and/or tribes (OCE 1988). Hence, catchment management was central to integrated land and water management for *Wadawurrung* people.

There are few written accounts of the Barwon River at the time of European settlement. Gellibrand provides most of them. However, there are more visual records, particularly those of John Helder Wedge in his illustrated accounts of travelling through *Wadawurrung* Country with William Buckley in 1835 (see Brown 1989 for further details).

After camping at Indented Head, on 5 February 1836, Gellibrand wrote (cited in Sayers 1983, p. 288):

> *When we quitted the natives, we directed our course to the head of the Barwon River. The river is about two miles* [3.2 km] *wide … We then proceeded through a fair country, near the margin of the river, until we arrived at a flat, where the river is at least eight* [12.8 km] *or nine* [14.5 km] *miles wide. At this flat there are some very good native wells, called Yan-Yan.*

Six months earlier, John Helder Wedge, in his explorations of the Geelong region with William Buckley, recorded that on 24 August 1835 the Barwon was little more than knee-deep just above its entry to Lake Connewarre (Brown 1989).

On 6 February 1836, Gellibrand (in Sayers 1983, p. 293):

> *... crossed over the cataract for the purpose of examining the Barrabool Hills, which had presented a most inviting appearance in our progress up the Barwon River ... a deep river, about 60 feet* [18 m] *wide for many miles up; in the winter a large body of water passes down it.*

In August 1836, on the first launch of a European boat on the Barwon River, Captain Edward Primrose Tregurtha described it as having 'a depth of $2^1/_2$–4 fathoms [4.5–7.3 m], with a moderate tide, and free of snags, and stumps, a most lovely country and the river itself alive with wild fowls of all descriptions'.

Wetlands

Victoria's wetlands are defined as areas subject to inundation. They can be classified as shallow and temporary, or deep and more permanent, and as freshwater or saline (Papas and Moloney 2012). By global standards, biodiversity in Australian wetlands is low. However, the flora in these wetlands was more diverse than that of the adjacent river (Boulton and Brock 1999). Connectivity between river channels and adjacent refuge wetlands was important in providing area to recolonise following long dry periods (Nielsen and Brock 2009).

Globally, wetlands occupy less than 9% of the Earth's land area (Zedler and Kercher 2005). At the time of European settlement, wetlands covered 3.5% of the Victorian landscape and ~8% of the land area in the Corangamite region. Today, wetlands in the Corangamite region cover 5% of the region (CCMA 2013). These wetlands are unevenly distributed throughout the Corangamite region, with the majority (84% of the wetland area) occurring in the Victorian Volcanic Plain bioregion, which covers 46% of the region. The second-highest number of wetlands are found in the Otway Plain bioregion, which includes the Barwon River and its tributaries and the Bellarine Peninsula (Sheldon 2006). Table 8.1 shows that the amount of wetlands increases as the Barwon River flows through the landscape.

Several wetlands in the Barwon catchment that are important to Aboriginal and contemporary Australian society are discussed below. The name in brackets refers to the Aboriginal Country in which the wetlands occurred.

Bermongo was a shallow freshwater marsh located around current- day Deans Marsh in the headwaters of the Barwon catchment on the northern foothills of the Otway Ranges. It was the home of the *Yan Yan Gurt* clan of the *Gadabanud* nation (Niewójt 2009). William

Table 8.1. Land area that comprises wetlands in the Barwon catchment (%)

Zone	% of land
Upper Barwon	1
Middle Barwon	1.8
Leigh	0.8
Bellarine	9.6

Source: Clarkson (2007).

Buckley, the 'Wild White Man', from the neighbouring *Wadawurrung* Nation, described exchanging tubers for eels (Flannery 2002).

Lake Murdeduke is a permanent saline lake ~1500 ha in size that supports large numbers of migratory shorebirds. Stone houses were found around Lake Murdeduke. It is considered *Wadawurrung* Country (Hale and Butcher 2011).

Lake Modewarre, situated 30 km south-west of Geelong, is a relatively shallow crater lake that in 2005 was around 2 m deep (EPA 2005). Water levels in Lake Modewarre fluctuate with climatic changes. About 1849–50, Lake Modewarre was nearly dry, but had only dried up in recent years and only once reached a very low level before that (*Geelong Advertiser* 1928). This trend has continued until today.

While Lake Modewarre is in the Otway Basin (Leahy *et al.* 2010), it is connected to the Barwon catchment in numerous ways.

William Buckley stayed at Lake Modewarre on numerous occasions. A *Wadawurrung* village was located at the site where the Buckley Railway Station would later stand (Geelong Advertiser, 5 May 1905). Around the time of European settlement Lake Modewarre was perfectly fresh. Eels were in great abundance, and the lake was a home for the Bunyip (Flannery 2002). In fact, the first record of the Bunyip was at Lake Modewarre. The *Wadawurrung* hunted for swans there during winter.

Jerringot Living Station was sheltered by considerable rises to the west and south-west, the direction of the prevailing cold winds. The site overlooked a 600 ha expanse of river alluvium, sand, gravel and clay, a vast river flat, the provident soil and permanently deep conditions of which supported many edible resources (Lane 1992). This was the location of the spring village of the *Wada wurrung balug*.

Reedy Lake, the largest area of deep-water marsh within the CCMA region (CEM 2005) possibly filtered large amounts of sediment carried by the Barwon River before entering Lake Connewarre (Billows and Gwyther 2007). It was also a late spring village of the *Wada wurrung balug*.

The 3000 ha coastal Lake Connewarre wetlands complex is found in the Barwon River estuary, one of the largest estuarine systems on the central Victorian coast (Billows and Gwyther 2007). During early European settlement, the lake was shallow with a black mud floor (Gill and Lane 1985). Situated ~9 km upstream of the mouth of the Barwon River, it provides summer feeding habitat for large numbers of migratory waders, shorebirds, waterfowl and other wetland birds (Billows and Gwyther 2007). Today it forms part of the Port Phillip (western shoreline) and Bellarine Peninsula Ramsar wetland site. The *Wadawurrung* lived here during the warmer months in permanent villages including at Campbell Point, Point Impossible and Breamlea.

Major ecological drivers of waterways in the Barwon catchment

It is generally understood that wetland formations are determined by the cumulative interactions of hydrology, landscape position, sediment dynamics, storm-driven processes, sea-level rise, subsidence and colonisation and disturbance by animals (Lee *et al.* 2006). The same applies for rivers and streams.

The two primary ecological drivers within waterways in the Barwon catchment are hydrology and water quality. Hydrology, or the timing, quantity and duration, of water flow is the key ecological driver. Most chemical and biological processes in standing and running waters are controlled and constrained by the water regime and its components

(Boulton and Brock 1999; Boulton *et al.* 2014). Arguably, the variability of the water regime is most important (Boulton *et al.* 2014).

The natural flow in a river continually changes, mainly due to the variability of rainfall (EPA 2004). Groundwater also plays a fundamental role, contributing baseline flows to many local waterways like the Barwon River and Kardinia Creek (Jones 1984). Such variability has existed for thousands of years (Nielsen and Brock 2009) with rainfall 18 000 years ago being as little as 50% present-day values (Gell *et al.* 2013; Finlayson *et al.* 2013). Variability also occurred during seasons and across years, decades and centuries (see Table 8.2). Major flooding events have occurred approximately every 20–30 years. A flood is a natural phenomenon that occurs when water covers land that is normally dry (DELWP 2016). A major flood occurs when the annual flood exceedance probability is 5% or less, or 20 years or greater (DELWP 2016).

Table 8.2. Hydrological description of FLOWS flow components

FLOWS flow component	Hydrological description	Relevant season	Frequency in the Barwon at Geelong
Cease-to-flow (also called zero flows)	Cease-to-flow is defined as periods where no flows are recorded in the channel	Not present in some streams, nearly always occurs in summer, but can also occur in winter	Several times each year in smaller waterways
Low flows	Low flows are the natural summer/autumn baseflows that maintain water flowing through the channel, keeping in-stream habitats wet and pools full	Summer	Annual
Low-flow freshes	Low-flow freshes, small and short-duration flow events that last one to several days as a result of localised rainfall during the low-flow period	Summer	2 events per year, 14 days
High baseflows	High baseflows refer to the persistent increase in baseflow that occurs with the onset of the wet season	Winter	Annual
High-flow freshes	High-flow freshes refer to sustained increases in the high-flow period as a result of sustained or heavy rainfall events.	Winter	4 per season, 6 days
Bankfull flows	Bankfull flows fill the channel but do not spill onto the floodplain	More common in winter, but occurs in summer	Every 1.2 years, 16 days
Overbank flows	Overbank flows are higher and less frequent than bankfull flows, and spill out of the channel onto the floodplain	More common in winter, but occurs in summer	Every 2–5 years

Source: Lloyd Environmental *et al.* (2005, 2006).

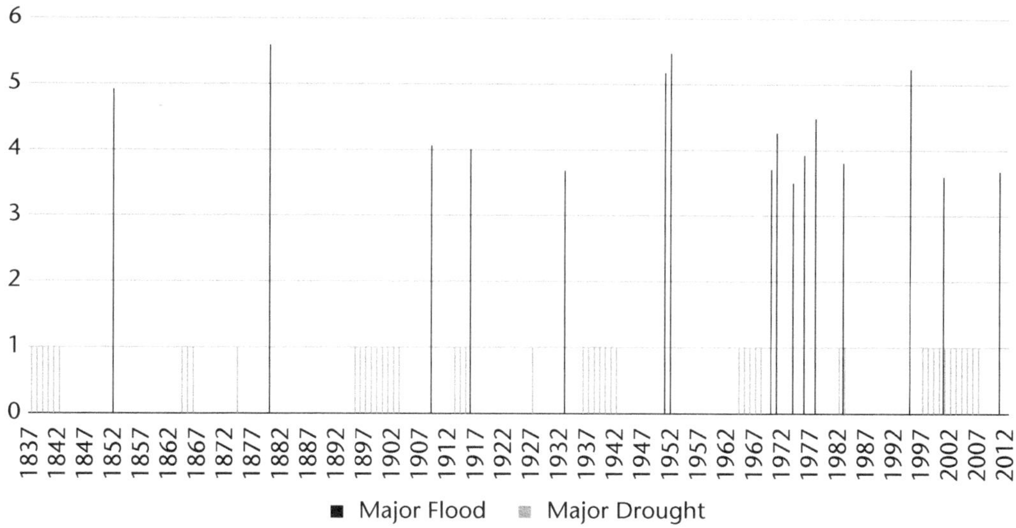

Fig. 8.1. Number of major climatic events in Geelong since European settlement. Source: Based on Figs. 1.7 and 2.4 in Smail (1998).

A drought is a prolonged abnormally dry period when the amount of water available is insufficient to meet normal human needs. It is generally measured by assessing rainfall deficiencies over three or more months. Major droughts are defined as 'serious' or 'severe'. A serious drought occurs when rainfall is in the lowest 5–10% for the period in question (BoM 2018). A severe drought means the lowest 5% of rainfall on record (Smail 1998) (Fig. 8.1).

This variability has caused Australian waterways to be characterised by a regular wetting and drying regime. Consequently, Australia supports a high number of species that show movements in response to boom and bust cycles (Prober *et al.* 2005; Prober and Thiele 2005; Robin *et al.* 2009) in terrestrial grasslands (Antos and Williams 2015) and waterways (Nielsen and Brock 2009). Biota of rivers and wetlands either tolerated these fluctuations or moved to more suitable habitats (Thomson *et al.* 2012). As a consequence, while such wetlands support a wide diversity of wildlife, the fauna was highly nomadic (OCE 1988). This was the case in Geelong where most wetlands were ephemeral, that is, not permanently holding water. There are numerous records of waterbird numbers fluctuating greatly in waterways in the Barwon catchment in the early days of European settlement.

Under natural conditions, much of the rainfall in riverine and wetland catchments was trapped and salts were drawn into groundwater aquifers, transpired by deep-rooted vegetation, evaporated or flushed out during the next high-flow events (Nielsen and Brock 2009). Floods transport the majority of nutrients in most riverine areas (Boulton and Brock 1999), particularly in wetlands, cut-off meanders, abandoned channels (Boulton *et al.* 2014) and riverine floodplains. Flooding releases a pulse of nutrients that, together with light and water, provide the resources for germination and growth of both micro and macro photosynthesisers (Boulton and Brock 1999). With the Barwon River being a lowland river in Geelong, flow would have been slow except in floods, when the water would have had a turbid appearance. Erosion of soil and sedimentation of wetlands is a natural process that occurs as slow removal by gentle sheet flows or rapid loss during torrential downpours or landslides. It comes from two main sources: catchment and channel (Boulton and Brock 1999). Hence, these lowland floodplain environments became more fertile with time (Romanowski 2009) and eventually become terrestrial ecosystems (Boulton *et al.* 2014).

In Geelong, the floodplain needs to be inundated every two to three years to promote breeding and recruitment of fish (Lloyd Environmental *et al.* 2005) and every two to five years to promote the growth of perennial flora and allow the growth and seed set of opportunistic floodplain flora (Lloyd Environmental *et al.* 2005). The natural hydrology of rivers such as the Barwon has several components, as outlined in Table 8.2. Each component plays a fundamental role in healthy functioning waterways. The mean annual natural yield of the Barwon catchment (excluding the Moorabool) is 270 GL (Lloyd Environmental *et al.* 2005). This is low compared to other rivers in Victoria.

Water quality

On a global scale, Australian waters contain low levels of nutrients. The three macronutrients phosphorus, nitrogen and carbon are the most important (Boulton *et al.* 2014). Other important water quality elements are salinity, pH and turbidity, an indicator of suspended sediment.

Under pre-European settlement conditions, just before pools in ephemeral wetlands and rivers dried up, water quality for salinity and nutrients was extremely poor (Boulton and Brock 1999). This natural drying of pools is critical in enabling bacteria to rapidly take up soluble phosphorus and making it unavailable until rewetting (Boulton and Brock 1999). This meant that algal blooms were a natural feature of Australian rivers and wetlands (Boulton and Brock 1999).

While water in rivers and wetlands at this time would have been very turbid, for most of the time water in the billabongs was black (Otto 2011) but clear, due to tannins from Eucalypts that surrounded them. Early descriptions of rivers such as the Yarra and Barwon were that the waters were clear (*The Argus*, 2 August 1859). Natural freshwater systems generally have a pH range of 6.5-8.0 and natural marine waters are generally close to 8.2 (EPA 2004).

Salinity levels in the Barwon catchment also varied greatly across time, and increased as water flowed down the river. A substantial salinisation event in nearby Lake Colac, evident at *c.*500 years BP, lasted several decades (Gell *et al.* 2013). Two saline shallow aquifers, a saline shallow aquifer (>3000 uS/cm) around Inverleigh and a brackish (900–3000 uS/cm) shallow aquifer (DWR 1989b) in Geelong were the dominant hydrological condition in summer and provided most of the low-level base flow in the Barwon at Geelong (Lloyd Environmental *et al.* 2005). This meant that water in the Barwon for 'three parts of the year, as during the summer months at Melbourne, is not in a fit state to drink' (*Sydney Monitor and Commercial Advertiser*, 19 April 1839). Such conditions were determinants for seasonal movement of *Wada wurrung balug* between their different living places, to places such as Reedy Lake and Bream Creek where fresh water was available. They were also needed for the creation of wells along the shore of Corio Bay and the Bellarine Peninsula.

Threats

Waterways have been identified as the ecosystem most impacted by European settlement. The threats impacting rivers and wetlands are largely the same. Sites not meeting one or more of the ecological objectives are identified as posing a potential risk to the health of the ecosystem (EPA 2004).

Threats come from the surrounding land (catchment) and within the waterway itself. The three major threats to Australian waterways – salt, nutrients and altered flows – come mostly from two land-based sources: point sources discharged via pipes, and diffuse sources, which come from a range of dispersed urban and rural land use activities. Point sources include sewage treatment plants, urban stormwater pipes, irrigation drainage and intensive animal industries. Diffuse sources of pollution include soil erosion from mining and agriculture, stormwater runoff, agricultural runoff and forestry activities. Diffuse pollution is more difficult to manage than point source pollution because it requires fundamental changes in land use activity and management.

At the catchment level, the most prominent impacts on waterway health in Geelong since European settlement have been grazing, gold mining, water regulation and extraction. Within the urban boundaries of Geelong, industry, residential living and roads have had the most impact. This is common for rivers across Victoria (VCMC 2017).

The interrelationship between salinisation, erosion and nutrient release (Prosser et al. 2001) make it difficult to attribute changes to particular drivers of change, and therefore particular changes in land use (Gell *et al.* 2009). A holistic approach needs to be taken when investigating risks to an ecosystem (EPA 2004).

Today, natural forests and woodlands remain in parts of the Otways and Brisbane Ranges headwaters areas, forming 13% of the catchment area (excluding the Moorabool catchment). Grassland used for grazing and cleared land used for grain cropping occupy 81% of the catchment (excluding the Moorabool catchment). The other major categories of land use are intensive agriculture (2%), plantation (1%) and urban areas (2%) (DWR 1989a) (Lloyd Environmental *et al.* 2005). As Table 8.3 identifies, market gardens contribute the highest total nitrogen and total phosphorus loads to waterways. Market gardens have not existed next to the Barwon River in Geelong since just after World War II (Wissing and Jones 2018). Urban areas have the next highest nutrient loads, followed by improved pasture. However, given the small proportion of the Barwon catchment that is urban compared to grazing and cropping, agricultural activities have a much larger impact on waterways.

Within the river channel of the Barwon River there are several threats to the condition of the waterway, including bed instability and degradation, change in flow regime and reduced riparian connectivity, degraded riparian vegetation and reduced vegetation width, barriers to fish passage and loss of instream woody habitat (CCMA 2014). For wetlands in the Connewarre subcatchment, the most commonly recorded management issues are altered water regimes, nutrient enrichment, grazing and encroachment (CEM 2005).

During the 19th century, the instream human activity that had the most impact was the practice of sheep washing before shearing, which was undertaken in local waterways before the establishment of wool scours along the Barwon River in the 1850s. Sheep scab was also managed in and around waterways. Treatment included the use of tobacco washes, acids and arsenic. Studies of arsenic contamination in Australia have found that it can persist and have impacts on aquatic life for more than 150 years (EPA 2017). For most of the 20th century, major instream impacts have resulted from drainage, removal of woody debris, installation of dams and weirs, and river straightening. The 21st century has focused on reducing the impacts of such works in waterways.

Table 8.3. South-east Australian average annual nutrient export data (kg/ha/yr)

	% land use per landscape zone (CCMA 2014)[1]						Total phosphorus		Total nitrogen	
Broad land use type	**Leigh**	**Murdeduke**	**Upper Barwon**	**Mid Barwon**	**Moorabool**	**Bellarine**	**Range**	**Typical**	**Range**	**Typical**
Urban	5			2		15	0.4–3.6	1	3.2–22.4	6.6
Peri-urban	5					5				
Improved pasture							0.1–0.7	0.3	0.6–4.6	3.3
Unimproved pasture	64	70	59	69	68	56	0.07	0.07	2.2	2.2
Cropping		25	4	20	5	8				
Market gardens					8		2.7–14.3	7.1	20–34.5	26
Conservation	5	3	6	2	4	12	0.03–0.1	0.06	0.9–1.5	1.1
Forestry	9		30		15					
Other	12	2	1	7		4				

Source: Modified from Young *et al.* (1996).

Conclusion

Australian waterways have been consciously and actively managed for tens of thousands years by Aboriginal people. These include the Barwon River and Geelong, place names that both originate from the language of the Traditional Owners, the *Wadawurrung* people. The Aboriginal management of waterways and their surrounding land catchments worked with the natural ecological boundaries and associated ecological processes. Central to the Aboriginal approach was recognition that most waterways in the region had boom and bust cycles in which water would be present for periods of time, attract wildlife and attract people, then decline in quantity and quality. When this happened, wildlife and people would move to the next place where suitable resources could be found. Permanent seasonal residential areas of at least 5000 years (Gill and Lane 1985) were associated with local rivers and wetlands.

European colonisers of what is now Geelong introduced new philosophies and land and water management approaches that had not evolved in Australia and were not environmentally successful even where the Europeans came from. Contemporary Australian approaches to land and water management are more aligned with traditional Aboriginal approaches, but the management of impacts to rivers and wetlands around places like Geelong is still not sustainable. To become sustainable, we must understand how our waterways function so that when we restore them we are more likely to be reinstating the original, pre-contact ecological processes (drivers) than treating the symptoms. Understanding the ecological processes of local waterways has been the key focus of this chapter.

References

Antos M, Williams NSG (2015) The wildlife of our grassy landscapes. In NSG Williams, A Marshall and JW Morgan (Eds) *Land of Sweeping Plains: Managing and Restoring the Native Grasslands of South-eastern Australia* (Chapter 4). CSIRO Publishing, Melbourne.

Barmuta LA (2003) Imperilled rivers of Australia: challenges for assessment and conservation. *Aquatic Ecosystem Health and Management* **6**(1), 55–68. doi:10.1080/14634980301476

Billows C, Gwyther J (2007) *Ecological Study of Lake Connewarre Wetlands Complex*. Deakin University, Geelong.

BoM (2018) *Climate Glossary*. Bureau of Meteorology, Canberra. http://www.bom.gov.au/climate/glossary/drought.shtml (accessed 15 May 2018).

Boulton AJ, Brock MA (1999) *Australian Freshwater Ecology Processes and Management*. Gleneagles Publishing, Devon.

Boulton A, Brock M, Robson B, Ryder D, Davis J (2014) *Australian Freshwater Ecology: Processes and Management*. 2nd edn. John Wiley & Sons, Brisbane.

Brown PL (Ed.) (1989) *Andrew alias William Todd (John Batman's Recorder) and his Indented Head Journal 1835 (Chief Illustrator, J.H. Wedge)*. Geelong Historical Society, Geelong.

Canning S (2009) Broad-scale paleo-environmental reconstructions of southern Victoria, Australia. *Artefact* **32**, 4–14.

CCMA (2005) *The Environmental Flow Determination for the Barwon River: Issues Paper*. Corangamite Catchment Management Authority, Colac, Victoria.

CCMA (2013) *Corangamite Catchment Management Strategy 2013–2017*. Corangamite Catchment Management Authority, Colac, Victoria.

CCMA (2014) *Corangamite Waterway Strategy 2014–2022*. Corangamite Catchment Management Authority, Colac, Victoria.

CCMA (2017) *Draft Seasonal Watering Proposal for the Lower Barwon Wetlands 2017–18, March 2017*. Vol. 2. Corangamite Catchment Management Authority, Colac, Victoria.

CEM (2005) *Corangamite Wetland Inventory.* Centre for Environmental Management, Corangamite Catchment Management Authority, Colac, Victoria.

Cheal D (2010) *Growth Stages and Tolerable Fire Intervals for Victoria's Native Vegetation Data Sets.* Fire and Adaptive Management Report no. 84. Department of Sustainability and Environment, Melbourne.

Clarkson T (2007) *Corangamite Soil Health Strategy* 2007. Corangamite Catchment Management Authority, Colac, Victoria.

Davis JR, Koop K (2006) Eutrophication in Australian rivers, reservoirs and estuaries: a southern hemisphere perspective on the science and its implications. *Hydrobiologia* **559**, 23–76. doi:10.1007/s10750-005-4429-2

DELWP (2016) *Victorian Floodplain Management Strategy.* Department of Environment, Land, Water and Planning, Melbourne.

Dobbie M, Green R (2013) Public perceptions of freshwater wetlands in Victoria, Australia. *Landscape and Urban Planning* **110**, 143–154. doi:10.1016/j.landurbplan.2012.11.003

DSE (2004) *EVC/Bioregion Benchmark for Vegetation Quality Assessment Victorian Volcanic Plain Bioregion: EVC 56 – Floodplain Riparian Woodland.* Department of Sustainability and Environment, Melbourne.

DWR (1989a) *Water Victoria: An Environmental Handbook.* Department of Water Resources, Melbourne.

DWR (1989b) *Water Victoria: A Resource Handbook.* Department of Water Resources, Melbourne.

Ede FJ, Hunt TD (2008) *Habitat Management Guide: Riparian. Weed Management in Riparian Areas: South-eastern Australia.* CRC for Australian Weed Management, Adelaide.

Edmonds L (2004) *Living by Water. A History of Barwon Water and its Predecessors.* Barwon Water, Geelong.

EPA (2004) *Guidelines for Environmental Management: Risk-based Assessment of Ecosystem Protection in Ambient Waters.* Publication 961. Environment Protection Authority, Melbourne.

EPA (2005) *Lake Modewarre Water Quality 2004–05.* Publication 1018. Environment Protection Authority, Melbourne.

EPA (2017) *Mercury and Arsenic in Victorian Waters: A Legacy of Historic Gold Mining.* Publication 1637.1. Environment Protection Authority, Melbourne.

Finlayson CM, Davis JA, Gell PA, Kingsford RT, Parton KA (2013) The status of wetlands and the predicted effects of global climate change: the situation in Australia. *Aquatic Sciences* **75**, 73–93. doi:10.1007/s00027-011-0232-5

Flannery T (2002) (Ed.) *The Life and Adventures of William Buckley: Thirty-two Years a Wanderer amongst the Aborigines of the then unexplored Country round Port Phillip.* Text Publishing, Melbourne.

Gammage B (2012) *The Biggest Estate on Earth: How Aborigines made Australia.* Allen & Unwin, Sydney.

Geelong Advertiser, Skeleton for the museum, Friday 5 May 1905, p. 2; Lake Modewarre: Ministerial Proposal – Geelong may need area for water storage basin, Tuesday 24 January 1928, p. 5.

Gell P, Fluin J, Tibby J, Hancock G, Harrison J, Zawadzki A, Haynes D, Khanum S, Little F, Walsh B (2009) Anthropogenic acceleration of sediment accretion in lowland floodplain wetlands, Murray–Darling Basin, Australia. *Geomorphology* **108**(1), 122–126. doi:10.1016/j.geomorph.2007.12.020

Gell P, Mills K, Grundell R (2013) A legacy of climate and catchment change: the real challenge for wetland management. *Hydrobiologia* **708**, 133–144. doi:10.1007/s10750-012-1163-4

GHD (1982) *Geelong Floodplain Management Study.* Report prepared by GHD Pty Ltd for State Rivers and Water Supply Commission of Victoria, Melbourne.

Gill ED, Lane LN (1985) Sea levels and Aboriginal oyster midden at Lake Connewarre, Victoria, Australia. *Proceedings of the Royal Society of Victoria* **97**(2), 95–100.

Hale J, Butcher R (2011) *Ecological Character Description for the Western District Lakes Ramsar Site.* Report prepared for the Department of Sustainability, Environment, Water, Population and Communities, Canberra.

Humphries P (2007) Historical Indigenous use of aquatic resources in Australia's Murray–Darling Basin, and its implications for river management. *Ecological Management and Restoration* **8**(2), 106–113.

Jones G (1984) *Growing Together. A Gardening History of Geelong – Extending to Colac and Camperdown.* George Jones.

Jones R (2010) *Green Harvest: A History of Organic Farming and Gardening in Australia.* CSIRO Publishing, Melbourne.

Lane LN (1991) Yollinko Park. Unpublished address to the Highton Rotary Club, 14 November 1991.

Lane LN (1992) Living Stations: Wathaurong tribe. The Jerringot Site. Unpublished paper.

Leahy P, Robinson D, Patten R, Kramer A (2010) *Lakes in the Western District of Victoria and Climate Change.* Publication 1359, November 2010. Environment Protection Authority, Melbourne.

Lee SY, Dunn RJK, Young RA, Connolly RM, Dale PER, Dehay R, Lemckert CJ, McKinnon S, Powell B, Teasdale PR, Welsh DT (2006) Impact of urbanization on coastal wetland structure and function. *Austral Ecology* **31**(20), 149–163.

Lindenmayer D (2007) *On Borrowed Time: Australia's Environmental Crisis and What We Must Do About It.* Penguin, Melbourne.

Lloyd Environmental, Ecological Associates, Fluvial Systems (2005) *Environmental Flow Determination for the Barwon River: Issues Paper.* Corangamite Catchment Management Authority, Colac, Victoria.

Lloyd Environmental, Ecological Associates, Fluvial Systems (2006) *Environmental Flow Determination for the Barwon River: Final Report – Flow Recommendations.* Corangamite Catchment Management Authority, Colac, Victoria.

Lovett S, Price P (1999) *Riparian Land Management Technical Guidelines. Vol. 1: Principles of Sound Management.* Land and Water Resources Research and Development Corporation, Canberra.

Luckin B (2000) Pollution in the city. In MJ Daunton (Ed.) *The Cambridge Urban History of Britain, Vol. 3 1840–1950.* Cambridge University Press, Cambridge.

Lunt ID, Bennett AF (1999) Temperate woodlands in Victoria: distribution, composition and conservation. In RJ Hobbs and CJ Yates (Eds) *Temperate Eucalypt Woodlands in Australia: Biology, Conservation, Management and Restoration* (Chapter3). Surrey Beatty & Sons, Sydney.

Lunt ID, Morgan JW (2002) The role of fire regimes in temperate grasslands of south-eastern Australia. In RA Bradstock (Ed.) *Flammable Australia: The Fire Regimes and Biodiversity of a Continent.* Cambridge University Press, Melbourne.

Lunt ID, Spooner PG (2005) Using historic ecology to understand patterns of biodiversity in fragmented agricultural landscapes. *Journal of Biogeography* **32**, 1859–1873. doi:10.1111/j.1365-2699.2005.01296.x

Nielsen DL, Brock MA (2009) Modified water regime and salinity as a consequence of climate change: prospects for wetlands in southern Australia. *Climatic Change* **95**, 523–533. doi:10.1007/s10584-009-9564-8

Niewójt L (2009) *Gadubanud* society in the Otway Ranges, Victoria: an environmental history. *Aboriginal History* **33**, 175–199.

OCE (1988) *State of the Environment Report 1988: Victoria's Inland Waters.* Office of the Commissioner for the Environment, Melbourne.

Otto K (2011) *Yarra: The History of Melbourne's Murky River.* Text Publishing, Melbourne.

Papas P, Moloney P (2012) *Victoria's Wetlands 2009–2011: Statewide Assessments and Condition Modelling.* Technical Report Series no. 229. Arthur Rylah Institute for Environmental Research, Department of Sustainability and Environment, Melbourne.

Pascoe B (2014) *Dark Emu Black Seeds: Agriculture or Accident?* Magabala Books, Broome, WA.

Presland G (2009) *The Place for a Village: How Nature shaped the City of Melbourne.* Museum Victoria, Melbourne.

Presland G (2014) A boggy question: differing views of wetlands in 19th century Melbourne. *Victorian Naturalist* **131**(4), 96–105.

Prober SM, Thiele KT (2005) Restoring Australia's temperate grasslands and grassy woodlands by integrating function and diversity. *Ecological Management and Restoration* **6**(1), doi:10.1111/j.1442-8903.2005.00215.x

Prober SM, Thiele KR, Lunt ID, Koe TB (2005) Restoring ecological function in temperate grassy woodlands: manipulating soil nutrients, exotic annual and native perennial grasses through carbon supplements and spring burns. *Journal of Applied Ecology* **42**, 1073–1085. doi:10.1111/j.1365-2664.2005.01095.x

Prosser IP, Young WJ, Rustomji P, Hughes A, Moran CJ (2001) A model of river sediment budgets as an element of river health assessment. In *Proceedings of the International Congress on Modelling and Simulation* (MODSIM 2001). 10–13 December, pp. 861-866. Modelling and Simulation Society of Australia and New Zealand, Canberra. https://www.mssanz.org.au/MODSIM01/Vol%202/Prosser.pdf (accessed 1 September 2018).

Robin L, Heinsohn R, Joseph L (Eds) (2009) Boom and Bust: Bird Stories for a Dry Country. CSIRO Publishing, Melbourne.

Romanowski N (2009) *Planting Wetlands and Dams.* CSIRO Publishing, Melbourne.

Sayers CE (Ed.) (1983) *Letters from Victorian Pioneers: A Series of Papers on the Early Occupation of the Colony, the Aborigines etc addressed by Victorian Pioneers to His Excellency Charles Joseph Latrobe Esq, Lieutenant Governor of the Colony of Victoria 1853.* Lloyd O'Neil for Currey O'Neil, Melbourne.

Sheldon RA (2006) *Corangamite Wetlands Strategy 2006–2011.* Corangamite Catchment Management Authority, Colac, Victoria.

Smail L (1998) *Weather Wonders of Geelong.* L. Smail, Grovedale, Victoria.

Sydney Monitor and Commercial Advertiser, Port Phillip: Geelong, Friday 19 April 1839, p. 2.

The Argus, Geelong, Tuesday 2 August 1859, p. 6; More about Chinese gardening, Friday 20 July 1866, p. 7; On the Barwon, Saturday 8 December 1866, p. 1.

The Australasian, The Naturalist. Bush notes concerning eels, Saturday 2 July 1821, p. 51.

Thomson J, Bond N, Cunningham S, Metzeling L, Reich P, Thompson R, Mac Nally R (2012) The influences of climatic variation and vegetation on stream biota: lessons from the Big Dry in south-eastern Australia. *Global Change Biology* **18**, 1582–1596.

VCMC (2017) *Catchment Condition and Management Report 2017.* Victorian Catchment Management Council, Melbourne.

White AK (2012) *Ecosystem Conceptual Models for Victorian Ecosystems.* Parks Victoria Technical Series no. 65. Parks Victoria, Melbourne.

Wijesuriya WS (1999) *Investigation into the Relationship between Biomass Reduction, Soil Disturbance and Weed Invasion in Basalt Plains Native Grassland Remnants in Victoria, Australia.* School of Life Science and Technology, Faculty of Engineering and Science, Victoria University, Melbourne.

Wilson AD (1990) The effect of grazing on Australian ecosystems. *Proceedings of the Ecological Society of Australia* **16**, 235–244.

Wissing RJ, Jones DS (2018) A tale of three cities: food in Aboriginal, European and Chinese Geelong. *Proceedings of the 14th Urban History Planning History Biennial Conference, 'Remaking Cities'.* 31 January–2 February 2018, RMIT, Melbourne.

Wong N, Morgan J (2007) *Review of Grassland Management in South-eastern Australia.* Parks Victoria, Melbourne.

Yates A (2010) *Victorian Infrastructure Report Card 2010.* Engineers Australia, Melbourne.

Young WJ, Marston FM, Davis JR (1996) Nutrient exports and land use in Australian catchments. *Journal of Environmental Management* **47**, 165–183. doi:10.1006/jema.1996.0043

Zedler JB, Kercher S (2005) Wetland resources: status, trends, ecosystem services, and restorability. *Annual Review of Environment and Resources* **30**, 39–74. doi:10.1146/annurev.energy.30.050504.144248

9

Marine and coastal environments

Chris Smyth

> *Australia faces very real risks from climate change. The future for south-eastern Australia is one of increased severe weather events, bushfires and, later this century, sea level rise. Geelong is exposed to many of these climate risks and we must start planning now to adapt to a future, changing climate.*
>
> *The challenge for communities all over the world is that the extent and the timing of the impacts of a changing climate remain uncertain. One of the great uncertainties is in the future emission levels of the greenhouse gases that cause climate change. We all hope that a global agreement to reduce greenhouse gas emissions can be achieved soon, but we cannot delay action on this hope.*
>
> *This is the reality the City of Greater Geelong must plan for.*
>
> – Mayor John Mitchell (cited in CoGG 2011, p. 4).

Since European settlement, Geelong's marine and coastal environments and ecological processes have influenced, and in turn been impacted by, the region's development and population growth. In this 21st century, the effects of climate change will be far-reaching.

A change of scene

If Matthew Flinders were alive today, scaled Arthurs Seat and looked west to the Bellarine Peninsula, the view would be starkly different from what he saw in April 1802. The Peninsula would be more sharply defined without its grassy woodland cover, the coast changed by urban development.

In writing of his few days exploring Port Phillip Bay while camped along the coast between what are now Portarlington and Indented Head, Flinders mentioned his contact with *Wadawurrung* people, whose traditional boundaries stretch from the Werribee River to Lorne. He observed rising smoke, 'deserted fireplaces and heaps of oysters', a hut, a 'party of the inhabitants' at a distance, and gave a 'shag and some trifling presents' to three camp visitors whom he later shared a meal with.

Before the arrival of Flinders, the *Wadawurrung* people had been living in the Geelong region for more than 25 000 years, gathering a bountiful harvest of food and fibre from the coastal wetlands, intertidal platforms, river estuaries and grassy woodlands, and developing strong spiritual and cultural connections to Sea Country.

Recent estimates put the population of the 25 *Wadawurrung* clans at 700 when Flinders made contact. By 1853 it may have been reduced to 30–40 (Everist 2004) due to shootings, poisonings, disease and the loss of hunting grounds.

That was not the end of the *Wadawurrung* story. Today, their descendants are maintaining their cultural identity and connections with Sea Country through the Wathaurong Aboriginal Co-operative in North Geelong and the *Wadawurrung* (Wathaurong Aboriginal Co-operative) in Geelong and Ballarat.

The *Wadawurrung*'s traditional lands included the Bellarine Peninsula, which is part of the Otway Plain bioregion, and the Victorian Volcanic Plain bioregion, where Geelong now sits. Changes to the bioregions' vegetation cover can be measured by comparing today's cover with pre-1750s estimates. Seven of eight ecological vegetation classes found on or near the Geelong region's coast are either vulnerable or endangered, with losses ranging from 16% to 94% (Table 9.1).

Spatial changes within the Central Victoria (open coast) and Victorian Embayments marine bioregions are not so easily defined – a historical ecological baseline is yet to be established for Victoria's marine environments. However, current information indicates significant changes to seagrasses, oyster and mussel reefs, fish populations, bottom sediments and beaches.

These changes to terrestrial habitats and species began in the late 1830s when squatters and their sheep arrived in the Geelong region, followed by fishers, timber cutters, shippers, miners, industrialists, rail and road builders, real estate developers, residents and holidaymakers. Each has left their mark on the region's marine and coastal environments.

Before the motor car, people made their way to, from and along the coast by foot, horse, wagon, ship and later train. Today the region's residents and its visitors are far more

Table 9.1. Threatened ecological vegetation classes of the Geelong region

	Otway Plain			Victorian Volcanic Plain		
Ecological Vegetation Class (EVC)	**Pre-1750 (ha)**	**% left**	**Status**	**Pre-1750 (ha)**	**% left**	**Status**
1 Coastal Dune Scrub/ Coastal Dune Grassland Mosaic	2049	67	Depleted	n/a	n/a	n/a
9 Coastal Saltmarsh	725	15	Endangered	2294	46	Vulnerable
161 Coastal Headland Scrub	799	84	Vulnerable	20	35	Vulnerable
163 Coastal Tussock Grassland	113	68	Vulnerable	127	28	Vulnerable
302 Coastal Saltmarsh/ Mangrove Shrubland Mosaic	4047	76	Endangered	1099	56	Endangered
858 Coastal Alkaline Scrub	4657	24	Endangered	377	10	Endangered
55 Plains Grassy Woodland	17193	6	Endangered	716941	8	Endangered
175 Grassy Woodland	79132	7	Endangered	39916	12	Endangered

Source: DELWP (nd); VNPA (2014).

connected to the coast, but connectivity for coastal wildlife has seriously declined. It would be far worse had governments in the 1870s not reserved a narrow strip of public land above the high-water mark along 96% of the coast. Although under intense pressure from coastal development and now climate change, this narrow band and several nature reserves offer refuge for the region's wildlife, as do the offshore marine national parks and sanctuaries created in 2002.

For the purpose of discussion, this chapter reviews changes in the region's marine and coastal environments along four coastal strips between Anglesea and Little River:

1. Anglesea to Point Lonsdale
2. Point Lonsdale to Point Henry
3. Corio Bay: Point Henry to Point Lillias
4. Geelong Arm: Northern Shore.

These four strips are defined not by jurisdictional boundaries but by physical geography (open coast or inside Port Phillip Bay) and urban and industrial development (Geelong waterfront).

As background to this chapter, Victoria's coastline from the high-water mark to 3 nautical miles offshore (the boundary between state and Commonwealth waters) is owned and managed by the Crown, along with land of varying widths above the high-water mark (along 4% of the coast, private land abuts the high-water mark). Although almost two-thirds of the coast is in national and other conservation parks managed by Parks Victoria, the remainder is managed in sections by various land managers, including minister-appointed committees of management, local councils as committees of management, and Parks Victoria. Their powers and responsibilities are derived from the *Crown Land (Reserves) Act 1978* (Vic.), while their management activities must be consistent with the strategies and coastal management plans prepared under the *Coastal Management Act 1995* (Vic). In 2018, this Act was replaced by the *Marine and Coastal Act*, which initiated a new state-wide marine and coastal policy, a marine spatial planning framework and five-yearly state of the marine and coastal environment reports.

Between Anglesea and Point Lonsdale (Strip 1) there are six separate land managers: Great Ocean Road Coast Committee, Parks Victoria, the City of Greater Geelong Council, the Department of Environment, Land, Water and Planning (DELWP), Barwon Coastal Committee of Management and the Borough of Queenscliffe. These five land managers essentially operate independently from one another, and the two committees also operate independently from local government. Such a fragmented coastal planning and management framework lacks integration and encourages duplication, limited visions and conflicting agendas. In the marine environment there is no overall plan, with various industry sectors such as fishing, petroleum and shipping managed under separate legislation and by different agencies with diverse aims. Such a discussion is outside the scope of this chapter; however, it is noted that the *ad hoc* nature of terrestrial and marine management and the mix of priorities has and will continue to add pressure on the vulnerable natural landscapes and seascapes identified in this chapter.

Anglesea to Point Lonsdale

Sheer cliffs, wide sandy beaches, high sand dunes, occasional headlands, intertidal and subtidal limestone reefs and bare sand are found along 50 km of high-energy open coast from Anglesea to Point Lonsdale.

The Coastal Headland Scrub on the clifftops between Anglesea and Bells Beach has been largely retained, kept as public land and protected in the Great Otway National Park. From

Jan Juc, the cliffs drop away and sand dunes, punctuated by the occasional headland, dominate the scenery to Point Lonsdale. Much of the hinterland's Grassy Woodland was cleared for agriculture, while coastal alkaline scrub has been hard-hit by the growth of Torquay, Barwon Heads, Ocean Grove and Point Lonsdale.

Vegetation cover between the towns has also been degraded by farms, golf courses and roads. To access the famed surf at Bells Beach, makeshift tracks and car parks were pushed through the coastal alkaline and headland scrubs in the 1960s, 100-year-old Moonah (*Melaleuca lanceolata*) trees have been lopped to improve spectator views, and weeds have invaded. The local group, Surfers Appreciating Natural Environment, have fenced and weeded remnant vegetation to help restore badly damaged areas.

Sand dune habitats between Torquay and Point Lonsdale have also been impacted by population growth and linear coastal development. Regional and local transport corridors provide multiple access points to these sensitive coastal areas, and the development of car parks, camping areas and other infrastructure have significantly added to the impact.

Torquay and Ocean Grove, along with the inland Armstrong Creek, a 30-minute drive from the seashore, are now designated growth areas as planners use the *G21 Geelong Regional Growth Plan* (G21 GRA 2013) to manage the City of Greater Geelong's projected 30% population growth between 2018 and 2036 (idcommunity 2018). Armstrong Creek will eventually house 50 000–65 000 residents in 22 000 homes; its 2016 population was a little over 4000. This rapid population growth across the Bellarine Peninsula (Clifton Springs is also designated as a growth area) will significantly increase visitor pressure on coastal habitats, with demands for more roads, car parks, access tracks, toilets and other facilities.

The coastal vegetation has also been undermined by invasive species, mostly spread from farms and gardens. Other species were planted in the 1970s to prevent sand drifts blocking Thirteenth Beach Road, which was built on the primary dune in the 1930s as a hoped-for Great Ocean Road gateway. Even so, there is high native plant diversity with 122 species, eight of state significance, including Coast Wirilda (*Acacia uncifolia*) and Coast Twinleaf (*Zygophyllum billardierei*) (Trengove 2013). The open-coast dunes are also home to the Swamp Wallaby (*Wallabia bicolor*) and Short-beaked Echidna (*Tachyglossus aculeatus*).

Environmental impacts on the open coast extend beyond the beaches and dunes into coastal wetlands. At Barwon Heads, the Lake Connewarre wetlands complex of five fresh and saline lakes and reedy marshes has formed on the floodplain of the Barwon River. Here the river discharges into Bass Strait at the end of its 160 km journey beginning in the Otway Ranges and crossing the Geelong region. The wetlands complex is one of 65 Australian sites listed on the Convention on Wetlands of International Importance (Ramsar Convention) and is important for migratory wader birds including the Marsh Sandpiper (*Tringa stagnatilis*), Sharp-tailed Sandpiper (*Calidris acuminata*) and Red-necked Stint (*Calidris ruficollis*).

On its cross-region journey, the Barwon River passes through agricultural, urban and industrial areas where it collects sediments, nutrients and chemicals and deposits them in the wetlands. Encroaching urban development, upstream water extraction, invasive species such as European Carp (*Cyprinus carpio*), Spartina Grass (*Spartina* x *townsendii*) and feral Cats (*Felis catus*), stormwater pollution and the trampling of shoreline vegetation are all major risks to the future of the Lake Connewarre complex.

As the coastal towns grew, so too did waste discharges from stormwater and sewage outfalls, creating local pollution hazards. From 1915 to the 1980s, Geelong's untreated sewage was discharged at Black Rock near Breamlea. Though now treated, the locals call the site Turd Rock and the Beachsafe website says, 'Polluted and hazardous beach with a rock dominated surf. Should be avoided' (SLSA 2018).

The beaches of the open coast are occasionally visited by the Little Penguin (*Eudyptula minor*), while the threatened Hooded Plover (*Thinornis rubricollis*) searches for amphipods in sea wrack. At the Barwon Bluff Marine Sanctuary, 27 conservation-listed birds have been recorded, including the Southern Giant Petrel (*Macronectes giganteus*), the Great Knot (*Calidris tenuirostris*) and Caspian Tern (*Hydroprogne caspia*) (Barton *et al.* 2012).

Below the high-water mark, intertidal and subtidal marine life was given greater protection in the Point Addis and Port Phillip Heads marine national parks and Point Danger and Barwon Bluff marine sanctuaries in 2002.

Intertidal reef flora is dominated by the brown algae, Neptune's Necklace (*Hormosira banksii*), which provides habitat for the Common Periwinkle (*Austrocochlea constricta*), Turban Shell (*Turbo undulatus*) and Cart-rut Shell (*Dicathais orbita*). Porter and Wescott (2010) found these molluscs were smaller in high-use areas and suggested that collection and trampling were changing community composition.

Crayweed (*Phyllospora comosa*) and common Kelp (*Ecklonia radiata*) often form dense canopies on subtidal reefs, providing habitat for Blacklip Abalone (*Haliotis rubra*) and the Southern Rock Lobster (*Jasus edwardsii*), once plentiful on intertidal reefs. Intensive harvesting caused their numbers to plummet and intertidal collection of abalone is now banned across Victorian waters. For rock lobster, the ban applies only to Port Phillip Bay.

Fish swimming above the subtidal reefs on the open coast include Bluethroat Wrasse (*Notolabrus tetricus*), Yellowtail Kingfish (*Seriola lalandi*), Southern Hulafish (*Trachinops caudimaculatus*), Sea Sweep (*Scorpis aequipinnis*), Scalyfin (*Parma victoriae*), Senator Wrasse (*Pictilabrus laticlavius*), Herring Cale (*Odax cyanomelas*) and the Western Blue Devilfish (*Paraplesiops meleagris*), for which Port Phillip Heads is considered an important stronghold for this colourful fish. Offshore, the Humpback Whale (*Megaptera novaeangliae*) and Australian Fur Seal (*Arctocephalus pusillus*) may be seen passing, while Bottlenose Dolphin (*Tursiops truncatus*) chase Southern Calamari Squid (*Sepioteuthis australis*).

At the entrance to Port Phillip Bay, a deep canyon is draped in sponge gardens containing more than 270 sponge species, with at least 115 that are endemic to The Heads (Barton *et al.* 2012). Dredging for the 2008 channel-deepening project 'exacerbated natural erosion processes, causing ongoing rubble formation and periodic rock falls into the canyon. Although this process is highly spatially confined, the canyon habitat is also highly spatially confined, making the threatening process more significant' (Edmunds *et al.* 2009, p. 53). The sponge gardens have since been listed as threatened under Victoria's *Flora and Fauna Guarantee Act 1988*.

Another threatened community, Giant Kelp (*Macrocystis pyrifera*), was once extensive but is now found only in small patches on subtidal reefs at Point Addis and Barwon Bluff, having disappeared from The Heads, Lonsdale Bight and off Point Lonsdale (TSSC 2012). Key factors in its decline are rising water temperatures and reduced nutrient loads, due to climate change, and reduced water quality.

Point Lonsdale to Point Henry

Rounding Point Lonsdale and entering Port Phillip Bay, the open-coast's high cliffs, tall sand dunes and limestone reefs are replaced by a large tidal delta, narrow beaches, low-lying land and soft clay, silt, sand and gravel shorelines. Nearshore there is bare sand, occasional rocky reefs and extensive seagrass meadows: large areas of *Amphibolis antarctica* grow in Lonsdale Bight, while Swan Bay and the Bellarine Bank off Clifton Springs are covered in *Zostera* species.

Almost all Coastal Alkaline Scrub, including large areas between Point Lonsdale and Queenscliff and at Portarlington, and plains grassy woodland and grassy woodland, is gone. The Coastal Saltmarsh/Mangrove Mosaic on the western side of Swan Bay is largely intact, as is the Coastal Dune Scrub/Grassland Mosaic between Point Lonsdale and Queenscliff (and a patchy foreshore sliver from St Leonards to Portarlington).

To the west of Point Lonsdale, the Lonsdale Lakes is another important coastal wetland complex suffering from development. The Point canal estate will result in 'the loss of ~2.3 ha of Coastal Saltmarsh, and the potential degradation of retained areas of saltmarsh adjoining the development, as well as the removal of ~32 ha of Saline Aquatic Meadow and ~7.5 ha of Coast Moonah Woodland' (Boon *et al.* 2011). There are concerns that nutrient discharges from the estate into Lakers Cutting, and ultimately Swan Bay, will lead to excessive algal growth and the loss of marine fauna.

Swan Bay, initially called Swan Pond by Matthew Flinders, can host close to 3000 Black Swan (*Cygnus atratus*) on its shallow and sheltered waters (SBEA 2017). Seagrass meadows are dominant, Long Eelgrass (*Zostera nigricaulis*) and Short Eelgrass (*Zostera muelleri*) the main species. Juvenile King George Whiting (*Sillaginodes punctatus*) shelter among the seagrasses, initially feeding on small invertebrates before heading into Port Phillip Bay's deeper waters. Other fish species in Swan Bay include the Spotted Pipefish (*Stigmatopora argus*), Green-back Flounder (*Rhombosolea tapirina*), Black Bream (*Acanthopagrus butcheri*), Australian Giant Cuttlefish (*Sepia apama*), Smooth Stingray (*Dasyatis brevicaudata*) and Yank Flathead (*Platycephalus speculator*).

Each summer, threatened international migratory waders and waterbirds, including the Marsh Sandpiper (*Tringa stagnatilis*) and Eastern Curlew (*Numenius madagascariensis*), visit this Ramsar site (and part of the Port Phillip Heads Marine National Park).

The Salt Lagoon State Nature Reserve is a hyper-saline lagoon cut off from Port Phillip Bay by the construction of the St Leonards–Portarlington Road. It was once considered for salt production, and used as a tip, but is of national significance for visits by the Common Greenshank (*Tringa nebularia*) and of state and regional significance for 15 other bird species including the Black-winged Stilt (*Himantopus himantopus*) (state) and the Hoary-headed Grebe (*Poliocephalus poliocephalus*) (regional) (Hewish 2003).

The Bellarine Peninsula's last remaining stand of Coastal Moonah Woodland (Parks Victoria 2011) is protected within Edwards Point Nature Conservation Reserve, while the only remnant of Damp Sand Herb-rich Woodland is in Portarlington's Point Richards Flora and Fauna Reserve, where the threatened Southern Brown Bandicoot (*Isoodon obesulus*) and Growling Grass Frog (*Litoria raniformis*) are found.

Queenscliff, Portarlington, Indented Head and St Leonards initially grew around small ports used by commercial fishers, by farmers who found them more accessible than Geelong's port, and by holiday-makers arriving on paddle-steamers.

Day-trippers flocked to what became known as Clifton Springs after the 1870 discovery of mineral springs. Regular ferries dropped their passengers off at a new pier where they could enjoy the hospitality of a nearby hotel and two-storey stone bathing house. But economic depressions, fire and septic tank pollution that made the water unfit to drink, saw the enterprise decline and little remains of the structures today.

Natural resource use along this coastal strip was intensifying. Seagrass, sand, guano and kelp were extracted, rocks were removed from the beach at Indented Head, leading to erosion (Bird 2011), and wattle trees were stripped of their bark, cut for firewood and cleared for agriculture. Catchment clearance caused the siltation and shallowing of Swan Bay (Duncan 2006).

Shell grit was mined for use in glass-making and local lime kilns, and as farm fertiliser, chicken feed and ships' ballast. At Clifton Springs a:

> *... 600 mm high deposit of shell grit at the high-water mark ... was in 1933 ... systematically bagged up, by teams of up to 50 men, and sent away to a Melbourne company for use in the manufacture of glass. This action plus the dredging of the Bay exposed the foreshore to erosion* (CSGC nd, p. 7).

As well as shell grit extraction, mussels and oysters were being dredged from the Geelong Arm (of Port Phillip Bay) for food and to fuel local lime kilns. Scallop dredging largely replaced the fishery in the 1960s but was closed in 1997 due to environmental concerns. The dredging decimated Port Phillip Bay's oyster and mussel reefs (Ford and Hamer 2016).

Reversing habitat damage was one of the arguments used in support of a proposed 4 km artificial reef and aquaculture project between Clifton Springs and Indented Head. Although the proposal did not progress, the HaeJoo company's managing director said it would provide protection for the Bay's eroding beaches and 'restore some of the habitat destroyed over the years' (Fowles 2015). Artificial reefs attract fish but they do not necessarily increase fish numbers and can make them more vulnerable to being caught (Grossman *et al.* 1997).

The Portarlington Harbour's recent redevelopment is in support of the aquaculture, angling, boating and cruise ship sectors. It is now the operational centre for Blue Mussel (*Mytilus edulis*) and Native Oyster (*Ostrea angasi*) farming at Clifton Springs, Grassy Point and Kirk Point in the Geelong Arm, and the Bay's scallop diving fishery, established in 2014. Although shellfish aquaculture can remove nutrients from coastal systems, and so reduce algal bloom risk, some scientists believe it may also reduce nutrients for other marine life (Rice 2008). Recent research has shown that the Bellarine Bank's seagrasses are ephemeral, their cover declining in response to reduced nutrient supply during drought (Jenkins and Keough 2015).

Other potential environmental impacts of aquaculture can include site habitat damage, the spread of disease among marine life, litter – mussel shells have accumulated on McDermotts Beach opposite the farms (Bird 2011) – and loss of public access to the area. While current operations are relatively small, these issues will require analysis before the expansion of aquaculture in the Geelong Arm.

Boating has dramatically increased across Port Phillip Bay and in the Geelong region. In 1962, an aerial survey of Port Phillip Bay identified just 1208 recreational fishing boats: 49 between Point Lillias and Point Wilson, 195 between Point Henry and Queenscliff and four in Corio Bay (Lynch 1966). By 2014 there were 172 000 boats registered in Victoria, with 117 000 being used on Port Phillip Bay and Westernport (BIAV 2015). The Boating Industry Association, whose members sell boats, wants 'more ramps, better quality ramps and more parking' that 'will encourage greater boat usage on our waterways'.

The Victorian government wishes to increase the state's angler numbers to 1 million (VFA 2018) although sales of recreational fishing licences in 2016–17 were only 270 000. However, for some species such as Snapper (*Pagrus aurata*) and Flathead (*Platycephalus* spp.), anglers have been catching far more than commercial fishers. With the closure of Port Phillip Bay to commercial netting, which is aimed at improving catches by anglers, significant catch data will be lost at a time when the impact of recreational fishing will require greater scrutiny – there is no rigorous monitoring or ecological sustainability assessment of angling and its associated boating activity and infrastructure.

Of the general impact of fishing on marine life, Edmunds *et al.* (2009) believe that it 'has the most profound influence on marine natural values in Victoria' and has led to collapses in Barracouta (*Thyrsites atun*), Southern Elephant Seal (*Mirounga leonina*), Flat Oyster (*Ostrea angasi*), Scallop (*Pecten fumatus*), Pilchard (*Sardinops sagax*) and Greenlip Abalone (*Haliotis laevigata*). More recently, angling pressure, along with drought, has been implicated in the significant decline of the Southern Sand Flathead (*Platycephalus bassensis*) in Port Phillip Bay (Hirst *et al.* 2014; Parry and Hirst 2016).

Corio Bay: Point Henry to Point Lillias

Point Henry and Point Lillias mark the entrance to Corio Bay and are linked by a submerged sand bank covered in seagrass and with small patches of Sea Squirts (*Pyura stolonifera*).

For many years, the shallowness of the sand bank – at times cattle were driven along it to cross Corio Bay – prevented the entry of ships with more than 3 m draught. As early as 1841, the *Geelong Advertiser* (26 June 1841) was campaigning for dredging to allow 'the largest class of merchantmen' to 'enter the compact basin of Corio Bay'. However, Point Henry remained the key landing site from where people and cargo were taken into Geelong on smaller boats until 1893, when the Hopetoun Channel was cut through the sand bank.

Although access to Corio Bay was restricted, Geelong was growing. Fellmongering, boiling-down works, wool-scouring and lime-burning sprung up along the foreshore, discharging waste into the Bay that, along with sewage discharges, was raising community concerns. From 1838, lime from the western cliffs of Limeburners Point, and gravel and shells from beaches, was used in kilns that burned local timber day and night. But as beachgoer numbers at Eastern Beach rose, and complaints about noise disturbance and foreshore damage grew, the industry shifted to the eastern side of Limeburners Point in the 1860s. It remained there until the 1880s.

Geelong's industrial development continued in the 20th century and pressures on the Bay increased. In 1979, Labor MLC Rod Mackenzie said that the former Bolte government's development vision was why 'the fish from Corio Bay taste like kerosene and why we cannot eat mussels because they are full of mercury and cadmium' (Mackenzie 1979, p. 3237).

Corio Bay's contamination was outlined in the *Port Phillip Bay Study* (CSIRO 1996), which stated that dieldrin levels in fish liver and lead in mussels were of concern to human health (land on the Bellarine Peninsula is also contaminated with dieldrin from agricultural pesticide use). Improved waste management has reduced pollution loads, but localised impacts persist near the petrochemical industry (Baker *et al.* 2016). In 2004, a Shell refinery oil spill coated 400 m of local beaches, shellfish, seagrass and mud flats. Ten years later, the refinery leaked 10 000 L of chemicals (Begg 2015).

As Geelong's industry developed, so did the port and, with it, the number of marine pests. A survey (Currie *et al.* 1998) of Corio Bay found 20 introduced marine pests, with the European Fan Worm (*Sabella spallanzanii*), Sea Squirts (*Styela clava*; *Ascidiella aspersa*), Asian Date Mussel (*Musculista senhousia*) and Asian Semele (*Theora lubrica*) in sufficient abundance to cause ecological damage by competing for space and food with native species.

The Asian Date Mussel was seen as a major risk (Currie *et al.* 1998) to Corio Bay's seagrass meadows, which are used for food and shelter by Yelloweye Mullet (*Aldrichetta forsteri*), Southern Garfish (*Hyporhamphus melanochir*), Black Bream (*Acanthopagrus butcheri*), Sixspine Leatherjacket (*Meuschenia freycineti*) and Goby, Toadfish and Pipefish species. Other fish species that can be found in this Bay, heavily compromised by contaminants and invasive pests, include Silver Trevally (*Pseudocaranx georgianus*), Blue Warehou (*Seriolella brama*), Eastern Australian Salmon (*Arripis trutta*) and Gummy Shark (*Mustelus antarcticus*).

One of Port Phillip Bay's largest seagrass meadows is found in Stingaree Bay on the western side of Point Henry. It sits offshore from the Moolap wetlands, which were converted to salt evaporation ponds in 1889 by the Cheetham Salt Co. Salt production has ceased, and the wetlands remain an important feeding and resting site for international migratory wader birds, including significant Australian populations of Red-necked Avocet (*Recurvirostra novaehollandiae*) and Banded Stilt (*Cladorhynchus leucocephalus*). They remain under threat from a proposed canal estate, inadequately managed water levels in the ponds, and residential and tourist development proposed in the draft *Moolap Coastal Strategic Framework Plan* (DELWP 2017).

North of Geelong, Limeburners Bay is 'the best-preserved estuary system in Port Phillip Bay' and 'contains the most extensive stand of mangroves in Port Phillip Bay and probably the largest intact salt marsh complex' (Parks Victoria 1999, p. 5). Its fauna is of national significance, with important habitat for large numbers of waders. The Yellow Sedge-skipper Butterfly (*Hesperilla flavescens*) is present and the critically endangered Orange-bellied Parrot (*Neophema chrysogaster*) is known to visit during winter.

Further east, the Cheetham Saltworks at Avalon produced salt from 1954 to 2002 in a series of evaporation ponds that replaced Coastal Saltmarsh, Coastal Tussock Grassland and Plains Grassland. Although water levels have not been maintained, it remains an important site for birds. Recorded sightings total 147 bird species (eBird Australia 2017), including the threatened Bar-tailed Godwit (*Limosa lapponica*) and Fairy Tern (*Sternula nereis*).

Geelong Arm: northern shore

One of Australia's richest saltmarshes stretches between the Avalon Saltworks and the Western Treatment Plant and is part of the Western Shoreline (Port Phillip Bay) and Bellarine Peninsula Ramsar site (which includes the coast from the Werribee River to Hovells Creek plus Swan Bay and the Connewarre Lake complex). Nearshore habitats are a mix of seagrass meadows, subtidal reefs and areas of bare sand. A large area of drift algae, a mat of red and brown algae detached from the seabed, is found off Kirk Point.

Half of Melbourne's sewage is treated at the Western Treatment Plant but it is also a haven for more than 280 bird species. These include the Red-kneed Dotterel (*Erythrogonys cinctus*), Freckled Duck (*Stictonetta naevosa*), Australia's rarest duck, and Pink-eared Duck (*Malacorhynchus membranaceus*).

The 1996 CSIRO study of Port Phillip Bay recommended a significant reduction in the plant's nutrient discharges. When implemented in 2004, there were concerns it could reduce the benthic biomass in the tidal flats used by feeding shorebirds. Although research (Rogers *et al.* 2013) did not draw a conclusive correlation between discharges and benthos densities, it did show that worm densities had declined and crustacean densities increased, yet edible biomass was largely the same. However, the research also revealed the critical importance to birds of the treatment plant's narrow tidal flats that are under threat of inundation by rising seas – sea level in Port Phillip Bay has risen 9 cm since the 1960s (Rogers *et al.* 2013).

In the 1990s, commercial harvesting of Sea Cucumber was considered for the Point Lillias and Point Wilson area (DNRE 1996), even though very little was known of the biology or ecology of *Australostichopus mollis*, one of the largest Sea Cucumbers in Victorian waters. There were also government plans to relocate the Coode Island bulk liquid chemical storage facility from the Maribyrnong River in Melbourne to Point Lillias, after a fire in 1991. This would have required the excision of 20 ha from the Ramsar site. The plan was abandoned after a concerted community campaign.

Restricted access by land to the Western Treatment Plant, the Commonwealth-owned Point Wilson ammunition storage site, and the tidal lagoon of the Spit Wildlife Reserve, has to date given protection to saltmarshes, wetlands and mudflats that would have otherwise been removed – half of the saltmarshes on the west coast of Port Phillip Bay have been lost since European settlement. Current threats to wildlife come via access from the water. Windsurfers, jet skis and recreational fishing boats can disturb roosting, feeding and breeding birds, and also damage shallow seabed habitats. The future expansion of nearby Avalon Airport and the proposed Bay West port project between Point Lillias and Point Cook, an option for Victoria's second container port, would also threaten this very important area.

Future drivers of change

Population growth and linear coastal development

The region's coastal towns were slow to grow but that changed with 1950s post-war immigration, the efforts of creative real estate developers in the 1960s, and Geelong's growth in the 1980s. Now, along the coast road between St Leonards and Portarlington, 'a fragile landscape is under intense pressure from linear development' and 'the foreshore has become degraded with vegetation removed' (Bellarine Bayside 2013, p. 3).

Settlement boundaries have been established in town structure plans for the smaller coastal towns, although there are pressures to expand them; existing subdivisions may only satisfy growth to the mid 2020s (CoGG 2008). They will likely become Geelong's dormitory towns, as have Ocean Grove, Torquay and Clifton Springs, which are now designated urban growth areas in the *G21 Regional Growth Plan* (G21 GRA 2013); Clifton Springs was recently extended west along its shoreline into Curlewis. Higher permanent populations will place greater pressure on natural resources throughout the year. Combined with proposals for major residential and tourist development for the Moolap and Point Henry area under the draft *Moolap Coastal Strategic Framework Plan* (DELWP 2017), and potential coastal development north of Geelong, linear coastal development could well continue and the conflicting nature of private and public property remain an ongoing challenge for the authorities.

Climate change

Projected sea level and temperature rises of up to 82 cm and 3°C respectively this century (Klemke and Arundel 2013) indicate the risks facing the region's marine and coastal environments and infrastructure.

Rising sea levels could erode 'soft' cliffs along the Bellarine Bank off Clifton Springs, as well as beaches and dunes between Torquay and Point Lonsdale. They could also reduce swells at Bells Beach and inundate low-lying intertidal reefs at Barwon Bluff, Point Danger and Point Lonsdale, and saltmarsh and mudflats in Swan Bay, Moolap and between Lime-burners Bay and Lake Borrie.

A rise in water temperatures could affect the larval stages for species like King George Whiting (*Sillaginodes punctatus*), and lead to the displacement of some endemic species by others, such as Blacklip Abalone by the Long-spined Urchin (*Centrostephanus rodgersii*) moving south on a strengthening East Australian Current (Hirst and Hamer 2013). Victoria's record water temperatures in 2016 may have encouraged more sharks along the coast, where they were following prey fish aggregating around food-rich river discharges after higher-than-average rainfall (Mannix 2017).

Close to 1800 houses on the Bellarine Peninsula's low-lying coastal land could be inundated by an 80 cm rise in sea level, according to climate coastal risk overviews carried out for local councils, management committees and the catchment management authority (Our Coast 2017). Properties between St Leonards and Portarlington have already been flooded by waves washing over low dunes and beaches.

There will be increasing pressure from the owners of at-risk properties for governments to defend their assets. Sand fencing, rock groynes, seawalls and groundwater pumps have been installed between Portarlington and Curlewis, and beaches renourished at Clifton Springs, Point Lonsdale, Portarlington and St Leonards, in part to provide for recreation but to also defend existing infrastructure. The building of defensive structures, as well as harbours and marinas, disrupts dynamic coastal processes, can lead to erosion (Point Lonsdale Beach) and deposition (Queenscliff Harbour), and initiate further infrastructure-building (Bird 2011).

Having allowed development too close to the shoreline, the challenge now for governments, along with land managers and the community, is to decide when and where it is appropriate to defend infrastructure, adapt to climate change and plan a retreat from the advancing sea. These decisions, and those for the BayWest port project, the Moolap–Point Henry area, aquaculture projects, coastal town expansions, recreational fishing and boating, and climate change responses, will shape the future of the Geelong region's marine and coastal environments.

References

Baker J, Long S, Hassell K, Pettigrove V, Gagnon M (2016) Health status of sand flathead (*Platycephalus bassensis*): inhabiting an industrialised and urbanised embayment, Port Phillip Bay, Victoria as measured by biomarkers of exposure and effects. *PLoS ONE* **11**(10), e0164257. https://doi:10.1371/journal.pone.0164257

Barton J, Pope A, Howe S (2012) *Marine Natural Values Study Vol. 2: Marine Protected Areas of the Central Victoria Bioregion*. Parks Victoria Technical Series No. 76. Parks Victoria, Melbourne.

Begg P (2015) Criminal probe into Viva Energy after operator of former Shell refinery leaks 10 000 litres into Corio Bay. *Geelong Advertiser*, 8 April 2015.

Bellarine Bayside (2013) *Northern Bellarine Foreshore Master Plan*. Bellarine Bayside, Victoria.

BIAV (2015) *Boating Industry Snapshot: Drivers of Growth in Victoria*. Boating Industry Association of Victoria, Melbourne.

Bird E (2011) *Changes on the Coastline of Port Phillip Bay*. Department of Sustainability and Environment, Melbourne.

Boon P, Allen T, Brook J, Carr G, Frood D, Hoye J, Harty C, McMahon A, Mathews S, Rosengren N, Sinclair S, White M, Yogovic J (2011) *Mangroves and Coastal Saltmarsh of Victoria: Distribution, Condition, Threats and Management.* Technical Report. Victoria University, Melbourne.

CoGG (2008) *Portarlington Structure Plan.* City of Greater Geelong, Geelong.

CoGG (2011) *Climate Change Adaptation Strategy.* City of Greater Geelong, Geelong. https://www.geelongaustralia.com.au/common/Public/Documents/8ce5879f7f85db8-Climate%20Change%20Adaptation%20Strategy.PDF (accessed 1 March 2018).

CSGC (nd) *A History of the Dell.* Clifton Springs Golf Club, Clifton Springs, Victoria.

CSIRO (1996) *Port Phillip Bay Environmental Study. Final Report.* CSIRO, Canberra.

Currie D, McArthur M, Cohen B (1998) *Exotic Marine Pests in the Port of Geelong, Victoria.* Report no. 8. Marine and Freshwater Resources Institute, Queenscliff, Victoria.

DELWP (2017) *Draft Moolap Coastal Strategic Framework Plan.* Department of Environment, Land, Water and Planning, Melbourne. https://www.coastsandmarine.vic.gov.au/__data/assets/pdf_file/0033/55986/Moolap-draftFrameworkPlan-April2017.pdf (accessed 1 March 2018).

DELWP (nd) *Characteristics of Threatened Communities.* Department of Environment, Land, Water and Planning. https://www.environment.vic.gov.au/__data/assets/pdf_file/0018/50418/Flora-and-Fauna-Guarantee-Characteristics-of-Threatened-Communities-.pdf (accessed 1 March 2018).

DNRE (1996) *Draft Fisheries Management Plan for Sea Cucumbers (Stichopus mollis) from Port Phillip Bay, Victoria.* Fisheries Division, Department of Natural Resources and Environment, Melbourne.

Duncan B (2006) The maritime archaeology and maritime cultural landscapes of Queenscliffe: nineteenth-century Australian coastal community. Unpublished PhD thesis, James Cook University, Townsville.

eBird Australia (2017) *Avalon Saltworks.* http://ebird.org/ebird/australia/hotspot/L2547648 (accessed 5 January 2017).

Edmunds M, Mustoe S, Stewart K, Sheedy E, Ong J (2009) *VNPA Nature Conservation Review: Marine Conservation Priorities and Issues for Victoria.* Australian Marine Ecology, Melbourne.

Everist R (2004) *Geelong and the Bellarine Peninsula.* BestShot, Geelong.

Ford J, Hamer P (2016) The forgotten shellfish reefs of coastal Victoria: documenting the loss of a marine ecosystem over 200 years since European settlement. *Proceedings of the Royal Society of Victoria* **128**, 87–105. doi:10.1071/RS16008

Fowles S (2015) $80m marine farm proposed for Bellarine Peninsula. *Geelong Advertiser*, 17 July.

G21 GRA (2013) *G21 Geelong Regional Growth Plan.* Geelong Regional Alliance, Geelong. http://www.g21.com.au/sites/default/files/resources/g21_regional_growth_plan_-_april_2013_-_low_res_0.pdf (accessed 5 January 2018).

Grossman G, Jones G, Seaman W Jr (1997) Do artificial reefs increase regional fish production? A review of existing data. *Fisheries*, April 1997.

Hewish M (2003) *Fauna Values of Three Sub-coastal Wetlands on the Bellarine Peninsula: Lake Victoria, Freshwater Lake and St Leonards Salt Lagoon.* Parks Victoria Technical Series no. 10. Parks Victoria, Melbourne.

Hirst A, Hamer P (2013) Implications of future climate for marine fish. In J Klemke and H Arundel (Eds) *Implications of Future Climate for Victoria's Marine Environment.* Glenelg Hopkins Catchment Management Authority, Hamilton, Victoria.

Hirst A, Rees C, Hamer P, Kemp J, Conron S (2014) *The Decline of Sand Flathead Stocks in Port Phillip Bay: Magnitude, Causes and Future Prospects.* Fisheries Victoria, Queenscliff.

idcommunity (2018) *City of Greater Geelong Population Forecast.* https://forecast.id.com.au/geelong (accessed 5 January 2018).

Jenkins G, Keough M (2015) *Seagrass Resilience in Port Phillip Bay: Final Report to the Seagrass and Reefs Program for Port Phillip Bay.* University of Melbourne, Melbourne.

Klemke J, Arundel H (2013) *Implications of Future Climate for Victoria's Marine Environment.* Glenelg Hopkins Catchment Management Authority, Hamilton, Victoria.

Lynch D (1966) Port Phillip survey 1957–1963: the fisheries. *Memoirs of the Natural Museum of Victoria* **27**, 7–18.

Mackenzie RA (1979) Speech by RA Mackenzie, 16 October 1979. *Hansard*, Victorian Parliament, Melbourne.

Mannix L (2017) More sharks than the last four years put together: be ready for a record shark season. *The Age*, 6 January.

Our Coast (2017) *Our Coast.* http://www.ourcoast.org.au/cb_pages/files/Indented%20Head%20Risks.pdf (accessed January 2017).

Parks Victoria (1999) *Ramsar Wetlands: Information Sheet.* Parks Victoria, Melbourne. http://www.environment.gov.au/water/topics/wetlands/database/pubs/18-ris.pdf (accessed 5 January 2018).

Parks Victoria (2011) *Edwards Point Nature Conservation Reserve: Park Notes.* Parks Victoria, Melbourne.

Parry G, Hirst A (2016) Decadal decline in demersal fish biomass coincident with a prolonged drought and the introduction of an exotic starfish. *Marine Ecology Progress Series* **544**, 37–52. doi:10.3354/meps11577

Porter C, Wescott G (2010) *Rocky Shores of Marine National Parks and Sanctuaries on the Surf Coast: Values, Uses and Impacts prior to Protection.* Parks Victoria Technical Series no. 22. Parks Victoria, Melbourne.

Rice M (2008) *Environmental Effects of Shellfish Aquaculture in the North-east.* North-eastern Regional Aquaculture Centre, University of Maryland, Maryland. https://agresearch.umd.edu/sites/agresearch.umd.edu/files/_docs/105-2008%20Environmental%20effects.pdf (accessed 5 January 2018).

Rogers D, Loyn R, Greer D (2013) *Factors influencing Shorebird Use of Tidal Flats adjacent to the Western Treatment Plant.* Technical Report Series no. 250. Arthur Rylah Institute for Environmental Research, Department of Sustainability and Environment, Melbourne.

SBEA (2017) *Swan Bay Environment Association.* http://www.sbea.org.au (accessed 5 January 2017).

SLSA (2018) *Surf Life Saving Australia.* http://www.beachsafe.org.au (accessed 5 February 2018).

Trengove M (2013) Barwon Coast vegetation management plan: final draft. Unpublished report prepared by Ecological Services for City of Greater Geelong, Geelong.

TSSC (2012) *Listing Advice for the Giant Kelp Marine Forests of South-east Australia Ecological Community.* Threatened Species Scientific Committee, Department of Sustainability, Environment, Water, Population and Communities, Canberra.

VFA (2018) *Target One Million.* Victorian Fisheries Authority, Melbourne. https://vfa.vic.gov.au/recreational-fishing/target-one-million (accessed 5 January 2018).

VNPA (2014) *EVC Analysis for the 2014 Nature Conservation Review.* Victorian National Parks Association, Melbourne.

PART 3

HUMANS AS AGENTS OF CHANGE IN THE GEELONG REGION

10

Key ecological principles adapted for regional green infrastructure

Dennis Williamson

> *Ecological design is heralded, and justly so, as being a primary way through the muddle we find ourselves facing ... By conscious reflection and intention we make whole what has been ripped apart, clean what has been contaminated, and use less more intelligently. The promise is beguiling.*
>
> – Higgs (2003, p. 272).

Introduction

This chapter reviews Key Ecological Principles (KEPs) adapted from Williamson (2017) that provide a useful framework to enhance the green infrastructure of Geelong's regional landscape. The KEPs would offer a pattern-oriented response to habitat fragmentation and climate change pressures that adversely affect indigenous species' survival in the region. The KEPs provide a landscape-scale approach that will require an evolution of ecology and cultural land use values for application to the Geelong region.

An ecological landscape shaped by its past

The Geelong region landscape is a product of its ecological evolution and cultural changes that have altered its original condition over time. The region's geological complexities are simplified in a surface view to that of a plains landform flanked by low mountain ranges and interspersed with isolated hills, river estuaries and coastal lowlands. The terrain is edged on the south and the east by Bass Strait and Port Phillip Bay.

The initial cultural shaping of the landscape was by clans of the *Wadawurrung* Aboriginal people to meet the needs of relatively small populations. They burnt small areas of open woodlands and indigenous grasslands to aid their hunting of kangaroos and other native fauna (Hateley 2010). From the late 1830s, European settlers came to graze sheep and cattle in the natural grasslands of the *Dijlang* region and establish the port city we now know as Geelong.

The grasslands and grassy woodlands with moderate- to low-density eucalypt, casuarina, acacia, banksia and other trees were associated with the basalt plains. More dense woodlands occurred on the granitic and sandstone hills and ranges of the You Yangs,

Table 10.1. General land use distribution in the regions' landscape zones (%)

	Landscape zones of the Geelong region				
Land use categories	**Moorabool**	**Hovells**	**Mid Barwon**	**Bellarine**	**Thompsons**
Conservation	4	8	2	12	27
Urban/peri-urban	0	19	2	20	7
Grazing	68	50	69	56	46
Cropping	5	7	20	8	0
Forestry	15	0	0	0	16
Horticulture	8	0	0	0	0
Other	0	16	7	4	4
Total	100	100	100	100	100

Source: CCMA (2014).

Brisbane Ranges, Barrabool Hills and Otway Ranges. Where the grasslands predominated, relatively narrow corridors of riparian trees grew along the plains, rivers and streams, including the Moorabool River and Barwon River. The general landscape character of the region was well depicted by Von Guérard in his 'View of Geelong', painted during 1856. In the estuary and coastal lowlands of Corio Bay, Lake Connewarre and elsewhere, tea-tree thickets (*Melaleuca* and *Leptospermum* spp.) and great swamps of rushes and heathlands existed, along with some mangrove forests.

The City of Greater Geelong has grown to a population of ~238 000 persons (30 June 2016) and is projected to expand at a rate of over 36% to 2036, reaching over 320 000 residents (idcommunity 2016a). Much of this will be urban-based growth, but the population of the surrounding G21 region, which contains more small towns and rural areas, is also expected to grow rapidly. Having a population of more than 302 000 during 2015 (idcommunity 2016b), it has already outstripped the 289 000 population projections for 2031 made during 2014 (Vaughn 2014).

The Corangamite Catchment Management Authority (CCMA 2014) shows that within the five landscape zones of the Geelong region, nature conservation or forest areas constitute a very small portion of the regional landscape and that agricultural, urban or mining uses predominate (refer to Table 10.1).

Superimposed on the previous landscape is a network of grey infrastructure, including roads and railways, fencelines and urban commercial, industrial and residential development. Agricultural uses provide a less hard, but nonetheless intensive, form of land use. These non-natural land use categories continue to intensify as the urban and regional population of the region increases.

The ecological predicament

Habitat fragmentation and species endangerment

Since the 1830s, the ecological structures and functions of the Geelong region have significantly changed and declined through vegetation clearing and habitat fragmentation. The Victorian Environmental Assessment Council (VEAC 2010) reported that the bioregions surrounding Geelong are very highly fragmented landscapes with only 4% of native vegetation remaining intact in the Victorian Volcanic Plains bioregion and only 15.6% remaining in the Otway Plains bioregion (DSE 2011). The intact areas of remnant vegetation and regrowth are found in the national parks, state forests and coastal reserves described previously.

Corangamite CMA (CCMA 2013) assessments paint the following picture of the region's ecological condition:

- The bioregional conservation status of the region indicates that vulnerable and threatened ecological vegetation classes (DELWP 2016a) are widely distributed throughout the region, but with higher concentrations around the estuaries and coastal fringes of the Bellarine Peninsula, in particular.
- The Moorabool Basin, central to the Geelong region, has the largest percentage of streams and waterways in poor to very poor condition and only 13% of streams within the entire Corangamite CMA region are in good condition, primarily those in the Otway Ranges outside our present focus region (CCMA 2013; Melbourne Water 2013).

Within the City of Greater Geelong, an extensive number of threatened fauna species are listed under the Victorian *Flora and Fauna Guarantee Act 1988* (SWIFFT 2016), including:

- 84 species of threatened birds
- nine species of threatened mammals
- five species of threatened fish
- three species of threatened reptiles
- two species of threatened amphibians
- four species of threatened invertebrates.

Similarly long lists of threatened species exist for Golden Plains Shire and Surf Coast Shire. Examples of critically endangered and endangered flora and fauna species include the Button Wrinklewort (*Rutidosis leptorhynchoides*), Spiny Rice-flower (*Pimelea spinescens* subsp. *spinescens*), Growling Grass Frog (*Litoria raniformis*) and Orange-bellied Parrot (*Neophema chrysogaster*).

Ecological restoration not achieved despite 40+ years of effort

Poor ecological habitat conditions have continued to occur despite a wide range of natural resource, ecological and biodiversity protection policies, strategies and legislation instigated at both state and federal government levels over more than 40 years, including:

- the *Victorian Conservation Trust Act 1972*
- the Commonwealth *Environment Protection and Biodiversity Conservation Act 1999*
- the Victorian *Wildlife Act 1975* and *Wildlife Regulations 2013*
- the Victorian *Flora and Fauna Guarantee Act 1988*
- *Protecting Victoria's Environment – Biodiversity 2037* (DELWP 2017)
- *Victoria's Native Vegetation Management – A Framework for Action (2002)*
- the Victorian *Wildlife Act 1975* and *Wildlife Regulations 2013*
- *Australia's Biodiversity Conservation Strategy 2010–2030.*

Although grassroots-level community and non-government environmental groups such as Landcare, Greening Australia and Australian Bush Heritage have been actively protecting and restoring many habitat areas, they have been largely dependent on available government or corporate funding and have not always been strategically focused. Continuing ecological decline has not been adequately reduced by these efforts.

Looming climate change effects and challenges

Climate change, resulting from global population and industrial effects on greenhouse gas emissions, has led to global warming (CSIRO and BoM 2016; NRMMC 2010; IPCC 2014; BoM and CSIRO 2016). The Australian Museum (2015) cautions that:

> *An average temperature rise of 2°C may result in the estimated extinction of 20–50 per cent of the world's species by 2030. If this current trend continues, the figure could reach 50–60 per cent by 2100.*

Others agree that climatic change is now evident in Australia and has the potential to cause mass displacement and extinctions of a very large number of flora and fauna species (MEA 2005; Steffen *et al.* 2009, BoM and CSIRO 2016). The cost of extreme weather events (severe storms, floods, bushfires, heatwaves and droughts) is now in the tens of billions dollar across Australia (BoM and CSIRO 2016). The added loss of homes, livelihoods and human lives is beyond estimation.

Ecological research on climate change adaptation solutions has been largely species-oriented, and effective research findings may be too little and too late to rescue most vulnerable species (Hughes 2003; Williamson 2017). Existing conservation reserves may be inadequate to provide the effective habitat refuges required by many indigenous species (Mackey *et al.* 2008). These areas will need to be supplemented by improved habitats within non-conservation public lands (i.e. road reserves and stream reserves), as well as through landscape restoration of private agricultural and urban lands (Fitzsimons and Wescott 2001).

Integrating cultural and natural landscape systems

KEPs to focus a more rapid response

Williamson (2017) identified 16 pattern-oriented KEPs for application within the Australian wheatbelt agricultural regions. The principles have general support within the Australian ecological and urban design literature. These principles are adapted in this chapter as a set of Rural–Urban KEPs for application to the Geelong region.

The principles provide a rapid response to climate change pressures on indigenous flora and fauna species habitats in general, without relying directly on longer-term climate research on specific species. The KEPs provide an opportunity and priorities for integrating the conservation of more natural and functional ecosystems within the overall landscape matrix, land use and infrastructure of the Geelong region (and many other regions of Australia). They focus our attention on many of the key elements and actions that are required if anything resembling natural ecosystems are to survive to 2050 and into the next century. The modified principles are listed in Table 10.2.

These principles are not intended to replace the range of excellent natural resource and biodiversity policies, strategies and plans already serving as useful guides for many of the threatened native flora/fauna species and ecosystems from the Commonwealth to local level for the region. Such plans include, for example, the *Corangamite Native Vegetation Plan 2003–2008* (CCMA 2003), *Flora and Fauna Guarantee Action Statements* (DELWP 2016b) and *Biodiversity Action Planning, Corangamite Region* (DELWP 2016c). However, the KEPs do distil and simplify many of the management priorities previously identified, bringing them together in an integrated format.

Table 10.2. KEPs for integration with the land use and green infrastructure of rural–urban regions

Actively manage the landscape as a healthy integrated whole
P1: Actively protect, restore and manage the entire regional landscape, its natural systems and ecological processes as a high-quality, healthy and integrated whole through the implementation of all KEPs. Ensure an ecologically healthy, biodiverse and balanced condition for locally appropriate indigenous flora and fauna communities (including the vegetative canopy, intermediate and ground cover layers)
P2: Actively control and eradicate feral pest fauna, invasive and noxious plants and weeds and the impacts of fire that threaten the survival of indigenous flora and fauna communities and species
Protect and enhance high quality and larger habitat patches and at-risk flora and fauna species
P3: Actively manage, protect and restore areas of highly significant natural ecosystems and flora/fauna habitats, including those of rare, endangered and threatened flora and fauna species
P4: Protect, restore and actively manage the larger parcels of natural habitats, remnant vegetation and ecosystems of 50 ha or greater area
P5: Regenerate and restore fragmented, degraded habitat areas and ecosystem adjacent to existing high-quality remnant vegetation patches
Protect and enhance critical natural water zones and flow regimes
P6: Actively protect, maintain and restore the natural vegetation structure, ecological functions and water flow regimes of riparian areas along rivers, streams, lakes, wetlands and natural drainage lines
P7: Maintain, establish or widen indigenous vegetation corridors per side along: a) smaller streams, lakes and wetlands to total widths of 50–100+ m, and b) arterial and rural road and railway reserves to widths of 30–150 m where possible (the wider, the better for habitats)
P8: Maintain, establish or widen indigenous vegetation corridors along larger rivers, lakes and wetlands to total widths of 75–250+ m per side, depending on circumstances (the wider, the better for flora and fauna habitats – as well as flood protection)
P9: Fence out grazing livestock, people and vehicles from sensitive water zones, habitat areas, water recharge zones and areas affected by soil erosion and soil salinity. Revegetate these sensitive areas with suitable locally indigenous trees, shrubs and groundcover plants
P10: Maximise the permeability of urban landcover to enhance natural water flows, water retention and reuse, flood control and water quality within urban areas and downstream. Create a 'sponge city' effect in which at least 40% of ground surface areas are permeable to rainwater infiltration and at least 70% of rainwater is infiltrated or re-used
Improve ecological connectivity by minimising gaps and creating stepping-stones
P11: Connect locally indigenous flora and fauna habitats by planting natural vegetation cover in patterns that enhance the ecological structures and functionality throughout the regional landscape matrix
P12: Actively protect, restore and create indigenous trees and remnant vegetation patches of 1–20 ha to maintain native habitat 'stepping-stones' through natural regeneration or planting for ecological connectivity and the movement of indigenous fauna through the landscape
P13: Close gaps to less than 1 km between native forest and woodland patches of 10–20 ha (the closer, the better for habitats)
P14: Close linear gaps between native vegetation corridors along streams, rivers, road reserves and fencelines, and between indigenous trees and shrubs in paddocks, to less than 100–200 m (the closer, the better for flora and fauna habitats) to create a 'stepping-stone' effect for faunal movement

Table 10.2. *(Continued)*

P15: Close gaps between individual and clusters of indigenous trees and shrubs in urban, peri-urban and rural areas to less than 50–200 m to serve as local habitat and as movement 'stepping-stones' for native flora and fauna (the closer, the better for fauna habitat)
P16: Create an effective, sustainable urban forest throughout all public and private urban lands. Ensure all new or redeveloped residential allotments establish a minimum of two to three large canopy/shade trees of at least 8–10 m + in height and 8–12 m in crown diameter. Ensure all nature strips, carparks and road/railway reserves include large canopy trees with a minimum of 20 m spacings wherever possible. Ensure at least 60% of urban forest plantings include locally indigenous tree, shrub and groundcover species. Establish and utilise green roof and green wall design where possible

Source: Adapted from Williamson (2017, Table 19, p. 347).

Actively manage the landscape as a healthy integrated whole

The first two KEPs (P1 and P2) highlight the priorities of:

- actively managing and protecting the regional landscape, its natural systems and ecological processes as an integrated whole to restore and maintain healthy conditions for indigenous flora and fauna communities and species
- actively controlling and eradicating threats to indigenous flora and fauna communities and species posed by feral pests (e.g. cats, foxes, rabbits), invasive and noxious plants and weeds, and the impacts of intensive and/or frequent bush and grass fires.

Active management and protection mean that the relevant ecosystems and ecological processes should be maintained where they exist, or restored to a healthy and resilient condition. This includes the ongoing maintenance of communities of locally indigenous (native) flora and fauna with appropriate species composition, diversity, and structural and functional organisation as compared with undisturbed habitats within the region. It does not mean simply setting aside conservation reserves and leaving them unmanaged.

Protect and enhance high-quality and larger habitat patches and at-risk flora and fauna species

KEPs P3–P5 focus on the ecosystems and indigenous flora and fauna communities that are of particularly high quality and significant (e.g. within portions of the Great Otways and Brisbane Ranges National Parks), or at-risk in terms of being rare, endangered or threatened. These occur throughout the region's agricultural landscapes, in relation to the Lake Connewarre wetlands and the coastal fringes of the region. P3–P5 also underscore the importance of protecting and managing larger parcels of remaining natural habitats and remnant vegetation of 50 ha or greater in area. They also focus on providing adequate buffer areas adjacent to the remaining fragmented but high-quality ecosystems and habitats.

Some of the existing protected areas relevant to these three KEPs include:

- Brisbane Ranges National Park
- Great Otway National Park and Anglesea Heath (which have among the richest biodiversity in the region)
- You Yangs Regional Park
- Lake Connewarre Wildlife Reserve (Ramsar Convention listed wetlands)

- Breamlea Flora and Fauna Reserve
- Serendip Wildlife Sanctuary
- Avalon Coastal Reserve
- The Spit Wildlife Reserve
- Limeburners Lagoon (Hovells Creek) Coastal Reserve
- state forests and other large regional parks.

Protect and enhance critical natural water zones and flow regimes

KEPs P6–P10 focus on the protection and enhancement of ecological structure and functions, as well as natural water flow regimes along smaller and larger streams, rivers, lakes and wetlands. They promote the maintenance and widening of native vegetation corridors and buffers along the edge of these water features and within floodplains. These corridor and buffer widths may vary considerably from rural to urban areas. However, restoration and management of such riparian and water zones can have particularly positive effects on the ecological health of the wider region, including improved biodiversity, reduced flood risks and reduction of water pollution (Bennett *et. al.* 2014; Lake 2007; Cullen 2007; Barton and Davies 1993).

Regarding appropriate stream corridor habitat reserve widths, government departments, municipal councils and land developers have tended to fall short of the ecological principles proposed. For example, open space buffers along extended sections of Armstrong Creek are less than 10 m per side through residential areas (Collie Pty Ltd 2010).

KEP P9 recommends the protection of sensitive water zones by fencing out grazing livestock, people and vehicles. The Corangamite Waterway Strategy (CCMA 2014) and various other plans have addressed this issue in a similar manner. This principle should be extended to urban land use encroachment over such key water zones.

KEP P10 addresses the issue of rainwater permeability within urban areas. Historically, urban development has tended to divert, channelise and bury natural stream systems in underground pipes which eventually flow to downstream outfalls on the banks of larger rivers. A high percentage of urban land cover (often 60–80% in residential areas) tends to be impermeable, leading to the requirement for underground stormwater systems to prevent flooding (DELWP 2015). This and the trend to larger footprint and higher-density residential dwellings has led to more expensive stormwater systems, pollution of streams and the additional expense of installing artificial wetlands, wetland swales and rain gardens. These remedial measures may help, but only remove a fraction of the polluting contaminants and sediments. Addressing the problem at its source, P10 recommends that at least 40% or more of urban ground surface areas should be permeable to rainwater infiltration and at least 70% of rainwater should be re-used, like the 'Sponge City' movement currently advocated in PR China (Zevenbergen *et al.* 2018).

Improving ecological connectivity

Ecological and native vegetation connectivity within the Geelong region is especially poor throughout the agricultural grazing and cropping areas and along the key river and stream corridors, as well in some coastal areas. Clearing of these small islands of vegetation is often exacerbated by modern-day mechanised agricultural practices, as well as by urban development. *Biodiversity Action Planning, Corangamite Region* (DELWP 2016c) does not always assist the maintenance of such vegetative islands when undermined by the flexible regulation of native vegetation clearing through the 'net gain' and 'biodiversity offsets' concepts of the *Permitted Clearing of Native Vegetation – Biodiversity Assessment Guidelines* (DELWP 2013).

KEPs P11–P16 focus on improving structural and functional connectivity between remnant vegetation and habitat patches. First, P11 recommends planting natural vegetation cover in patterns that enhance the ecological structures, functionality and connectivity of indigenous flora and fauna habitats (Bennett *et al.* 2006).

Some ecologists have indicated from their research findings that agricultural uses should not occupy more than 30% of the landscape, or that at least 30% of the total landscape should have native vegetation cover (McIntyre *et al.* 2004; Lindenmayer *et al.* 2006; Cullen 2007; Doerr *et al.* 2013). However, Williamson (2017) found in his survey of Australian wheatbelt region farmers and natural resource managers that they are not comfortable with such exacting percentages. Therefore, the KEPs were amended to indicate a more general balance of ecological conservation and agricultural production within a regional landscape.

P12 encourages the active protection and establishment of indigenous vegetation patches of 1–20 ha to maintain native habitat 'stepping-stones'. P13–P16 seek to close excessive gaps between indigenous vegetation patches to improve species movement. Assessment of areas that require gap reduction can be done using aerial photos, Google Earth satellite imagery, and DELWP Connectivity mapping available via Victoria Resources Online (DEWLP 2016d).

P15 and P16 address urban circumstances where a combination of roads, overhead electricity and communications wiring and higher-density development have severely reduced tree canopies and indigenous vegetation cover and habitats. The latter issue is especially pronounced in residential areas where larger 'McMansions' and multi-dwelling developments on previous single-dwelling allotments occupy up to 80% of many urban blocks, making it impossible to retain or establish more than one large canopy tree per allotment.

Green infrastructure and environmental services

On face value, the 16 KEPs are primarily aimed at increasing the area of and improving the condition and connectivity of regional tree cover, indigenous flora and fauna habitats, and ecosystem functions. At a deeper level, however, implementation of the KEPs is also about providing a reasonable balance in land use between commercial agriculture and urban uses and non-commercial environmental land uses. Converting as much of the existing and future grey (built) infrastructure to green infrastructure through nature-based solutions (NBS) is a major goal (Hansen *et al.* 2011).

CSIRO (2015) defines green infrastructure (GI) as:

> *... all of the vegetation that provides environmental, economic and social benefits such as clean air and water, climate regulation, food provision, erosion control and places for recreation. GI includes urban parks and reserves, wetlands and stream corridors, street trees and roadside verges, gardens and vegetable patches, bikeways and pedestrian trails, wall and rooftop gardens, orchards and farms, cemeteries and derelict land.*

The related concepts of environmental services or ecological benefits and green infrastructure have become increasingly referred to and applied to varying degrees in the fields

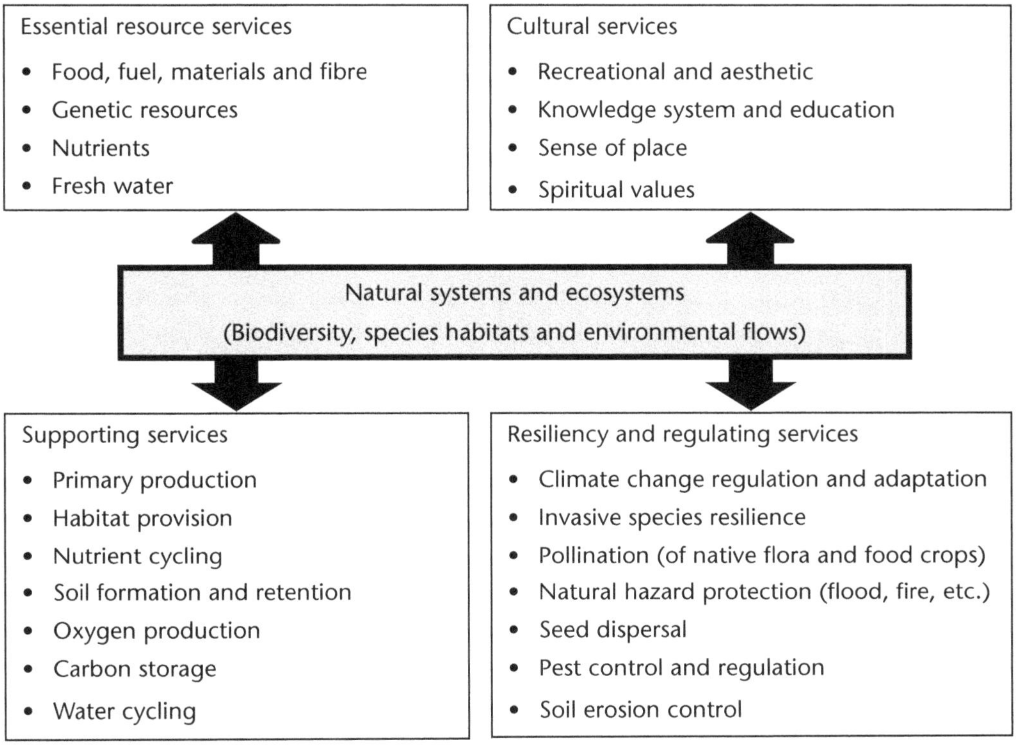

Fig. 10.1. Ecosystem services provided by biodiversity and natural systems. Source: Adapted from DEWHA (2009, p. 9).

of ecology and economics over the past 20 years. The wide range of services and potential economic value of ecological services are illustrated in Fig. 10.1 and have been evaluated on a global scale by Costanza *et al.* (2014).

Biodiversity generates ecosystem services, and it will provide the resilience that ecosystems need to stay in balance 'in the face of impacts or pressures that arise from natural or human-made interactions or events' (DEWHA 2009, p. 6).

Despite general misconceptions that green infrastructure may not have commercial or economic value, Williamson (2017) has identified several Australian studies that quantified significant economic benefits in the provision of environmental services in rural agricultural landscapes. Regionally, the Australian Bureau of Statistics (ABS 2010) found that the rivers, wetlands and floodplains of the Murray–Darling Basin generate AU$187 billion in ecosystem services annually, while the terrestrial ecosystems of that region provide up to AU$325 billion in ecosystem services per year. The New South Wales Office of Environment and Heritage (OEH 2017) found that native vegetation on farms and in rural catchments can 'improve economic outcomes for farmers by improving land value, increasing production outcomes, and reducing operating costs'.

The City of Greater Geelong has undertaken the application of an urban green infrastructure framework primarily focused on reducing urban heat mitigation through the planting of trees along streets and in open space areas (Norton *et al.* 2015). The City has also added *Sustainable Infrastructure Guidelines* to its *Infrastructure Design Manual* to further the use of recycled materials and reduce the carbon footprint of road and other infrastructure construction projects (Sheldon 2015) and encourage green roof and other green building construction. However, a more comprehensive green infrastructure framework that addresses the broader natural systems and landscape, similar to that advocated by the KEPs, would significantly enhance the City's current actions in this realm.

One of the aims of the KEPs is simply to increase indigenous tree and vegetation cover. Trees provide a simple, but representative, example of the environmental service values provided by one component of green infrastructure. Beecham and Lucke (2015) provide a useful review of the economic values estimated by a range of other researchers regarding the value of trees within urban settings for such services as reduced stormwater management costs, air pollution savings, carbon sequestration and annual energy saving. Overall, the annual net benefit per tree ranged from AU$45 to $245, depending on the research study cited. They found that the overall economic benefits of urban trees varied with the size of the tree, with the economic value multipliers for small, medium and large trees being 1.0, 2.9 and 4.1, respectively. An Australian Government (2017) survey of urban home purchasers in Australia found that they show a willingness to pay AU$36 000–54 000 per property in residential precincts that have rain gardens. Other studies have shown that a large quality tree in a residential front garden can add 5% to the property value and that houses on tree-lined streets have sale values 30% higher than those on streets without trees (Gonzalez 2007; Plant et al. 2016; Moore 2015).

As promoted by Norton *et al.* (2015), increasing the area of urban forest can also have a significant effect on the reduction of adverse urban heat island effects. The US Environmental Protection Agency (US EPA 2018) states that:

> *Trees and vegetation lower surface and air temperatures by providing shade and through evapotranspiration. Shaded surfaces, for example, may be 20–45°F* [11–25°C] *cooler than the peak temperatures of unshaded materials. Evapotranspiration, alone or in combination with shading, can help reduce peak summer temperatures by 2–9°F* [1–5°C].

It notes that a study of five US cities indicates that the accrued annual benefits of urban trees outweighed investment and maintenance costs.

In rural north central Victoria, Polyakov and Pannell (2016) found that added native vegetation cover can increase the value of rural properties, but that the optimal percentage of vegetative cover may range from 20% for a property of 5000 ha or greater to as much as 45% for properties of 1 ha or less. In addition, land degradation resulting from native tree clearing, habitat fragmentation and related issues of soil erosion, soil salinity and reduced water resources can have a significant negative economic effect on rural landholders and economies. Lefroy *et al.* (1993) pointed out that, in Western Australia, an estimated AU$615 million/year in environmental costs occurred due to land degradation – approximately 17% of the state's total agricultural production.

All of the above benefits would enhance the Geelong region and should be seen as incentives for application of the KEPs.

A regional concept plan for ecological restoration

A concept plan that integrates application of KEPs at the landscape scale is suggested in Plate 10.1. Major riparian vegetation corridors are required to improve connectivity along the major rivers, such as the Moorabool and Barwon rivers, and along smaller streams, such as Armstrong Creek south of Geelong. Additional protection and connectivity of indigenous grasslands are needed in the rural areas, and biolinks between larger reserves in the You Yangs, Brisbane Ranges, Otway Ranges and coastal reserves.

Applying the principles would provide significantly improved ecosystem conservation and connectivity. This would ensure:

- ecologically functional riparian biolinks along major rivers, streams and wetland complexes
- overland connections between indigenous grasslands, grassy woodlands and forests through the northern region of high value and vulnerable ecological vegetation classes
- protection and management of the entire coastal shoreline and major river estuaries
- a more balanced network of urban forests and permeable ground within the urban areas.

Summary and discussion

Key ecological principles have been drawn from a wide body of ecological and green infrastructure literature to provide an integrating framework for the conservation of natural ecosystems and cultural land use in the regional landscape. This framework also provides a pattern-oriented (not a species-oriented) response to the effects of climate change. A first-tier plan for the Geelong region based on these principles is presented in Plate 10.1. This provides a broad ecological design and structure that will encourage healthy functionality in the natural systems of the region.

The City of Greater Geelong *Biodiversity Strategy* (CoGG 2003) and other regional strategies reflect many of these principles and the suggested plan. However, these principles have not been implemented well over past years, due to a lack of focus and various socio-economic barriers.

Second-tier application of the KEPs and green infrastructure plan will require assistance from ecologists and natural resource management specialists to identify appropriate plant communities and species for future restoration works. Application of the principles will need to rely on historical ecological records but should also recognise that static ecosystem equilibrium is not possible, and that ecosystems and species are constantly subject to disturbances and will change and adapt over time. These are concepts outlined by such ecologists and environmental syndicates as Holling (1973), Gunderson (2000), the Sustainable Scale Project (SSP 2013) and the Resilience Alliance (2016). Climate change and our cultural needs may influence landscape evolution and hybrid or novel ecosystems of the future (Higgs 2003; Hobbs *et al.* 2013). As such, the proposed ecological principles and model are aimed firmly at the future, not the past. The ecological principles and the model they create seek to avoid an imbalance in natural systems, abrupt ecological changes and losses of essential ecosystem services that could seriously undermine the sustainability and survival of our economic and social fabric.

The required social will and community co-operation must be driven by a cultural acceptance of and demand for the environmental services described in Fig. 10.1. For all the reasons outlined in this chapter, it is essential that we integrate ecology into our regional landscapes through the application of key ecological principles and green infrastructure to achieve a sustainable balance in our natural and cultural systems.

References

ABS (2010) *Year Book Australia, 2009–10: Australia's Biodiversity.* Cat. no. 1301.0. Australian Bureau of Statistics, Canberra. http://www.abs.gov.au/ausstats/abs@.nsf/ Previousproducts/301.0Feature%20 Article12009%E2%80%9310?opendocument&tabn (accessed 25 April 2017).

Australian Government (2017) *Stakeholder Annual Report.* Cooperative Research Centre for Water Sensitive Cities, Melbourne.

Australian Museum (2015) *What are the Impacts of Climate Change?* Australian Museum, Canberra. http://australianmuseum.net.au/what-are-the-impacts-of-climate-change (accessed 9 December 2016).

Barton JL, Davies PE (1993) Buffer strips and streamwater contamination by atrazine and pyrethroids aerially applied to *Eucalyptus nitens* plantations. *Australian Forestry* **56**, 201–210. doi:10.1080/00049 158.1993.10674608

Beecham S, Lucke T (2015) Street trees in paved urban environments: the benefits and challenges. In G Williams (Ed.), *Proceeding of the 16th National Street Tree Symposium*, pp. 24–36. ARRIS, Highgate, SA.

Bennett AF, Radford JQ, Haslem A (2006) Properties of land mosaics: implications for nature conservation in agricultural landscapes. *Biological Conservation* **133**, 250–264. doi:10.1016/j.biocon.2006.06.008

Bennett AF, Nimmo DG, Radford JQ (2014) Riparian vegetation has disproportionate benefits for landscape-scale conservation of woodland birds in highly modified environments. *Journal of Applied Ecology* **51**, 514–523. doi:10.1111/1365-2664.12200

BoM and CSIRO (2016) *State of the Climate 2016.* Bureau of Meteorology and Commonwealth Scientific and Industrial Research Organisation, Canberra.

CCMA (2003) *Corangamite Native Vegetation Plan 2003–2008.* Corangamite Catchment Management Authority, Colac, Victoria.

CCMA (2013) *Corangamite Region Catchment Management Strategy 2013–2019.* Corangamite Catchment Management Authority, Colac, Victoria.

CCMA (2014) *Corangamite Waterway Strategy 2014–2022.* Corangamite Catchment Management Authority, Colac, Victoria. http://www.ccma.vic.gov.au/admin/file/content2/c7/Public%20consultation%20summary%20report_CCMA%20style.pdf (accessed October 2016).

CoGG (2003) *Biodiversity Strategy.* City of Greater Geelong, Geelong. https://www.geelongaustralia.com.au/common/Public/Documents/8cbc8661b5926ee-BiodiversityStrategy.pdf

Collie Pty Ltd (2010) *Armstrong Creek East Precinct Structure Plan, Plan 3.* Report prepared for the Department of Planning and Community Development and the City of Greater Geelong.

Costanza R, de Groot R, Sutton P, van der Ploeg S, Anderson SJ, Kubiszewski I, Farber S, Turner RK (2014) Changes in the global value of ecosystem services. *Global Environmental Change* **26**, 152–158. doi:10.1016/j.gloenvcha.2014.04.002

CSIRO (2015) *Establishing a National Agenda for Urban Green Infrastructure.* Commonwealth Scientific and Industrial Research Organisation, Canberra. https://www.csiro.au/en/Research/LWF/Areas/Resilient-cities-21C/Green-infrastructure (accessed 12 February 2018).

CSIRO and Bureau of Meteorology (2016) *Australia's Changing Climate.* Australian Climate Change Science Programme, CSIRO, Canberra.

Cullen P (2007) Water in the landscape: the coupling of aquatic ecosystems and their catchments. In DB Lindenmayer and RJ Hobbs (Eds) *Managing and Designing Landscapes for Conservation: Moving from Perspectives to Principles. Conservation* (pp. 458–472). Blackwell Publishing in association with the Zoological Society of London, London.

DELWP (2013) *Permitted Clearing of Native Vegetation: Biodiversity Assessment Guidelines.* Department of Environment, Land, Water and Planning, Melbourne. http://www.depi.vic.gov.au/ data/assets/pdf_file/0011/198758/Permitted-clearing-of-native-vegation-Biodiversity-assessment-guidelines.pdf (accessed October 2013).

DELWP (2015) *Understanding the Residential Development Standards (ResCode).* Planning Practice Note 27. Department of Environment, Land, Water and Planning, Melbourne.

DELWP (2016a) *Ecological Vegetation Classes, Bioregional Conservation Status and Benchmarks.* Department of Environment, Land, Water and Planning, Melbourne. http://www.depi.vic.gov.au/ environment-and-wildlife/biodiversity/evc-benchmarks (accessed 11 December 2016.

DELWP (2016b) *Biodiversity Action Statements.* Department of Environment, Land, Water and Planning, Melbourne. http://delwp.vic.gov.au/environment-and-wildlife/conserving-threatened-species-and-communities/flora-and-fauna-guarantee-act-1988/action-statements (accessed October 2016).

DELWP (2016c) *Biodiversity Action Planning, Corangamite Region.* Department of Environment, Land, Water and Planning, Melbourne. http://vro.agriculture.vic.gov.au/dpi/vro/coranregn.nsf/ pages/ corangamite_bio_action_planning (accessed October 2016).

DELWP (2016d) *Modelled Native Vegetation and Connectivity Mapping, Victoria Resources Online.* Department of Environment, Land, Water and Planning, Melbourne. http://mapshare2.dse.vic.gov.au/ MapShare2EXT/imf.jsp?site=bim (accessed October 2016).

DELWP (2017) *Protecting Victoria's Environment – Biodiversity 2037.* Department of Environment, Land, Water and Planning, Melbourne.

DEWHA (2009) *Ecosystem Services: Key Concepts and Applications.* Occasional Paper No. 1. Department of Environment, Water, Heritage and the Arts, Canberra.

Doerr VAJ, Williams KJ, Drielsma M, Doerr ED, Davies M, Love J, Langston A, Low Choy S, Manion G, Cawsey EM, McGinness HM, Jovanovic T, Crawford D, Austin M, Ferrier S (2013) *Designing Landscapes for Biodiversity under Climate Change: Final Report.* National Climate Change Adaptation Research Facility and CSIRO, Gold Coast, Queensland.

DSE (2011) *Victoria's Native Vegetation Management: A Framework for Action.* Department of Sustainability and Environment, Melbourne.

Fitzsimons J, Wescott G (2001) The role and contribution of private land in Victoria to biodiversity conservation and the protected area system. *Australian Journal of Environmental Management* **8**(3), 142–157. doi:10.1080/14486563.2001.10648524

Gonzalez C (2007) Why a tree-lined nature strip can add 30% to your property value. *Sydney Morning Herald*, 9 March 2007, Sydney.

Gunderson LH (2000) Ecological resilience: in theory and application. *Annual Review of Ecology and Systematics* **31**, 425–439. doi:10.1146/annurev.ecolsys.31.1.425

Hansen H, Veerkamp C, Nordin A, Lazarova T, Hedlund K, Olsson P, Schipper A (2017) Assessment of biophysical and ecological services provided by urban nature-based solutions: a review. *Naturvation: Cities, Nature, Innovation.* Deliverable 1.3 Part II. https://naturvation.eu/result/assessment-biophysical-and-ecological-services-provided-urban-nature-based-solutions

Hateley R (2010) Fire-stick farming in Victorian forests. In *The Victorian Bush: Its 'Original and Natural' Condition.* Polybractea Press, Melbourne.

Higgs E (2003) *Nature by Design: People, Natural Processes and Ecological Restoration.* MIT Press, Cambridge, Mass.

Hobbs RJ, Higgs ES, Hall C (2013) *Novel Ecosystems: Intervening in the New Ecological World Order.* John Wiley & Sons, New York.

Holling CS (1973) Resilience and stability of ecological systems. *Annual Review of Ecology and Systematics* **4**, 1–23.

Hughes L (2003) Climate change and Australia: trends, projections and impacts. *Austral Ecology* **28**(4), 423–443. doi:10.1046/j.1442-9993.2003.01300.x

idcommunity (2016a) *City of Greater Geelong: Population Forecasts.* http://forecast.id. com.au/geelong (accessed 9 December 2016).

idcommunity (2016b) *G21 Region: Community Profile, Population Estimates.* http://profile.id.com.au/ g21-region/population-estimate (accessed 9 December 2016).

IPCC (2014) *Climate Change 2014: Synthesis Report. Contribution of Working Groups I, II and III to the Fifth Assessment Report of the Intergovernmental Panel on Climate Change*. (Eds RK Pachauri, LA Meyer). Intergovernmental Panel on Climate Change, Geneva.

Lake PS (2007) Flowing waters in the landscape. In DB Lindenmayer and RJ Hobbs (Eds) *Managing and Designing Landscapes for Conservation: Moving from Perspectives to Principles* (pp. 445–457). Blackwell Publishing in association with the Zoological Society of London, London.

Lefroy EC, Hobbs RJ, Scheltema M (1993) Reconciling agriculture and nature conservation: toward a restoration strategy for the Western Australian wheatbelt. In DA Saunders, RJ Hobbs and PR Erlich (Eds) *Nature Conservation 3: Reconstruction of Fragmented Ecosystems* (pp. 243–257). Surrey Beatty & Sons, Sydney.

Lindenmayer DB, Franklin JF, Fischer J (2006) General management principles and a checklist of strategies to guide forest biodiversity conservation. *Biological Conservation* **131**, 433–445. doi:10.1016/j.biocon.2006.02.019

Mackey BG, Watson JEM, Hope G, Gilmore S (2008) Climate change, biodiversity conservation, and the role of protected areas: an Australian perspective. *Biodiversity (Nepean)* **9**(3–4), 11–18. doi:10.1080/14888386.2008.9712902

McIntyre S, McIvor JG, Heard KM (Eds) (2004) *Managing and Conserving Grassy Woodlands*. CSIRO Publishing, Melbourne.

MEA (2005) *Ecosystems and Human Well-Being: Our Human*. Millennium Ecosystem Assessment.

Melbourne Water (2013) *Waterway Corridors Guidelines for Greenfield Development Areas within the Port Phillip and Westernport Region*. Melbourne Water, Melbourne. http://www.melbournewater.com.au/Planning-and-building/Forms-guidelines-and-standard-drawings/Documents/Waterway-corridors-Greenfield-development-guidelines.pdf

Moore GM (2015) The winds of climate change are already blowing in the direction of Australian arboriculture. In G Williams (Ed.) *Proceedings of the 16th National Street Tree Symposium* (pp. 21–23). TREENET, Arris, Highgate.

Norton BA, Coutts AM, Livesleya SJ, Harris RJ, Hunter AM, Williams NSG (2015) Planning for cooler cities: a framework to prioritise green infrastructure to mitigate high temperatures in urban landscapes. *Landscape and Urban Planning* **134**, 127–138. doi:10.1016/j.landurbplan.2014.10.018

NRMMC (2010) *Australia's Biodiversity Conservation Strategy 2010–2030*. Natural Resource Management Ministerial Council, Canberra. http://www.environment.gov.au/topics/biodiversity/biodiversity-conservation>

OEH (2017) *Why is Native Vegetation Important?* New South Wales Office of Environment and Heritage. http://www.environment.nsw.gov.au/vegetation/importance.htm#10 (accessed 29 April 2017).

Plant LJ, Morrison T, Rambaldi A (2016) Street trees: paying their way in property value benefits. *Acta Horticulturae* (1108), 13–24. doi:10.17660/ActaHortic.2016.1108.2

Polyakov M, Pannell D (2016) Accounting for private benefits in ecological restoration planning. In D Ansell, F Gibson and D Salt (Eds) *Learning from Agric.-Environment Schemes in Australia: Investing in Biodiversity and Other Ecosystem Services on Farms* (pp. 181–190). Australian National University Press, Canberra.

Resilience Alliance (2016) i http://www.resalliance.org/ index.php/key_concepts (accessed October 2016.

Sheldon V (2015) *City of Greater Geelong: Changing Construction – Using Recycled Materials in Infrastructure*. 2015 IPWEA IFME Conference, Institute of Public Works Engineering. https://www.ipwea.org/publications/library (accessed 22 March 2018).

SSP (2016) *The Sustainable Scale Project*. http://www.sustainablescale.org/ Advisorypanel.aspx (accessed October 2016).

Steffen W, Burbidge AA, Hughes L, Kitching R, Lindenmayer D, Musgrave W, Stafford Smith M, Werner PA (2009) *Australia's Biodiversity and Climate Change: A Strategic Assessment of the Vulnerability of Australia's Biodiversity to Climate Change*. CSIRO Publishing, Melbourne.

SWIFFT (2016) *State Wide Integration of Flora and Fauna Teams, Threatened Fauna Greater Geelong City*. https://www.swifft.net.au/explore.php?q=threatened+fauna+greater+geelong&submit=search

US EPA (2018) *Using Trees and Vegetation to Reduce Heat Islands. US Environmental Protection Agency*. https://www.epa.gov/heat-islands/using-trees-and-vegetation-reduce-heat-islands (accessed 18 February 2018).

Vaughn N (2014) *Greater Geelong Region Profile*. G21 Geelong Regional Alliance. City of Greater Geelong, Colac-Otway Shire, Golden Plains Shire and Surf Coast Shire, Victoria.

VEAC (2010) *Remnant Native Vegetation Investigation: Discussion Paper*. Victorian Environmental Assessment Council, Melbourne.

Williamson DN (2017) Australian wheatbelt farmers' and NRM responses to ecological principles, climate change, agricultural production and biodiversity effects at a landscape scale. Unpublished PhD thesis. Deakin University, Geelong.

Zevenbergen C, Fu D, Pathirana A (Eds) (2018) Sponge cities: emerging approaches, challenges and opportunities. Special Issue in *Water*, 2017/18/. http://www.mdpi.com/journal/water/special_issues/Sponge-Cities (accessed February 2018).

11

A landscape at risk

Phillip B. Roös

> *Many years ago, this land ... extended right out to the ocean. Port Phillip Bay was then a large flat plain where Boon Wurrung hunted kangaroos and cultivated their yam daisy. But one day there came a time of chaos and crises. The Boon Wurrung and the other Kulin nations were in conflict. They argued and fought. They neglected their children. They neglected their land. The native yam was neglected. The animals were killed but not always eaten. The fish were caught during their spawning season. As this chaos grew the sea became angry and began to rise until it covered their plain and threatened to flood the whole of their country.*
>
> *The people went to Bunjil, their creator and spiritual leader. They asked Bunjil to stop the sea from rising. Bunjil told his people that they would have to change their ways if they wanted to save their land. The people thought about what they had been doing and made a promise to follow Bunjil. Bunjil walked out to the sea, raised his spear and directed the sea to stop rising. Bunjil then made the Boon Wurrung promise that they would respect the laws.*
>
> – Briggs cited in Low Choy and Jones (2013, p. 183).

Our climate is changing – rapidly. This is due to anthropogenic factors; the rapid changes are caused by human activities from our industrial-mechanistic society, which is human-centric and less concerned with the environment and natural habitats. In contrast, the Indigenous peoples across the world understand the law of the land and respect the environment, demonstrated in the story of *Bunjil*. The Geelong region is facing major impacts due to climate change, and we need to critically review the issues of a built environment that do not show capacity or the appetite for keeping up with the changing natural environment to which it is connected.

The state of the planet

Climate change impacts, adaptation and resilience

Climate change and the degradation of our ecosystems are having major impacts upon our human settlements. These changes, occurring on an unprecedented scale never before imagined, are a major challenge for the human race. Climate change is and will continue to have many significant and possible severe impacts upon human society and the landscapes we reside upon and live within as well as on ecological systems generally

(Alley *et al.* 2003; IPCC 2013; IPCC SREX 2012; Robertson 2006; Staudinger 2013). While these impacts are numerous and interconnected in scope, they are also spatial, diverse and temporally variable in scale and location (Kenny *et al.* 1995). This is indeed a complex issue that is global in scale. Such global issues are equally affecting and influencing the local context of the Geelong region, Victoria as a whole and the Australian continent.

Built environment's impact on the climate

The built environment constitutes only 2.8% of the global land area of Earth (Ruth and Coelho 2007). This area hosts the main location of human social, cultural and economic life, and more than 50% of the world's population live in urban environments in and around cities (UN 2007). The UN estimates that in the next 30 years the current 3.5 billion city-dwellers will nearly double, adding another 2.5–3 billion people (UN Habitat 2013). The world is urbanising quickly, and under the business-as-usual scenario it is likely that more than 70% of the world's population will live in cities by 2050; a scenario where greenhouse gas emissions will increase dramatically (Angel *et al.* 2005). This contribution to greenhouse gas emissions is adding to scientific discourses about climate change, and will add to the physical impacts on climate. The building construction sector in Australia significantly negatively impacts upon biodiversity loss, ecosystem degradation and suburban sprawl, especially in coastal areas as most of Australia's cities and regional towns are located on the coast. Undoubtedly, one of the coastal regions that are part of this discourse is the Geelong region.

Impacts of climate change on the built environment

The impacts of climate change upon the built environment are many and diverse, and both direct and indirect. They vary greatly, depending upon the location and local quality of the existing built environment. Climate change poses serious threats to urban infrastructure, quality of life and entire urban systems, especially settlements and cities that are located next to the coast (Hoornweg *et al.* 2010). In Australia, 86% of the population lives in the coastal zone, which is not surprising, as two-thirds of the Australian continent is classified as an arid zone. Up to 35% of buildings in the Australian coastal zone are at risk of inundation, and many items of significant community infrastructure are situated in vulnerable coastal locations (DCC 2009).

One very difficult issue, that is rarely mentioned in the literature, is the possibility that the built environment in these low-lying areas will be abandoned due to relocation of people necessitated by inundation and flooding (Roös 2017). This would represent one of the largest losses of value in land and infrastructure, and the largest transfer of economic wealth in human history. What will happen with the buildings and infrastructure left behind? This will provide an enormous economic challenge as well as an environmental challenge for the Geelong coastal region, if a derelict built environment is left to its own devices. Tensions will grow as specific landowners and residents demand increasingly costly infrastructure construction to mediate the effects and protect their assets while others push for less costly shifts in habitation (Hoornweg *et al.* 2010). Sea-level rise modelling and studies indicate that most of the towns along the Bellarine Peninsula and the greater Geelong region coast would be impacted by inundation due to sea-level rise. Towns that will be affected include Geelong, Portarlington, Indented Head, Queenscliff, Point Lonsdale, Barwon Heads, Breamlea, Torquay and Anglesea (Fig. 11.1). Other communities along the Great Ocean Road coast will also be affected (Roös 2017; GORCC 2012). The complexity of solving future retreat and settlement establishment will provide major land

Fig. 11.1. Barwon Heads is at risk due to climate change impacts and sea-level rise. Photo: Donna Squire.

use and social planning challenges. This is especially so when considering that many sensitive ecological systems and natural environments are dependent on and located next to existing settlements. This scenario provides an opportunity to reconsider contemporary planning and design methods for urban development, and acknowledge the need to design and plan with nature, respecting the importance of ecological systems as an integrated part of our human species' future on this planet.

Climate change and the Geelong region

A landscape at risk

Climate change in the Geelong region is unlikely to cause new inland and coastal hazards, but at many locations it will make existing hazards worse. The most significant hazards identified by the *Climate Resilient Communities of the Barwon South West* project (CRC 2014) include:

- an increase in extreme heat days
- a decrease in mean annual precipitation/rainfall
- an increase in the severity of precipitation events
- an increase in sea level (CRC 2014).

The Geelong region has already experienced extreme weather events, such as the extreme heat on Christmas Day 2015 that caused a bushfire in Wye River that destroyed one in three homes (Minear 2015), and extreme and recurrent rain events in Geelong, such as the January 2016 event that flooded many main roads and was recorded as a 1 in 50-year storm, the unprecedented thunderstorm and flash flooding on 24 April 2017 that resulted in 20 cars being trapped, and the fierce winds and flash flooding in February 2018 that involved roofs being ripped off as part of a damaging hailstorm (Dmytryshchak 2018). Climate change effects are already happening, and are not a threat to ignore.

The climate risk assessment conducted by the *Climate Resilient Communities of the Barwon South West* project documented more than 150 risks for the larger Geelong and South West Victoria region. This assessment concluded that temperature increase and variability was the highest risk, followed by precipitation variability then sea-level rise. Other categories included secondary climate effects such as increased bushfire risk and increased wind speeds associated with increased storms (CRC 2014).

For the Geelong region, the regional sectors most at risk include natural assets, infrastructure, industry, community well-being, government assets and services, as well as buildings. The highest at risk are the natural assets, followed by infrastructure, then community well-being (CRC 2014).

Sea-level rise is considered to be the biggest threat to the Geelong coast and towns located along the coast (most of the towns in the region are located along the coast), resulting in increased rates of coastal erosion, more extensive and frequent coastal flooding, increased intrusion of seawater into estuaries and coastal aquifers, changing water quality, changing groundwater characteristics and sedimentation, and increased seawater temperatures that may affect ecosystems (Roös 2017; DSE 2012).

Fig. 11.2. Geelong City is at risk due to climate change impacts and sea-level rise. Photo: Donna Squire.

The exposure of key natural assets along the Geelong region coast, mostly the sandy areas, estuaries, bays and inlets, makes them vulnerable to permanent and temporary inundation, including coastal recession. Coastal erosion due to higher sea levels is a key risk for coastal areas within the Geelong region. The region has ~750 residential buildings that lie within 110 m of 'soft' erodible shorelines, and most of the coastal caravan parks along the Bellarine Peninsula would be affected by sea-level rise (Roös 2017; CoGG 2011). Issues relating to coastal storms are already evident, prompting local councils to attempt to restore some of the region's coastal dune systems due to sea-level inundation and storm surges. In the absence of coastal protection measures or other adaptation responses, most of the buildings on the shoreline and primary dunes will be at risk of increased erosion from sea-level rise and storm surges, with devastating effects for both built and natural assets.

The *Surf Coastal Climate Change Vulnerability and Adaptation Project* (GORCC 2012) report identified that significantly increased numbers of assets are located in areas that could be affected by coastal recession when allowing for 100 m of recession for each metre of sea level rise in susceptible areas, in accordance with the Bruun Rule (GORCC 2012). The risk areas identified were based upon sea-level rises of 0.2–0.8 m, projected from 2030 to 2100. Taking into account sea-level rises of up to 1.9 m by 2100 under a worst-case scenario (RCP8.5), the risks will increase substantially (DCC 2009). Assets vulnerable to inundation and coastal recession are more prevalent along the coast of the Geelong region.

In many areas, the Victorian coast is quite resistant to and has historically witnessed the potential impacts of sea-level rise and storm surges due to its geomorphological formations, but in other areas, especially those with sand formations and river inlets, the coast is very vulnerable. It is thus important to identify and understand the biophysical characteristics of the Geelong region coast.

The biophysical characteristics of the Geelong region coast

The biophysical characteristics of the coast differ in shape and form around Australia. The Australian coast has four major coastal regions, each with its own geomorphology and climate characteristics. The dynamic interactions of land and sea over time shaped the coastline as it is today, with over 10 000 sandy beaches making up half the coastline, and the remainder comprising mostly rocky and cliff shorelines (McInnes *et al.* 2009). The Victorian coast can mostly be classified as a rocky limestone coastline, with sandy beaches mostly located at river inlets. The Geelong region coast matches these Victorian coastal characteristics. The calcareous beach sands are 'perched' on the calcarenite ridges, a typical feature of the coast of the Geelong and Great Ocean Road region. In the coastal region from Torquay towards the Bellarine, the coastal area is mostly low-lying areas with coastal dunes, wetlands and salt lakes (Fig. 11.3). This geomorphological arrangement is key in identifying which areas of the Geelong region coast are vulnerable to climate change impacts, as well as determining the natural attributes that contribute to the characteristics and values of coastal settlements in these coastal zones.

The *Victorian Coastal Hazard Guide* (DSE 2012) identified the relative sensitivity of coastal landforms to changes in different climate drivers for open and re-entrant coastal areas. The *Guide* states that coasts open to the sea are characterised as being exposed to swells, they are mainly rocky and sandy coasts, and that they are subject to high-energy erosion. The biophysical features of the greater Geelong region and the Great Ocean Road region coast are, in most parts, highly resistant to coastal erosion due to the geological

Fig. 11.3. Coastal dunes with wetlands and salt lakes at Breamlea. Photo: Donna Squire.

formations (GORCC 2012). But the inlets, beaches, sand dunes and low-lying areas are susceptible to coastal erosion. These areas are more vulnerable to sea-level rise, and are in easily accessible areas where human settlement is concentrated.

A future climate

To be able to identify the future climate for the Geelong region, climate change factors need to be considered, based on climate models. Climate change projections for the Geelong region were developed, and represented in this chapter, using the climate change factors derived using the *Climate Futures Tool* (Clarke *et al.* 2011). The three Global Climate Models considered to best represent the general pattern of climate change projections for 2030–90 were based on the high emissions scenario according to RCP8.5 (business-as-usual growth in emissions), including changes in rainfall and temperature as well as sea-level rise. This will ensure that climate change projections over each timeframe and for various climate characteristics are consistent. This scenario also reflects a plausible upper limit to climate change up to and at 2090. Due to the uncertainty of the changing climate, projections from 2090 to 2100 and beyond are mostly uncertain.

Temperature

The average temperatures will continue to increase in all seasons. For the near future (2030), the annually averaged warming is projected to be around 0.6–1.2°C above the climate of 1986–2005. By 2070, the projected range of warming is 1.1–2.9°C, depending on future emissions. There will be more hot days and extreme heat events. There will be a substantial increase in the temperatures reached on hot days. There will be more hot days (>35°C), and warm spells will last longer. Under the high emissions scenario, the projection for Geelong is 13 average hot days per year over 35°C. This directly impacts fire weather conditions; fire weather is projected to be much harsher in the future. Fire weather is a measure of fuel dryness and hot, dry, windy conditions. An increase in frequency of very high and extreme fire danger days is projected.

Rainfall

Natural climate variability will dominate any projected changes in the near future (2030), and year-to-year changes in rainfall will dominate trends caused by greenhouse gases. There will be less rainfall in winter and spring. Despite an overall decline in rainfall, the rain that will fall will occur in short intense downpours and in extreme storm events. Time spent in drought will increase and the predicted average rainfall will reduce by -8% by 2050, and up to -21% by 2070.

Sea level

Sea level will continue to rise. Heights of extreme sea-level events will increase, and the ocean temperature will get warmer. Conservatively, by 2030, the projected sea-level rise for the Geelong coastal region is 0.08–0.18 m above the 1986–2005 level. By 2070, sea level is projected to rise 0.20–0.46 m under the lower emissions scenario and 0.25–0.56 m under the high emissions scenario. However, other studies indicate these levels may be higher, up to 0.8–1.8 m. It must be noted that these levels are also applicable to the areas of Port Phillip Bay. Additionally, storm surges will increase the risks to coastal towns.

Geelong region risk profile

A warmer and drier climate, as well as sea-level rise predicted for the future, will impact the Geelong region in different ways. Noting that vulnerability is a function of exposure and includes the capacity to adapt, key vulnerable areas for social, environmental and economic sensitivities identified include:

- communities in coastal towns
- primary industry across the region
- industrial activity in Geelong
- ecosystems vulnerable to heat, drier climate and coastal inundation are a significant risk across the entire coastline.

Key risks identified for the Geelong region are listed in Table 11.1.

Adapting to an uncertain future

To a large extent, climate impacts will be experienced and managed at local and regional scales within the Geelong region. These impacts will interact with other non-climatic forces, which may result in amplification of negative consequences for local communities (CoGG 2011). Adaptive capacity is thus important.

The term 'adaptive capacity' is used to describe how well a community is able to respond to the impacts of climate change, and offers a measure by which the resilience of a community can be determined. Adaptive capacity is influenced by factors such as community diversity, flexibility, ability to learn and general levels of disadvantage. This cuts across

Table 11.1. Geelong regional risks profile

Risk exposure	Vulnerability indicator: Social	Economic	Environmental
Temperature An increase in mean temperature and/or extreme heat events	• Health impacts on vulnerable people • Increased recreational use of waterways • Increased costs from needing to build resilience • Potentially reduced economic growth	• Increased water needs/demand (mainly irrigation and outdoor) • People and businesses lose access to water and energy (load shedding, inadequate backup/generators) • Increased risk of bushfire	• Impacts on ecological functions, reducing ecosystem services – habitat loss and more favourable habitat for pests
Rainfall A decrease in precipitation and less frequent 50 mm rainfall events	• Increased water restrictions • Inability to water gardens/playing fields • Water Catchment Management Authority – loss of on-ground works	• Impacts on road infrastructure with more linear cracking, drainage infrastructure and increased landslips • Increased bushfires • Reduced water quality	• Increased flooding events increase fish deaths (estuaries) • Increasing bushfire risk increases the impacts on biodiversity • Increased acid sulphate soils • Increased demand for natural resources
Sea-level rise Increased storm surge events and inundation due to sea-level rise impacts	• Higher demand for access to coast • Private and government land and amenities impacted • Policy uncertainty and changes result in community pressures • Tourism impacts – loss of amenities and resultant income	• Increased loss of businesses and community amenities • Loss of residential properties • Loss of infrastructure including roads, energy and water	• Increased salinisation of river systems and freshwater lakes • Coastal retreat of vegetation • Retreat of coastline • Erosion due to storm surge and higher sea levels • Loss of estuaries and salt marshes

Source: Adapted from CRC (2014).

indicators relating to community well-being, the economy, the built environment and the natural environment.

It must be noted that ecological systems in the natural environment have a limit to their adaptive capacities, very dependent on the overall ecology and constant climate conditions of the place. The climate models indicate that the Geelong region will face a different and harsher climate. If any other influences reduce the Geelong region's performance and resilience, the overall adaptive capacity may be diminished, and the communities of both human and natural environments may be more vulnerable to the impacts of climate change.

Climate change adaptation and resilience

Adaptation refers to 'adjustments in social, environmental and economic systems in response to actual or expected climatic stimuli and the effects or impacts' (McCarthy 2001). Adaptation involves changes in processes, practices and structures 'to moderate potential damages or benefits associated with climate change' (Smit 2003; IPCC 2014a). In the context of climate change, several definitions of adaptation are found in climate change literature, similar to the following example:

> *Adaptation is the process of adjustment to actual or expected climate and its effects. In human systems, adaptation seeks to moderate harm or exploit beneficial opportunities. In natural systems, human intervention may facilitate adjustment to expected climate and its effects* (IPCC 2014b, p. 40).

Adaptation to climate change necessarily includes adaptation to variability. Variability requires that adaptation include both the process of adapting as well as appreciation of the condition being adapted. Variability is evident in autonomous and spontaneous adaptation, and it is argued that human and natural systems will, to some extent, adapt autonomously and that planned adaptation can supplement autonomous adaptation. However, 'options and incentives are greater for adaptation of human systems than for adaptation to protect natural systems' (IPCC 2001, pp. 6–8). In unmanaged natural systems, adaptation is reactive – it is the means by which species and communities respond to changed conditions. With predicted accelerated climate changes, it is a recognised concern that many species will not be able to adapt to fast-changing environments (IPCC SREX 2012). Planned adaptation within managed ecosystems could be a possible solution for the changing future. The patterns of nature need to be considered in any process, implemented or managed by humans, to achieve sustainable outcomes. Unfortunately, this is not the case in the current practice of climate change risk assessments and adaptation planning. The predominant approach for planned adaptation is the risk management adaptation approach.

Planned anticipatory adaptation based on the risk management approach is aimed at reducing a system's vulnerability by diminishing risk or improving adaptive capacity. Numerous reasons have been given for pursuing planned adaptation, and it can yield benefits regardless of the uncertainty and nature of climate change (Ali and Minura 1999). Planned adaptation initiatives are regarded not as a substitute to the mitigation of climate change, but as a necessary strategy to manage its impacts. Risk assessment and adaptation planning is best undertaken as a participative exercise that engages the community and draws on their knowledge of the 'system' (human systems) under consideration.

Risk is defined as 'the effect of uncertainty on objectives' (AGO 2006). Risk frameworks need to be developed to support planning and decision-making in the context of uncertainty and the objectives to be achieved. Given the long-term nature of climate change and the considerable uncertainty about the timing and severity of some potential impacts, it is important that maladaptation does not occur, and that actions are taken to avoid 'over or under responses' to adaptation (GORCC 2012). The avoidance of maladaptation is specifically applicable to the Geelong region and is known as the 'precautionary approach' to climate change adaptation (IPCC 2007).

Unfortunately, human adaptation considerations, including the use of the precautionary approach, are self-centred – adaptation or mitigation considerations are framed and structured for human benefit only (Berke 1995; Munasinghe 2000; Campbell *et al.* 2009). A holistic integrated adaptation discourse, where ecosystems (vegetation, animals, etc.)

and human systems are in equilibrium, needs to be addressed. It is in this discourse of climate change impacts on the Geelong region that we want to find a way to work with nature, to work with our environment to adapt to future challenges. The Geelong region is in essence a coastal region – the question is how the government deals with coastal and hinterland issues. The process of coastal management needs to be reviewed.

Coastal management and planning

For many centuries, humans have deliberately exploited the resources of the coast and modified its environment. Throughout the ancient world, civilisations built sea walls and ports, diverted rivers, used coastal soils for agriculture, and exploited marine life (Kay and Alder 2005). Human activity to preserve the coastal character and ecological integrity is a very recent activity; it was in the 1930s that the first protective coastal park was established, under the concept of integrated coastal management (Wescott 1988; Sorensen 1997). Since then the concept of coastal management and planning has evolved and it is now generally accepted that coastal resources can only be effectively evaluated and managed in the context of a total ecosystem associated with human cultural and social environments (Kay and Alder 2005).

Unfortunately, this is not always the case. In Victoria, the decision-making and authority for approvals in the scenario where developments traverse public and private land, is a major problem in terms of effective implementation, especially in the context of the environment. Macintosh (2012, p. 6) concludes in *Coastal Adaptation Planning: A Case Study on Victoria, Australia* that:

> *the Victorian framework is based on an awkward distribution of powers and responsibilities. It provides broad principles that are intended to guide decision-making by planning and responsible authorities. However, it does not dictate outcomes and is vague about what is expected in the treatment of coastal issues.*

Part of this problem is that the autonomy of local councils in the planning system is tightly constrained. This is due to the multiplicity of powers of other governmental bodies such as the state Planning Minister, the referral authorities, the Victorian Coastal Council including Regional Coastal Boards, and the Victorian Civil and Administrative Tribunal.

Another constraint is that an urban/growth boundary is not set, and the continued movement towards and across this notional boundary into the natural landscape of coastal areas results in coastal suburban sprawl (Wescott 1998; VCC 2013). This is best demonstrated in the town of Torquay with the approval of new greenfield developments on the edge of the urban boundary in the Spring Creek catchment. The state Planning Minister approved the rezoning of 240 ha of agricultural land west of Torquay on the town boundary and declared Spring Creek an 'urban growth' zone (Fig. 11.4). This sparked fear from local residents and environmental groups who condemned the proposed sprawl, which occurred despite a state government-approved structure plan that had previously set town boundaries (Dowling 2014). Notwithstanding all the issues around policy, even with good intentions, the main problem of future sustainable development that needs to address minimising human impacts on the coastal environment has not been resolved.

The literature on coastal management and coastal environmental planning reveals that relationships between social and ecological systems are fundamental, and that the protection of significant environmental features is important (VCC 2013). The *Victorian Coastal*

Fig. 11.4. Coastal sprawl in the Spring Creek area of Torquay: Photo: Donna Squire.

Strategy (VCC 2013) notes that, to take into account environmental considerations, development, recreation and tourism and economic opportunities need to be restricted to activity nodes and recreation nodes. Activity nodes include community recreation facilities and tourism activities, located within settlements that are adjacent to the activity centres identified in planning schemes, and include public and private land (VCC 2013). The City of Greater Geelong has indicated its interest in the management of coastal reserves, both along the Bay and open coastline. The intent is to manage these coastal reserves and development, taking into account the improvement of regional open space status, natural environments and cultural values. However, it is evident that, with the pressure of population growth and the need to develop residential areas and supportive infrastructure in our coastal areas, environmental considerations will be put on the back burner. The question is, when will government (at all levels) acknowledge the crisis we face and change policy for the better?

Conclusion

Changes in the climate and rising sea levels make the Geelong region coast more vulnerable to the dynamic forces of nature. Social and economic forces, including new developments that result in coastal sprawl, also bring stress to the coastal and hinterland areas. Combine the impacts of anthropogenic-induced climate effects and pressure from peak tourism, and we are looking at a more vulnerable Geelong region into the future.

Governments adopt various practices to deal with these issues. Decisions are made that include climate change adaptation actions based on risk assessments, economic assessments, coastal management and adaptation planning policies. But what is lacking in the various government and coastal management strategies and policies, is that an understanding of the complex patterns of human–ecological systems is important to inform efforts to manage and plan for a sustainable future.

Facing unprecedented changes in the environment in the Geelong region due to future climate effects, the way forward is to reconnect humans with nature and to discover the patterns for the co-evolution of both, to establish resilience for the future. It is time to understand the natural law of the land, and learn lessons from the past – applying Indigenous knowledge to today's practice. It is time to respect the environment, and honour the promise by *Boon Wurrung* to *Bunjil* that they would respect the laws of the land.

References

AGO (2006) *Factors and Methods Workbook, December 2006: For Use in Australian Greenhouse Gas Emissions Reporting.* Australian Greenhouse Office, Canberra.

Ali A, Minura N (Eds) (1999) Climate change impacts and adaptation assessment in Bangladesh. *National Assessment Results of Climate Change: Impacts and Responses: CR Special* 6, 109–116.

Alley RB, Marotzke J, Nordhaus WD, Overpeck JT, Peteet DM, Pielke RA Jr, Pierrehumbert RT, Rhines PB, Stocker TF, Talley LD, Wallace JM (2003) Abrupt climate. *Science* **299**(5615), 2005–2010. doi:10.1126/science.1081056

Angel S, Sheppard S, Civco D (2005). *The Dynamics of Global Urban Expansion.* World Bank, Washington, DC.

Berke PR (1995) Natural-hazard reduction and sustainable development: a global assessment. *Journal of Planning Literature* **9**(4), 370–382. doi:10.1177/088541229500900404

Campbell AM, Kapos V, Sharlemann JPW, Bupp P, Chenery A, Coad L, Dickson B, Doswald N, Saiful M, Khan I, Kershaw F, Rashid M (2009) *Review of the Literature on the Links between Biodiversity and Climate Change: Impacts, Adaptation and Mitigation.* Technical Series no. 42, 124. Montreal. https://www.cbd.int/doc/publications/cbd-ts-42-en.pdf (accessed 1 March 2018).

Clarke JM, Whetton PH, Hennessy KJ (2011) *Providing Application-specific Climate Projections Datasets: CSIRO's Climate Futures Framework.* Climate Change Australia, Climate Adaptation Flagship, CSIRO Marine and Atmospheric Research, Melbourne.

CoGG (2011) *Climate Change Adaptation Study.* City of Greater Geelong, Geelong.

CRC (2014) *Regional Report Climate Resilient Communities of the Barwon South West.* Victorian Adaptation and Sustainability Partnership Co-operative Research Centre, Melbourne.

DCC (2009) *Climate Change Risks to Australia's Coast: A First Pass National Assessment.* Department of Climate Change, Canberra.

Dmytryshchak G (2018) Fierce winds, flash flooding and hail as Geelong hit by brief storm. *The Age*, 10 February 2018. https://www.theage.com.au/national/victoria/fierce-winds-flash-flooding-and-hail-as-geelong-hit-by-brief-storm-20180210-p4yzwx.html (accessed 30 May 2018).

Dowling J (2014) Planning Minister Matthew Guy sparks sprawl fears as Torquay land is rezoned. *The Age.* http://www.theage.com.au/victoria/planning-minister-matthew-guy-sparks-sprawl-fears-as-torquay-land-is-rezoned-20140320-355wx.html (accessed 20 March 2018).

DSE (2012) *Victorian Coastal Hazard Guide.* Department of Sustainability and the Environment, Melbourne.

GORCC (2012) *Surf Coastal Climate Change Vulnerability and Adaptation Project.* Great Ocean Road Coast Committee, Melbourne._https://www.gorcc.com.au/projects/great-ocean-road-renewal-project/ (accessed 1 March 2018).

Hoornweg D, Sugar L, Anderson C, Brada P, Trejos CC, Dave R, Lee M (2010) *Cities and Climate Change: An Urgent Agenda*. International Bank for Reconstruction and Development, Washington DC.

IPCC (2001) *Climatic Change 2001: Working Group II: Impacts, Adaptation and Vulnerability – Enhancing Adaptive Capacity*. Intergovernmental Panel on Climate Change, Geneva.

IPCC (2007) *Climate Change 2007: Impacts, Adaptation and Vulnerability. Contribution of Working Groups I, II and III to the Fourth Assessment Report*. Intergovernmental Panel on Climate Change, Geneva.

IPCC (2013) *Climate Change 2013: The Physical Science Basis. Contribution of Working Group I to the Fifth Assessment Report of the Intergovernmental Panel on Climate Change*. Intergovernmental Panel on Climate Change, Geneva.

IPCC (2014a) *Working Group III Contribution to the IPCC 5th Assessment Report – Mitigation to Climate Change Technical Summary Final Draft*. Intergovernmental Panel on Climate Change, Working Group III, Geneva.

IPCC (2014b) *Working Group II Contribution to the IPCC 5th Assessment Report: Climate Change 2014: Impacts, Adaptation, and Vulnerability AR5 Final Draft Report*. Intergovernmental Panel on Climate Change, Working Group II, Geneva.

IPCC SREX (2012) *Managing the Risks of Extreme Events and Disasters to Advance Climate Change Adaptation: A Special Report of Working Groups I and II*. Intergovernmental Panel on Climate Change, SREX. Cambridge University Press, Cambridge.

Kay R, Alder J (2005) *A Brief History of Coastal Management and Planning*. Taylor & Francis, London.

Kenny GJ, Warrick RA, Mitchell ND, Mullan AB, Salinger MJ (1995) Climate impacts: an integrated model for assessment of the effects of climate change on the New Zealand environment. *Journal of Biogeography* **22**(4/5), 883–895. doi:10.2307/2845989

Low Choy D, Jones D (2013) Words, oral libraries and environmental responses: an Australian Indigenous perspective to climate change adaptation. In *IFLA50 Shared Wisdom in an Age of Change* (pp. 176–185). International Federation of Landscape Architects and New Zealand Institute for Landscape Architects, Wellington.

Macintosh A (2012) *Coastal Adaptation Planning: A Case Study on Victoria, Australia*. CCLP Working Paper Series 2012/2. Centre for Climate Law and Policy, Australian National University, Canberra.

McCarthy J (2001) *Climate Change 2001: Impacts, Adaptation and Vulnerability. Working Group II to the Third Assesment Report to the Intergovernmental Panel on Climate Change*. Cambridge University Press, Cambridge.

Minear T (2015) Great Ocean Road fires: how a bushfire burnt Christmas Day, *Herald Sun*, 26 December 2015. https://www.heraldsun.com.au/news/victoria/great-ocean-road-fires-how-a-bushfire-burnt-christmas-day/news-story/ (accessed 25 March 2018).

Munasinghe M (2000) *Climate Change and Sustainable Development Linkages: Points of Departure from the IPCC TAR*. https://www.researchgate.net/publication/228790602 (accessed 1 September 2018).

Robertson D (2006) Health effects of increase in concentration of carbon in the atmosphere. *Current Science* **90**(12), 1607–1609.

Roös PB (2017) *Regenerative-adaptive Design for Coastal Settlements: A Pattern Language Approach to Future Resilience*. PhD thesis. School of Architecture and Built Environment, Deakin University, Geelong. www.dro.deakin.edu.au/eserv/DU:30103436/roos-regenerativeadaptive-2017.pdf

Ruth M, Coelho D (2007) Understanding and managing the complexity of urban systems under climate change. *Climate Policy* 7(4), 317–336. doi:10.1080/14693062.2007.9685659

Smit B (2003) *Adaptation to Climate Change in the Context of Sustainable Development and Equity*. Quirk Press, Toronto.

Sorensen J (1997) National and international efforts at integrated coastal management: definitions, achievements, and lessons. *Coastal Management* **25**, 3–41. doi:10.1080/08920759709362308

Staudinger MG (2013) *Impacts of Climate Change on Biodiversity, Ecosystems, and Ecosystem Services: Technical Input to the 2013 National Climate Assessment.* US Geological Survey, Palo Alto.
UN (2007) *World Urbanization Prospects: The 2007 Revision Population Database.* Population Division, UN Department of Economic and Social Affairs, New York.
UN Habitat (2013) *The State of European Cities in Transition 2013.* UN Human Settlements Programme, Denmark.
VCC (2013) *Victorian Coastal Strategy 2013: Draft Report.* Victorian Coastal Council, Melbourne.
Wescott G (1988) *A Study in Environmental Planning and Management: The Coast of Victoria, Australia.* Deakin University, Melbourne.
Wescott G (1998) *Reforming Coastal Management to Improve Community Participation and Integration in Victoria, Australia.* Taylor & Francis, London.

12

The Geelong suburban dream: origins, history and future

Louise Johnson

The great Australian backyard is part of the larger psychological package called the Australian dream. It's a detached house on a quarter acre block with its own back and, ideally, front garden; it's the castle

– Dr Tony Mathews, in Brook (2018).

Introduction

Geelong suburbs share many characteristics with those in other western cities, including low density, location beyond the central city, social homogeneity and an association with nature, family living, health, open space and safety. The 'suburban dream' as it has been manifested in Australia also has a commitment to particular forms of housing – single-storey and detached – and tenure – home ownership. In addition, we must include in the Australian suburbs story their location on Aboriginal lands and ongoing Indigenous presence. The Geelong pattern does diverge from its metropolitan counterparts in being smaller in scale, lower in density and physically more diverse. After some further consideration of what makes a suburb, internationally and in Australia, this chapter will focus on the Geelong experience: the origins, history and current pattern of Geelong's suburbanisation. Finally, two very different suburbs will be singled out for closer attention – Warralily, a very new suburban development in the southern growth corridor, and Corio, in the older, industrial northern sector of the city. These suburbs represent two sides of Geelong's current economic, social and spatial order. Along with a newly emergent knowledge economy, affluence and population inflow, there is a darker story of social polarisation, economic restructuring and spatial renovation which perhaps prefigures a bifurcated social and spatial suburban future for this regional city.

What is a suburb?

Internationally, suburbs are diverse and changing, as peripheral and transitional urban development assumes different forms in the cities of the south as well as north (Harris and Lehrer 2017). For Robert Silverstone, suburbia has enormous variability and a long history, 'from the tight semi-detached estates of the unplanned suburbs of 1930s Britain to the more expansive unfenced tract housing of middle America, from the elegant villas of early

Victorian London to the clapboard and brick of 1950s Sydney' (Silverstone 1997, p. 4). What is common, though, is a search for the perfect marriage of nature and culture, a desire to create a balanced life, the best of both worlds (Silverstone 1997). Suburbia is an essential product of urban expansion, yet at the same time an escape from it and protest against it (Fishman 1987).

Elements which distinguish the suburb, especially for Australia, England, Canada and the US are peripheral location, as suburbs, at least initially, are located at the edge of the urban core and then on the edge of the city; a relationship to the central business district of functional dependence; a positive relationship to the countryside, with 19th-century suburbanisation linked to a Romantic ideal of country living; a lower density relative to the city; a housing type of single-family dwellings with gardens; social segregation, be it in class, race and/or ethnic terms; a commuting relationship to the core; and a particular cultural formation, as suburbs form a utopian middle landscape and private romantic paradise in contrast to the dysfunctional inner-city high-rise (McManus and Ethington 2007). There are strong parallels in these urban histories to developments in transport technologies, with the 18th and early 19th-century mixed-use, walking town giving way to the 19th-century more class-zoned, public transport city which then exploded spatially with mass private car ownership from the mid 20th century.

In Australia, suburbia has come to represent the quintessential national way of life – clean, low-density, healthy family living (Allon 2008). Despite the growth in central city apartment living in many of the metropoles, suburbs remain the dominant form of living – comprising 80% of the housing stock in Australia in 2011 (ABS 2011). In addition to these elements shared with other western countries, there are characteristics and developments which are peculiar to Australian suburbs. Australian cities and suburbs are built on Indigenous lands, which were taken without consultation or compensation, helping to elevate the role of surveyors and town planners in the appropriation process to produce a land market based on free land, sold and resold for higher and higher profits within a speculative market. This process of appropriation was not a one-off or completely successful, leading to a continuing Indigenous presence in Australian suburbs, be it physical, symbolic or a set of historical regulations, attachments and meanings.

In addition to the colonial foundation of Australian cities there is a strong association of Australian suburbia with home ownership. The drive for home ownership has led to Australia being recognised internationally as a special property-owning democracy with one of the highest rates of home ownership – peaking at 71.4% in 1966 – in the world. However, it fell to 67% by 2011 and is still going down, especially for the young (ABS 1966, 2011). The desire for and long-term reality of acquiring home ownership has been one of the main drivers for suburbanisation in Australia (Kemeny 1981; Greig 1995). The obverse to the high rates of private ownership are low levels of government-owned social housing. There has also been a wind-back in the role of the state as housing provider, from a high in the 1940s when the Commonwealth built close to 40% of all housing stock, to public housing now comprising a residual tenure form of around 4% of the national housing stock (Groenhart and Burke 2014; Troy 2012). Despite the withdrawal of governments from the actual provision of social housing because of our governance structures, Australian suburbs in comparison to those in the US, for example, have far greater social and spatial equity, primarily because they are serviced by state and federal government tax takes rather than being locally administered and financed.

What has further distinguished the Australian post-war suburb has been its growing social diversity. Australian cities echo changes in national immigration patterns which have led to increased population diversity since World War II, so that now many who live in suburbs are either born overseas or are the children of migrants (Burnley *et al.* 1997; ABS 2014). The 2016 Census shows that two-thirds (67%) of the Australian population were born in Australia. Nearly half (49%) of Australians had either been born overseas (first-generation Australian) or one or both parents had been born overseas (second-generation Australian) (ABS 2016). Since the 1980s, there has also been increased design diversity – with different households and 'lifestyle groups' represented in a range of house and block sizes – as well as growing medium- and high-rise dwelling units, especially in the major cities, though 88% are still separate houses (Johnson 2006). Despite a fall in the size of households, there has been an increase in house sizes which grew by around 40% in average floor area from 1984 to 2003. In NSW, the ACT and Queensland, the growth was over 50% (Allon 2008) so that Australians now have the largest houses on average in the world, at 245 m^2. Along with the growth in the size of houses there has been the emergence of the Master Planned Estate as the major form of suburban housing provision (UPR 2010). At least some of these estates are what Gleeson calls 'privatopias' – providing private services and protected by covenants and cost from social mixing (Gleeson 2006).

How, then, does the history and character of Geelong's suburbs mirror or diverge from these national patterns?

Geelong's suburbanisation

Geelong echoes many of these elements which characterise Australia's urban history, in its initial location on Indigenous lands and the roles of surveyors and land speculators in their appropriation. Occupied by around 25 clans of the *Wadawurrung*, each with 60–120 members, the original population may have been between 1600 and 3250, extending and moving across the coastal, wetland, volcanic plain and woodland areas of the central Victorian dividing range (Clark 1990). Thus clans congregated at Mirraewuae Swamp – east of the Grampians/*Gariwerd* – to hunt and conduct ceremonies with others and gathered at Lake Bolac for the annual eel (*Anguilla australis*) migration. Local Aboriginal peoples also attended major meetings at what became Melbourne's cricket ground, collected food from the rivers and bay and conducted vital meetings at what is now the Geelong football ground and Eastern Gardens (Clark 1990; CoGG 2014). They also moved across the region to the wetlands and coastal bounty of the Bellarine Peninsula, and made use of the greenstone available from Barrabool Hill quarries.

The area was initially mapped by Surveyor General Charles Grimes in 1803, who reported it as unsuitable for 'settlement'. Subsequent contacts with Europeans in search of convict outposts – David Collins' ill-fated adventure at Sullivan's Cove near Sorrento – were associated with Aboriginal deaths but also the adoption of Indigenous names for the area – *Jillong* for the white cliffs abutting *Corayo* bay (DPCD 2018).

Having evacuated the lands of their actual and symbolic Indigenous presence, through the impact of two waves of smallpox in the late 1780s and 1820s, and the occupancy by squatters and their sheep, in 1838 surveyor Henry Smythe marked out a grid-like pattern of urban and suburban blocks onto a blank canvas. The rectilinear plan for Geelong was anchored by shipping on Corio Bay and the fresh water as well as industrial possibilities offered by the Barwon River. Smythe designed five main streets running roughly north to south, intersected by 10 running west to east, providing 36 blocks, each of which was divided into 20 allotments. As with all other towns laid out across Australia and the British empire, this grid pattern affirmed imperial power while also facilitating land disposal (Jacobs 1996; Kain and Baigent 1992). The northern part developed at the expense of the southern; it became the main commercial and port centre of the emerging township, especially as the shops were moved back from the narrow Corio Terrace to the area around Market Square and Ryrie Street (Wynd 1979). The location of what became the central business district has barely changed since.

At the same time as the town blocks were surveyed and offered for sale – although few were actually sold before the gold rush – much of the surrounding land was divided into suburban lots of 25 acres (10.12 ha). The intent was that they were to be used by small settlers, to provide opportunities for the much-favoured yeoman farmer who would both supply the town with foodstuffs and anchor the social order against the power and illegal occupancy of the squatters. However, these blocks were soon subdivided and resold, beginning the tradition of suburban division, sale and resale which has underpinned much of the Australian urban political economy (Berry 1994). For example, PW Welsh & Co. soon sold suburban lots which cost around £8 per acre, for £45 and £50 an acre (Wynd 1979).

Near but separate from the town, the *Geelong Advertiser* extolled buyers in January 1841 to consider New Town:

> *One of the most contiguous SUBURBS to the general business sites in this Township ... it commands a noble view of the Bay ... and the flocks and herds upon the adjacent hills* (quoted in Wynd 1979, p. 9).

With an influx of people quadrupling Geelong's population over the 1840s, suburban subdivision and promotion proceeded apace. As a result, the village of Ashby was established, with 20 buildings in 1841, followed by Irishtown, Kildare, South Geelong, Little Scotland, Chilwell and New Town (Seaton 1983). These suburbs were differentiated by ethnicity and class. Artisans, mechanics and labourers occupied mixed-use, flatter areas, in crowded small cottages close to industry and the river; the wealthy lived in a more exclusive residential area with large blocks and substantial houses commanding the views and heights of New Town hill. South of the Barwon River, John Highett established Highton and Alexander Thompson developed Belmont in 1850, hoping to capitalise on land sales through subdivisions (Wynd 1971).

What was offered here and elsewhere is the earliest form of the suburban dream in Australia – a combination of deeply rooted values derived from Britain that Davison (1994, 2016) describes as evangelicalism, sanitarianism, romanticism and class distinction. Evangelicalism is anchored in the mid 18th-century religious revival, where life was a constant battle between God and Satan and the 'home was the Christian's safest earthly stronghold', preferably located 'within sight of the city, yet safely beyond its corrupting influence' (Davison 2016, pp. 76–77). If the suburban home was the hallowed respite from the city, it was also the centre of a gendered family, with its privatised, domestic space and open streets 'coded feminine' (Davison 2016, p. 79). But the suburb also arose from concerns with cleanliness, purity, health and safety compared to the darkness, sin, disease, crowds and dirt of the city. It was these ideas which led to the valuing of elevated sites, open to sea breezes (Davison 2016). These ideas also informed the development of water reticulation and waste removal systems as the first acts of city-wide planning at the turn of the 20th century as well as Aboriginal exclusion from Australian towns – they were forcibly relocated to fringe reserves (Jackson *et al.* 2018).

But if suburbanisation was an escape from the grime and disorder of the city, it was also a positive quest for connection with nature. Romanticism is an aesthetic ideal that, in the suburb, offers 'the town-dweller a glimpse of the peace and beauty of the countryside' (Davison 2016, p. 83). On the ground, such ideas were manifested in the design of houses in their separate lots with small, private gardens. Finally, as already shown for the early patterns in Geelong, the suburb is associated with class distinction.

Many of the earliest suburban estates were meant only for the wealthy and then the middle classes, in part to escape physical proximity with the lower orders as well as the sordid city environment. The democratic impulse of this house form was enabled in Australia by 'high wages, cheap land, generous credit and supportive governments' which made the suburb available to the working classes (Davison 2016, p. 90). There is a further element in the economics of suburban housing: namely, the vital role the suburban home plays in the accumulation of wealth by many in Australia (Davison 1994). Integral to the dream is the quest for home ownership, a drive that has fuelled suburbanisation and social mobility for over a century in this country and one which is now under considerable threat, though perhaps less in Geelong than in the metropolitan centres of Melbourne and Sydney.

If the early pattern of Geelong comprised a town centre anchored by the port and river, surrounded by an array of class and ethnically differentiated suburbs, the gold rushes of the 1850s saw the population of the town treble – from 8300 in 1851 to 23 300 in 1857 – and the original core extended to the boundaries of the town reserve. Along the Barwon River from the 1860s to the 1890s noxious industries were established, especially wool-washing and woollen mills, tanneries and fellmongers, all using large amounts of water for processing and the river as a drain for wastes (Wynd 1971, 1979). These industries were serviced by workers who lived in the city core and increasingly in the suburbs surrounding it, in Chilwell and South Geelong.

If many of the earliest suburbs developed along the river and were contiguous to the township, there was an ongoing drive to subdivide in advance of need, with Geelong joining Melbourne and Sydney in a major speculative boom in the 1880s. Thus 'estates' at Richmond, Norwood, Shirley and Roseneath were never developed beyond lines on a map. The attempt to capitalise on the Geelong–Melbourne railway in the 1850s produced two pubs, a store, a station and a school but little else until the boom of the 1880s when North Shore and St Helens were established, offering retreats from industrial Geelong (Wynd 1979).

It was ironically to these northern suburbs, with their large areas of flat land and good rail as well as port connections, that industry was attracted early in the 20th century. Thus Geelong became a major industrial centre as the nation moved from being primarily an agricultural exporter to being a manufacturer – the early woollen and flour mills as well as tanneries, rope works and food processors were joined by the Commonwealth Woollen Mills (1915), Cresco Fertilisers (1922), the Ford Motor Co. (1925) and the Corio Distillery (1928), all of which moved away from the river and central areas to north Geelong. They were joined soon after by Pilkington Brothers (1937), International Harvester and the Grain Elevators Board terminal (1940) (Wynd 1979). Combined with the development of an electric tram system that connected north to west and east Geelong as well as across the river to Belmont, the influx of industry and migrants created whole new suburbs – including North Geelong, Corio and North Shore – while also boosting older ones.

After World War II such industrial and suburban development continued, with the expansion of car and truck making, and the establishment of the Shell Oil refinery and Alcoa Aluminium spurring the creation of new suburbs to Geelong's north and west

(Norlane, Bell Park, Bell Post Hill, Hamlyn Heights and Montpellier) as well as in the south (Highton, Grovedale) and east (Thomson, Whittington, Newcomb and Moolap). As with the earlier wave of suburban development, these suburbs were differentiated by class and ethnicity, but also by the quality, size and tenurial form of their housing. Bell Post Hill became strongly identified with the Balts and southern Europeans escaping the Displaced Persons' Camps after World War II. They initially built their own houses on modest affordable blocks of land while securing jobs in the industries of the northern suburbs (Gibson 1988). The long-term decline in Geelong's industrial base has impacted on the suburbs built to support it but it has also been associated with newer areas that have burgeoned to serve the growing numbers of knowledge and service workers. It is a complex story of economic, social and spatial change that is best told through the history of two suburban areas – Armstrong Creek and Corio–Norlane.

Current experiences: Corio–Norlane and Armstrong Creek

Corio has its origins in the idealism and land speculation that characterised Geelong in the 1920s. This was the time when major industries were moving into the city, particularly the Ford Motor Co. If there was an emerging market for land, there was also a town planning movement that was systematising the best ways to design and build – as a 'Garden Suburb'. The Melbourne Subdivisions Co. created the up to date Corio Garden Suburb, publishing a 46-page booklet extolling its virtues to prospective buyers. This booklet shows the affirmation of the 1920s suburban dream. Its opening pitch assures prospective buyers:

> *It was recognised that to have a successful and a complete Garden Suburb, it was essential to secure about 600 acres* [242 ha] *of land within 6 to 7 miles* [9.6–11.2 km] *of an established city. For a suburb to be successful, the city must be progressive, and must be an educational, social and industrial centre, with increasing manufacturing industries, thus providing justification for increased expansion* (MSC 1924, p. 3).

The booklet extols Geelong as the 'metropolis of the Western District', the 'Bradford of Australia', the 'Naples of the South', the 'Birmingham of Australia' and the third city of the Commonwealth – labels and claims all indicative of its economic base and the buoyant optimism of the land speculator (MSC 1924). It also promoted the existence of the Melbourne–Geelong railway station and roadway, beach and plantation frontages and the growth of the city towards the metropolis along with affordable land, to attract investors:

> *These allotments are available on small deposit and on very small easy terms, and if purchased now and paid for in monthly instalments over the next five years, the cost of them will not be felt, but the increase in their value at the end of that time will be enormous* (MSC 1924, p. 11).

Like so many inflated claims for new suburban areas, less than a quarter of the 2555 allotments were sold by 1931 and only six houses had been built. It took until 1979 for some parts to be developed as the suburb of Rosewall (Wynd 1979).

By then Corio was joined by its neighbour Norlane as an area less of salubrious suburban dreams, and more as an area registering Geelong's most recent economic history. Corio had been conceived as a Garden Suburb in the 1920s; Norlane was planned and allocated to the new migrants recruited from England and Europe by Commonwealth and state

Fig. 12.1. Some of the original Norlane houses. Photo: Louise Johnson.

governments to serve the expanding post-war industries. The pre- and post-war housing shortage, along with the optimism for building a different society after World War II, had led to the formation of various State Housing Commissions. Funded federally, the Victorian Housing Commission constructed close to 2500 homes for low-income renters and migrant workers coming into Geelong for the Ford, International Harvester and related factories. Of these houses, 1200 were imported as prefabricated units from the Netherlands and France, with most modest weatherboard dwellings on large blocks of 600–800 m^2. The peak in building activity occurred from 1951 to 1957 and was associated with the creation of local strip shopping centres, such as Labuan Square, opened in 1954 to serve a dynamic, prosperous working-class suburb (Figs 12.1 and 12.2). At the time, there was full employment and 30% of Geelong's workforce was employed in manufacturing. In Corio, close to 40% of the workforce was in local industry (ABS 1961). In 1986 Norlane still had 44% of its workforce in manufacturing, though this had fallen to 25% by 2006, as part of a massive increase in unemployment (DPCD 2010).

The 1950s and 1960s were boom times, but the early 1970s heralded the beginning of a major national shift away from manufacturing. In Geelong, first the textile industry and then the truck and car industries came under sustained pressure with the dismantling of the tariff wall that had long protected them. In addition, the shift towards public housing being a welfare and residual tenure rather than one available to low-income workers stigmatised the area and heightened its association with poverty. A survey of residents in 2000 recorded that this area was perceived by those outside as the 'Bronx of Geelong', the ghetto. As one respondent noted, 'nobody wants to be here and we haven't got no choice. Like in LA, they've got their trailer park trash and in Geelong you've got Corio and Norlane trash' (Warr 2005, p. 299).

The northern suburbs of Geelong have therefore become associated with industry and with social disadvantage. While currently Geelong has 11% of its workforce engaged in

Fig. 12.2. Labuan Square shopping centre today. Photo: Louise Johnson.

manufacturing, in Corio and Norlane close to 20% work in industry (Corio 18.3%, Norlane 19.4%). The area is further distinguished by its quantity of public housing (22% in Norlane, 10.6% in Corio) and low household incomes. In Corio, 30% of the adult population earn below AU$600 per week and in Norlane 42%, compared to the City of Greater Geelong rate of 22.8% (DOTE 2015). In what used to be the industrial powerhouse of the city, there are now high levels of unemployment, limited educational qualifications and a concentration of low-income households. This is not just the etching of class onto the city, but a class divide exacerbated by economic restructuring and social stigmatisation, despite the many efforts to improve the social and urban fabric of the area with initiatives such as Neighbourhood Renewal and New Norlane.

On the other side of the economic, social and suburban ledger, there is a new economy emerging in Geelong. While there was a fall from 2006 to 2011 of over 14% in manufacturing employment in the Greater Geelong area (from 11 890 to 10 207), over the same time period there were increases in employment in financial and insurance services of 47.5% (from 1589 to 2339), in health and social assistance of 25.5% (from 10 025 to 12 578), in professional, scientific and technical services of 12.4%, construction of 11.4%, public administration and safety of 10.5% and accommodation and food services (8.7%) (Enterprise Geelong 2016). Significantly, the workers in these industries, especially those who are highly educated, tend to live in the expanding southern suburbs of the city. Those working in what is the 'knowledge economy' –information and media, telecommunications, financial and insurance services as well as professional, scientific and technical services – have a residential pattern that is the inverse of those in manufacturing. These service workers inhabit an arc from the inner core of the city to Highton and Wandana Heights (DPCD 2010). It is in these areas, and the central business district, that growth has been concentrated, with the areas of Grovedale, Highton and St Albans Park showing the most recent dramatic increases in population (DPCD 2010).

Alongside the growth in the service and knowledge economy has been growth in areas such as Armstrong Creek, which represents the newest suburban trends in the city. Directing the growth of the city away from Melbourne was a planning decision made in the 1990s, and the location of 60 000 people south along the Waurn Ponds valley towards Torquay and the coast is stimulating a series of Master Planned Estates to emerge from the rolling farmland. One of these – Warralily, named after the Aboriginal word for water lily – is set to accommodate 14 000 residents. The Aboriginal past has been recognised in a series of street and place names and, in consultation with the *Wadawurrung*, the restoration of Armstrong Creek (Pollock 2016). As the Heritage Study for the area affirmed, there are nine registered Aboriginal sites, six stone artefact scatters and three scarred trees. The study also noted that *Wadawurrung* people were forcibly removed from the city of Geelong from the 1850s at sunset, the downside of the sanitarian impulse which associated the Indigenous population with health and moral threats. One of the places they retreated to was a marshy block of land that a local landowner, John Stewart, carved from his property 'where the dispossessed could camp unmolested' (Turnbull 2009). This area is now one of a series of parks along the waterways within the development, ensuring both an abundance of open space and some recognition of its Indigenous past.

Warralily echoes the ongoing values that drive suburbanisation in contemporary Australia – glossy brochures extol the high level of coordinated master planning, environmental sustainability (via a density of 15 houses per hectare, energy-efficient homes, recycled water to all homes, lots of bike and walking tracks, fitness stations and gym), the orderly roll-out of retail, education and child-care services and accessibility to the city, coast and countryside. Significantly, it also has a full-time community development officer, employed since 2001 when there were only 300 residents, whose job it is to oversee the local drop-in centres, organise social events and produce newsletters. Actively building a 'community' is part of creating the 'Warralily way of life' for the downsizers, new families and the few singles who are moving into the area, primarily from the adjacent suburbs of Belmont and Grovedale but also from other parts of Geelong, Melbourne and Victoria (Pollock 2016). This will be done through activities and through the design of 'Warralily villages' or neighbourhood centres 'to suit the lifestyle you want'.

Virtually all of the homes are detached – though there are a few medium-density units close to the shopping centres – and the assumption is that all will be owner-occupied. There is no social or subsidised housing and, while there is much talk about diversity, this is primarily in terms of house and block sizes rather than income, ethnicity or religion. There is much emphasis on the possibility of a rail connection to Geelong – with an easement set aside and currently used as parkland – but so far there are only private buses along the main outer boundary roads. As with most suburbs built since the 1960s, the assumption is that car ownership and use are ubiquitous. An assessment in 2016 notes that there has been a delay in the timely delivery of facilities and services to the area, limited employment options on site, and a lack of housing diversity with small lot sizes and large houses with limited open space, despite many positive elements associated with its master planning (Montgomery 2016).

While the Warralily development is a handsome and considered one, it does not deliver on all promised fronts and it echoes the wider socio-economic polarisation that is occurring in Geelong. As the figures shown in Table 12.1 make clear, the affluence and suburban gloss of one part of the city occur while another part is suffering. The suburban dream may be alive for all involved, but only some can really access all of its positive dimensions.

Table 12.1. Indicators of economic and social structure: Corio–Norlane, Waurn Ponds, Greater Geelong and Melbourne (2011)

Indicators	Corio–Norlane	Waurn Ponds	Greater Geelong	Melbourne
% employed in manufacturing	17.7	9.8	10.9	10.8
% labourers	20.7	7.1	10.3	8.0
% managers and professionals	13.5	32.7	21.2	36.4
% public housing	10 Corio 22 Norlane	1.2	3	3.4
Households on high income (>$2500 pw)	4.9 Corio 2.79 Norlane	22.1	12.2	19.4
Households on low income (<$600 pw)	30.1 Corio 42.1 Norlane	8.8	24.8	19.2
% unemployed	10.9	4.7	5.3	5.5

Source: DOTE (2015).

Table 12.2. SEIFA Index 2001 and 2011 (no disadvantage = 1000)

Area	2001	2011
Corio (ABS state suburb)	864.86	856
Norlane–North Shore (ABS state suburb)	770.01	754.4
Norlane (ABS state suburb)	759	745.4

Source: ABS (2001, 2011).

These figures do not just confirm the class structure of the city, but illustrate its social polarisation. There are two major measures of socio-spatial divergence in Australia: the Australian Bureau of Statistics SEIFA index of Social Disadvantage and that calculated by Tony Vinson (Vinson 2007; DOTE 2015) and others for the Brotherhood of St Laurence. The results for Corio–Norlane confirm the worsening situation of this area as it moved from Band 3 to Band 1 in the *Dropping off the Edge* report over the period 2007–15 (DOTE 2015). It is currently ranked as one of the most disadvantaged localities in Australia. The bands combine a host of indicators of social dislocation and disadvantage: unemployment levels, criminal convictions, disability rates, child maltreatment, low family income, rental assistance, family violence and psychiatric hospital admissions. The SEIFA Index (detailed in Table 12.2) also shows that Corio and Norlane, and proximate areas in the northern part of Geelong, became more disadvantaged from 2001 to 2011 as their scores moved further and further away from 1000 or the Australian average.

Conclusion

Geelong suburbs conform in many ways to others across Australia, though their particular character also echoes the economic and social history of the city. Geelong and its suburban extensions are built on *Wadawurrung* lands and this Indigenous presence is still registered in them if consultation and care are taken to access it. Good planning practice in existing or new suburban developments should indeed acknowledge and work with this past as well as with the *Wadawurrung* peoples on how best to recognise this past and present.

The suburbs of Geelong were both contiguous and separate from the city, connected by walking or tramways, the fantasies of land speculators as well as the employment needs of residents. The suburbs have always been differentiated by class and ethnicity, and their placement, idealisation and reality for many has been derived from the values of evangelicalism, sanitarianism, romanticism and class distinction. Most recently Geelong's suburbs have registered contemporary developments in the social economy of the region, as manufacturing has declined and services have expanded, with their planning increasingly informed by commitments to sustainability and community through master planning. The result in this regional city has been far less diversity than in the metropolis, in terms of ethnicity but also tenure, as Geelong continues to value the separate house over medium and high density, and home ownership over other tenures, living out where possible the ongoing suburban dream.

References

ABS (1961) *Census of the Commonwealth of Australia*. Australian Bureau of Statistics, Canberra.

ABS (1966) *Census of the Commonwealth of Australia*. Australian Bureau of Statistics, Canberra.

ABS (2001) *2001 Census Data*. Australian Bureau of Statistics, Canberra. http://www.abs.gov.au/websitedbs/censushome.nsf/home/historicaldata2001?opendocument (accessed 1 September 2018).

ABS (2011) *Census of the Commonwealth of Australia*. Australian Bureau of Statistics, Canberra.

ABS (2014) *Australian Social Trends*. Cat. no. 4102.0. Australian Bureau of Statistics, Canberra.

ABS (2016) *2016 Census*. Australian Bureau of Statistics, Canberra. http://www.abs.gov.au/websitedbs/censushome.nsf/home/2016 (accessed 1 September 2018).

Allon F (2008) *Renovation Nation: Our Obsession with Home*. University of New South Wales Press, Sydney.

Berry M (1994) The political economy of Australian cities. In LC Johnson (Ed.) *Suburban Dreaming: An Interdisciplinary Approach to Australian Cities* (pp. 33–50). Deakin University Press, Geelong.

Brook B (2018) *As our Cities Become Denser, the Great Australian Backyard is Paying the Price*. News.com.au. https://www.news.com.au/lifestyle/home/outdoors/as-our-cities-become-denser-the-great-australian-backyard-is-paying-the-price/news-story/a6ea0576f9781a0869d811d3048156be (accessed 1 October 2018).

Burnley I, Murphy P, Fagan R (1997) *Immigration and Australian Cities*. Federation Press, Sydney.

Clark I (1990) *Aboriginal Languages and Clans: An Historical Atlas of Western and Central Victoria*. Department of Geography and Environmental Science, Monash University, Melbourne.

CoGG (2014) *Karreenga Aboriginal Action Plan 2014–2017*. City of Greater Geelong, Geelong.

Davison G (1994) The past and future of the Australian suburb. In LC Johnson (Ed.) *Suburban Dreaming: An Interdisciplinary Approach to Australian Cities* (pp. 99–113). Deakin University Press, Geelong.

Davison G (2016) *City Dreamers: The Urban Imagination in Australia*. New South Publishing, Sydney.

DOTE (2015) *Most Disadvantaged Postcodes in Victoria: Comparison with Previous Years*. Fact Sheet. Dropping off the Edge, Melbourne.

DPCD (2010) *Victoria's Regional Centres: A Generation of Change – Geelong*. Department of Planning and Community Development, Melbourne.

DPCD (2018) *Greater Geelong Planning Scheme*. Department of Planning and Community Development, Melbourne.

Enterprise Geelong (2017) *Enterprise Geelong Review Edition 3 – 2017*. City of Greater Geelong, Geelong. https://www.geelongaustralia.com.au/common/Public/Documents/8d535b269a70f14-finaleg3review-2017.pdf (accessed 1 September 2018).

Fishman R (1987) *Bourgeois Utopias: The Rise and Fall of Suburbia*. Basic Books, New York.

Gibson N (1988) *A Fresh Start in a New Land.* Shire of Corio, Geelong.

Gleeson B (2006) *Australian Heartlands: Making Space for Hope in the Suburbs.* Allen & Unwin, Sydney.

Greig A (1995) *The Stuff Dreams are made of: Housing Provision in Australia 1945–1960.* Melbourne University Press, Melbourne.

Groenhart L, Burke T (2014) *Thirty-five Years of Public Housing Supply and Consumption, 1981–2011.* AHURI Final Report no. 231. Australian Housing and Urban Research Institute, Melbourne.

Harris C, Lehrer G (Eds) (2017) *The Suburban Land Question.* University of Toronto Press, Toronto.

Jackson S, Porter L, Johnson LC (2018) *Planning in Indigenous Australia: From Imperial Foundations to Postcolonial Futures.* Taylor & Francis, London.

Jacobs J (1996) *Edge of Empire: Postcolonialism and the City.* Routledge, London.

Johnson LC (2006) Style wars: revolution in the suburbs? *Australian Geographer* **37**(2), 259–277. doi:10.1080/00049180600701935

Kain RJP, Baigent E (1992) *The Cadastral Map in the Service of the State: A History of Property Mapping.* University of Chicago Press, Chicago.

Kemeny J (1981) *The Myth of Home Ownership.* Routledge & Kegan Paul, London.

McManus R, Ethington PJ (2007) Suburbs in transition: new approaches to suburban history. *Urban History* **34**(2), 317–337. doi:10.1017/S096392680700466X

Montgomery D (2016) The price of growth. *Geelong Business News.*

MSC (1924) *'Corio': Garden Suburb of Geelong.* Melbourne Subdivision Co., Melbourne.

Pollock K (2016) Interview with the Warralily Community Development Officer, 12 December.

Seaton G (1983) *Model Borough: An Introduction to Newtown and Chilwell.* City of Newtown, Geelong.

Silverstone R (1997) Introduction. In R Silverstone (Ed.) *Visions of Suburbia* (pp. 1–25). Routledge, London.

Troy P (2012) *Accommodating Australians: Commonwealth Government Involvement in Housing.* Federation Press, Sydney.

Turnbull J (2009) *Armstrong Creek: Desktop Cultural Heritage Assessment.* Report prepared for Villawood Properties. Ochre Imprints, Melbourne.

UPR (2010) Special issue on master planned estates. *Urban Policy and Research.*

Vinson T, Rawsthorne M, Cooper B (2007) *Dropping off the Edge: The Distribution of Disadvantage in Australia.* Jesuit Social Services; Catholic Social Services Australia, Melbourne.

Warr D (2005) Social networks in a 'discredited' neighbourhood. *Journal of Sociology* **41**(3), 285–308. doi:10.1177/1440783305057081

Wynd I (1971) *Geelong: The Pivot.* Cypress Books, Melbourne.

Wynd I (1979) *Case Study of Geelong. Chapter 15. Economic Geography and Australian Regionalism: Part 2.* Deakin University Open Campus Program, Geelong.

13

Greater Geelong's planning future to 2050: determining spatial outcomes through agricultural land planning

Beau B. Beza, Josh Zeunert and Murray Herron

Most planners are far from being advocates, and are more typically soothsayers of informed complacency ... reiterating the conventional wisdom.

– Downton (2009, p. 481).

Introduction

Consistent with upward population growth trends in major urban areas across Australia, at the time of writing this chapter and with increases in population, the City of Greater Geelong (CoGG) is the fastest-growing regional centre in Victoria (ABS 2011). To provide for the city's substantially growing and projected population, low-density detached housing and associated developments are being built on agricultural land, in particular, high-quality productive land. This is occurring despite planning measures put in place to protect CoGG's further expansion into agricultural land. The state of Victoria's planning framework oversees development of land across the state, and is responsible for regulation of land use in the CoGG. This arrangement provides municipalities, like the CoGG, with semi-autonomous planning decision-making by using municipal Planning Schemes.

A key feature of the CoGG, and what distinguishes it from other urban settings throughout Victoria, is its Municipal Strategic Statement (CoGG 2016b) and other related context and visioning documents such as the *Northern and Western Geelong Growth Areas Project* (CoGG 2016a). These 'development-influencing' documents have important roles to play in guiding planning and development-related decision-making in regards to the duality of development/protection of land throughout the CoGG. Each document can be as prescriptive or ill-defined as its decision-makers desire. Unfortunately, documents such as the above demonstrate key shortcomings, made particularly apparent through the numerous and ongoing low-density sprawl outcomes consuming quality agricultural land in the municipality.

This phenomenon is in part due to the documents':

- use of vague language to effect or manage change
- lack of spatial specificity at appropriate scales.

Perhaps most revealing about these documents is that they fail to articulate a clear set of values within the development/conservation hierarchy relating to prioritising land use types. Consequently, and consistent with most other growing urban regions in Australian, market-lobbied and -led greenfield low-density detached housing development is the dominant typology accommodating population growth (Soule 2006; Weller 2009; Spencer *et al.* 2015; CoGG 2016b). Thus, rather than effecting an ecologically sustainable long-term outcome, they demonstrate a market-led approach to governance and planning; which is typical of western societies under neoliberal regimes since the late 1970s (Hackworth 2007; Leitner *et al.* 2007; Allmendinger 2016). In contrast, Downton's (2009) application of Richard Register's approach of 'Shadow Planning' for Adelaide, for example, utilises broad timescales (of hundreds of years) to allow much greater consideration of crucial factors in land use planning, namely, the intergenerational sustainability of natural resource planning. The authors of this chapter are unaware of any Australian municipalities conducting and implementing visioning at this level of foresight – accordingly, 'business as usual' and 'status quo' outcomes are the norm (Downton 2009, p. 481).

This chapter discusses the planning and development of land across the CoGG as a means to identify and highlight three pillars of land use planning to guide decision-making to 2050 and beyond. These pillars include:

1. dedicating land for the sole production and use of agriculture to facilitate local food systems in perpetuity
2. dedicating land for the conservation and restoration of ecosystems (which can be used to meet and exceed the UN Convention on Biological Diversity targets, or, for example, to pursue Wilson's vision of half of the Earth being designated a human-free natural reserve to preserve biodiversity (Wilson 2016)
3. planning for human factors and activities such as demographic, infrastructure (re)development and dynamic social elements.

Currently, the third factor, driven by population growth, dominates urban planning across Australia and the CoGG. This results in relegation of the other two pillars to piecemeal planning, incremental loss and reactive rather than proactive governance and planning. In terms of this chapter, the focus is on the first pillar because, as noted by Carey and James (2018), while food needs are increasing with population growth, urban sprawl is decreasing the capacity of cities' peri-urban fringes to meet these needs.

Kunstler (2006, p. 239) articulates the food planning challenge more bluntly:

> *To put it simply,* [we] *have been eating oil and natural gas for the past century, at an ever-accelerating pace. Without the massive 'inputs' of cheap gasoline and diesel fuel for machines, irrigation, and trucking, or petroleum-based herbicides and pesticides, or fertilizers made out of natural gas, [we] will be compelled to radically reorganize the way food is produced, or starve.*

Geelong

Geelong is the second largest city in Victoria (Infrastructure Victoria 2017) and is ~75 km south-west of Melbourne. Geelong (i.e. Greater Geelong, when referred to by the Australian Bureau of Statistics) covers an area of 1248 km^2 and has a population density of 187.04

residents per km^2 (ABS 2016). Australian cities are some of the lowest-density in the world (Spencer *et al.* 2015; Downton 2009) and the CoGG is no exception. Its regional setting is primarily urban with the majority of the population living in the Geelong urban area. Other smaller settlements within this local government area include Anakie, Balliang, Barwon Heads, Batesford, Ceres, Clifton Springs, Drysdale, Lara, Ocean Grove, Portarlington and St Leonards. The council governing the city formed in 1993 from the amalgamation of the Bellarine Rural City Council, Corio Shire Council, Geelong City Council, Geelong West City Council, Newtown City Council and South Barwon City Council along with parts of the Shire of Barrabool (*History of Council,* CoGG 2017). This chapter focuses on the CoGG rather than the overall region, to enable quantification consistent with municipal and ABS data.

Currently, the CoGG uses numerous distinct zones and overlays to regulate land use and development within its area boundaries. The largest zone by area is the Farming Zone (T2 in Geelong) which represents ~59.4% of the total area of the Geelong area (DPI 2013). CoGG's (2007) *Rural Land Use Strategy* describes this zone as characterised by 'agricultural industries [including]: broad acre cropping and grazing; vineyards; aquaculture and fishing; local food processors of poultry, rabbits and pigs (sourced from neighbouring municipalities); olive oil production and processing; cut flowers and turf cultivation' (CoGG 2007, p. 110). The *Rural Land Use Strategy* (CoGG 2007) shows a productive soil land system quality assessment for the CoGG, and part of this area is examined later in this chapter in relation to urban expansion (see Plate 13.1).

The visioning associated with the CoGG and that which complements the above planning framework is reflected in documents such as the *Geelong Region Plan: Sustainable Growth Strategy* (G21 GRA 2007) and *Plan Melbourne* (a document guiding Melbourne's growth; which refers to Geelong as a regional growth city in the state of Victoria) (Victoria 2017). Development of the *Geelong Region Plan* (now superseded) was created through 'the alliance of ... independent organisations with a shared vision for the Geelong region' and 'delivers a strategic framework upon which the [groups' stated] pillars and other stakeholders can prioritise and measure the success of their projects in moving towards the agreed vision for the Geelong region' (G21 GRA 2007, p. 5). The crux of the document's argument is the presentation of two scenarios based on substantial levels of population growth from the recent estimated population of 234 999 to 320 791 by 2036 (idcommunity 2016). The first scenario identifies a 'business as usual' approach to urban (re)development and the second focuses on 'partnerships with all levels of government, business and community' (G21 GRA 2007, p. 13) to achieve a desirable and 'sustainable' outcome.

To effect this desirable second outcome, the document employs a language of sustainability (in all its contemporary uses), collaboration and implementation. These are supported by several policy statements and objectives, and referenced to a range of short- and long-term timeframes. Achievement of both spatial clarity and (scientific) data, however, to guide and determine land use values and priorities and subsequent decision-making is largely absent in the G21 documents. This is despite numerous documents, datasets and other forms of deducible knowledge generation being referenced and listed in the evidence section of the plan (G21 GRA 2007). Its major reference to data appears under Policy 5.3 ('Use of data to plan and measure') where reference to greenhouse emissions, demographics and forecast systems and GIS datasets is presented (G21 GRA 2007); development-related targets are stated but not deducible or conveyed in a manner that demonstrates spatial land use scenarios and outcomes.

Importance of peri-urban and agricultural land protection

Contemporary scholarly planning discourses throughout the world are focusing on both the preservation of agricultural land in close proximity to urban areas and the nurturing of local agricultural industries (i.e. agricultural land planning). While this contemporary planning discourse gains momentum (APA 2007; Raja *et al.* 2008; Cabannes and Dubbeling 2000; Dubbeling *et al.* 2010; Hodgson 2010; Ilieva 2016; GLA 2010; Roggema 2016; Viljoen and Wiskerke 2012; Morgan 2013; Zeeuw and Drechsel 2015; Zeunert and Waterman 2018) it is not adequately embedded in day-to-day planning practice in Australia and continues to be a crucial and undervalued concern (Jones and Beza 2018; Sheridan *et al.* 2015; Sheridan *et al.* 2016). In fact, post-World War II planning in Australia has largely deferred to economic concerns (e.g. market-led responses to change) and logistical concerns (e.g. road transport) rather than prioritising land use planning elements essential for sustainability, such as ecosystem resilience and regeneration and production of food in and within close proximity to urban areas. While the CoGG *Rural Land Use Strategy* lists the state policy of 'Protect[ing] productive agricultural land' and ensuring that 'productive agricultural land is not lost due to permanent changes to land use' (CoGG 2007, p. 105), this chapter will demonstrate that not only has this policy direction been flouted, but it is continuing to be disregarded in growth area planning (Plate 13.1).

Visioning in Australia

Despite much of Australia being understood to be infertile and arid, with skeletal and nutrient-deficient soils, the expansive mainland landmass can lull inhabitants into a false sense of abundance in regards to productive agriculture land to provide for its growing population. Yet only 6% of land area is arable (World Bank 2016) and Australia is particularly vulnerable to increasing agricultural challenges presented by climate change (Flannery 2005).

As referred to previously, state governments in Australia hold primary responsibility for land use planning. While most planning systems in Australia zone for agriculture as a regulated land use, various loopholes and intense pressure from developers frequently result in rezoning these parcels to non-agricultural uses (primarily low-density detached housing). Similar outcomes also routinely result in loss of land and remnant native ecosystems with ecological value (Weller 2009). Lack of effective economic mechanisms and food-planning measures routinely results in peri-urban agricultural land laying fallow, which can result in the subsequent perception that this type of land is of low productive quality. Yet such potential development-related land, lying in wait, should not be confused with its productive capacity. Furthermore, much agricultural land, having already been cleared of substantial vegetation and rocks, allows for an easy and cost-effective transition to building development and housing; this helps to underscore its appeal to greenfield developers. Loss of such land is almost always permanent, with soil becoming degraded/eroded/contaminated and/or its recovery too expensive to enable agricultural reinstatement.

Australian cities, like many worldwide, have typically been built on productive agricultural land such as alluvial river valleys, plains and fertile pockets, especially near milder coastal fringes; Geelong included (CoGG 2007). The productive land capacity of Australia's urban regions is evidenced in a study by Houston (2005) stating that in the mid 1990s, the peri-urban regions of Australia's five largest mainland states produced almost 25% of the nation's gross agricultural value from less than 3% of the agricultural land. Despite this notable financial return, peri-urban agriculture cannot financially compete with urban land development for housing. Agricultural land preservation requires long-term vision and understanding from decision-makers to recognise its crucial role for ongoing local food supply security. It also requires strong ongoing political will to uphold such legislative measures in perpetuity against persistent pressure for urban fringe development (Buxton and Carey 2014).

Successful examples in Australia of long-term and large-scale agricultural land preservation are rare. In contrast, some global regions have more successfully implemented agricultural land protection and the ongoing measures required to make it work. British Columbia in Canada, for example, has had an agricultural land freeze in place since 1972. Agricultural land protection exists throughout the European Union and in New Zealand, and the *Resources Management Act* enforces agricultural land protection. 'Green belts' in England have stopped continuous sprawl of around 25 cities since 1935, and Oregon in the US has largely upheld urban growth boundaries since their implementation in 1973 (Zeunert 2017).

Food planning in the CoGG

The CoGG has recently shown positive signs in taking food planning more seriously through commissioning a *Food Policy Discussion Paper* (CoGG 2014), adopting a Food Policy (CoGG 2015) and commissioning a 'food hub' feasibility study (Jackson 2016). The Food Policy makes specific reference to land use planning, listing its first aim as:

> ***Land Use Planning*** *– using measures in the Local Planning Policy Framework to protect farmland and support community food production, and advocating to State Government for additional measures to assist farmland protection and increase water security; advocating for the establishment of intensive farming options (land and water based) that are identified as appropriate and sustainable for the region.*

Yet as noted in the *Food Policy Discussion Paper* (CoGG 2014), referring to Cust *et al.* (2013, p. 9), 'a significant amount of farmland has been lost in the region over the last 20 years, and there is pressure on farmland from continued population growth'. Plate 13.1 demonstrates that CoGG's historic, current and future land use planning has not considered preservation of productive land, which is described as 'high agricultural quality' or 'Class 2' (CoGG 2007, p. 13). The *Geelong Public Health and Wellbeing Plan 2013–17*, for example, notes that in the last 20 years the number of outdoor fruit and vegetable farmers decreased by 25%, with sheep and cattle farmers decreasing by 33%; the latter are identified in the G21 GRA documents as the main agricultural activities in the area (G21 GRA 2012). Trestrail *et al.* (2013) notes that many farms in the region are small and have a turnover of less than AU$50 000 per annum – low farming incomes makes selling agricultural land to developers an appealing proposition.

Current key areas in the CoGG accommodating population growth are numerous and occur in addition to the substantial areas of the Armstrong Creek Growth Area (the largest contiguous growth area in Victoria: CoGG 2018) and the earmarked Northern and Western Growth Areas Project (CoGG 2016a). These include but are not limited to Bannockburn, Clifton Springs, Colac, Lara, Leopold, Ocean Grove and Drysdale. All of these population growth areas provide low-density (around 12–15 dwellings per hectare) and almost exclusively detached housing, and many of the growth zones are located on former agricultural land of Class 2 and Class 3 quality ('high' and 'moderate', respectively (CoGG

2007) (Plate 13.1). This is despite the G21 alliance aspiration of 'reducing pressures on our agricultural areas and natural assets' (G21 GRA 2013, p. iv). In essence, the CoGG is building on its food basket, which will make it challenging to achieve the aims stated in the *Food Policy Discussion Paper*.

Unfortunately, the land use/development-related aspirations of local governments can be bypassed through a loophole in Victoria whereby the state Planning Minister can overrule local planning frameworks. An example of this in the CoGG was the Victorian state government rezoning of Lovely Banks (Northern Growth Area) on 6 November 2014 (Ministerial Amendment C322). This amendment resulted in rezoning the land from a farming zone (Class 3; see CoGG 2007) to an urban growth zone.

To better consider the agricultural and ecological land use planning pillars, the CoGG's population growth should occur through infill development (accommodating population growth within existing urban boundaries). If greenfield areas are developed, land of both low agricultural productivity (Class 4 and Class 5; CoGG 2007) and of low conservation and ecological restoration value should be targeted. As visible in Plate 13.1, two of the CoGG's substantial recent and planned growth areas occur on high-value (Class 2) agricultural land, namely, much of Waurn Ponds and part of Armstrong Creek (Urban Growth 2017). Infill development is an aspiration of G21 councils with 'data suggest[ing] that ~40% of housing activity is within established urban areas' (G21 GRA 2012, p. 33). While some infill is occurring in Geelong, the CoGG is currently not achieving this percentage of infill and has to date demonstrated minimal achievement of medium-density development compared with detached low-density housing. As noted by Bolleter (2016), much infill in Australia, especially outside of Sydney and Melbourne's inner residential towers, is initiated by private landowner subdivision (referred to as background infill), not council-led. Infill development can be encouraged through adhering to existing urban growth boundaries, stricter greenfield planning requirements and economic incentives. It is also important that infill and medium-density typologies incorporate a variety of living options as well as affordable and social housing. The CoGG's *Central Geelong Action Plan* (CoGG 2013) has recognised some of these factors and is attempting to address such challenges. Meanwhile, the irreversible loss of agricultural land continues.

Ecological and food planning

In addition to the loss of agricultural land, if land planning is primarily driven by population growth instead of being framed around a robust and broad- scale ecological land vision, the ecology of a region can be lost over time. Views about incorporating ecology as a key dimension to planning are demonstrated by Kilbane (2013), who presents the concept of a 'national green network for Australia', an ecological network spanning the continent. Such big-picture land planning that combines protection of remnant ecologies and interconnecting vegetated corridors – created through ecological restoration and strengthening existing linkages – could ideally be conducted across the CoGG and the G21 region. Accordingly, using a combination of ecological and agricultural factors at regional scales (including hydrological) would ideally determine where population growth may or may not occur.

In Australia, perhaps due to the dominant industrial methods of broad-scale and monoagriculture (Zeunert 2018a), ecology and agriculture are usually treated as separate and conflicting land uses. This is evident in the CoGG, where farming practices in the region have been linked with loss of biodiversity and damage to terrestrial and aquatic systems

(CoGG 2006). As noted in the *Food Policy Discussion Paper* (CoGG 2014), through encouraging more sustainable and ecologically sensitive forms of food production, such as regenerative agriculture (Zeunert 2018a), planning, policy and incentivisation can encourage rather than deter biodiversity and reduce conflict with nearby residential zones.

The CoGG can even look to older examples of late 19th- and 20th-century visions which recognised the need to include agriculture as a significant component in land use planning. Some examples include Howard's Garden City, Le Corbusier's Radiant Farm, Migge's 'everybody self-sufficient' models (Haney 2010; Zeunert 2017) and Wright's Broad Acre City (Zeunert 2017). These can be used as reference points from which to draw inspiration – in essence, there is no need to reinvent the wheel! These planning visions, however, were not assisted through the mid 20th-century's misleadingly named 'green revolution' in agriculture (Zeunert 2018a) which led the trend (and was assisted by western 20th-century planning) of decoupling agriculture from urban and peri-urban areas (Carey and James 2018; Zeunert 2018b). Many urban citizens, for example, do not know of or understand their fundamental reliance on the agriculture land parcel, and its expansive scale. Yet, as noted by Kunstler (2006) and Zeunert (2018a), this separation of agriculture from urban populations is likely to be temporary in the medium and long term, due to exhaustion of the fossil fuels and finite primary resources essential for green revolution agricultural production, mechanisation and transportation.

Greater Geelong in 2050

With the historical legacy of 'business-as-usual' low-density detached housing and unprecedented levels of population growth, what will the Greater Geelong region look like in 2050? While the G21 GRA (2007) document presents desirable and undesirable cartoon-like scenarios to address this question, a long-term vision that incorporates the pillars of agriculture and ecology, before that of population growth, is key to increasing the CoGG's sustainability and resilience to 2050 and beyond. Such an overarching vision needs to be executed at increasingly detailed spatial scales (e.g. 1:50 000), be informed and accompanied by keen use of relevant and tangible data, and communicated through clear and direct language to guide development.

To accommodate future residential development in the CoGG and, importantly, to help achieve the city's current and increased infill targets, conservation of Class 2 and arguably Class 3 agricultural land should be a primary concern. This can only be achieved through ceasing any further development on Class 2 land and planning for protection of Class 3 land (Plate 13.1; CoGG 2007). Such an approach refers back to visions for cities previously mentioned in this chapter (Howard, Le Corbusier and Migge along with Lloyd-Wright, Weller and Kilbane) that place value on key components that cannot be resurrected once lost (e.g. agricultural land, ecological conservation and restoration). The alternative for the CoGG is that development continues to be favoured over highly productive agricultural land and the city remains a 'status quo' setting in Australia.

References

ABS (2011) Australian Bureau of Statistics, Canberra. http://www.abs.gov.au/ausstats/abs@.nsf/Products/3218.0~2010-11~Main+Features~Victoria?OpenDocument (accessed on 1 July 2018).

ABS (2016) *General Community Profile, Australia*. Australian Bureau of Statistics, Canberra. http://www.censusdata.abs.gov.au/census_services/getproduct/census/2016/communityprofile/LGA22750?opendocument (accessed on 1 July 2018)

Allmendinger P (2016) *Neoliberal Spatial Governance*. Routledge, New York.

APA (2007) *Policy Guide on Community and Regional Food Planning*. American Planning Association, Chicago. https://www.planning.org/policy/guides/adopted/food.htm (accessed on 1 July 2018).

Bolleter J (2016) Background noise: a review of the effects of background infill on urban liveability in Perth. *Australian Plants* **53**(4), 1–14.

Buxton M, Carey R (2014) The use of planning provisions and legislation to protect peri-urban agricultural land. *Australian Environment Review* **September**, 191–195.

Cabannes Y, Dubbeling M (2000) Urban agriculture and urban planning: what should be taken into consideration to plan the city of the 21st century. Presented at *Urban Agriculture and Horticulture: The Linkage with Urban Planning*. 7–9 July 2000. Berlin, Germany.

Carey R, James S (2018) Peri-urban agriculture in Australia: pressure on the urban fringe. In J Zeunert and T Waterman (Eds) *Routledge Handbook of Landscape and Food* (pp. 213–228). Routledge, London.

CoGG (2006) *Bellarine Peninsula Strategic Plan 2006–2016*. City of Greater Geelong Council, Geelong. https://www.geelongaustralia.com.au/common/public/documents/8cc118122839958-Bellarine%20 Peninsula%20Strategic%20Plan%20Summary%202006.pdf (accessed on 1 July 2018).

CoGG (2007) *City of Greater Geelong: Rural Land Use Strategy, Geelong*. Report prepared by Parsons Brinckerhoff Australia, Bendigo for City of Greater Geelong, Geelong. https://www.geelongaustralia.com.au/common/public/documents/8cbc14eb82a8923-Rural%20Land%20Use%20Strategy%20 -%20Adopted%20Full.pdf (accessed on 1 July 2018).

CoGG (2013). *Central Geelong Action Plan*. City of Greater Geelong, Geelong. https://www.geelongaustralia.com.au/common/public/documents/amendments/8d2d0901dd9c6bb-CentralGeelongAction-Plan.pdf (accessed on 1 July 2018).

CoGG (2014) *Food Policy Discussion Pape*. Healthy Together, City of Greater Geelong, Geelong. http://www.geelongaustralia.com.au/common/Public/Documents/8d193df3e8e310f-Healthy%20Food%20 Policy%20Discussion_Paper.pdf (accessed on 1 July 2018).

CoGG (2015) *Food Policy*. City of Greater Geelong, Geelong. https://www.geelongaustralia.com.au/ct/documents/item/8d2a2f2d6881466.aspx (accessed on 1 July 2018).

CoGG (2016a) *Northern and Western Geelong Growth Areas Report: Context Report*. City of Greater Geelong, Geelong. http://www.geelongaustralia.com.au/nwgga/article/item/8d39501af448fc9.aspx> (accessed on 1 July 2018).

CoGG (2016b) *Greater Geelong Planning Scheme*. City of Greater Geelong and Department of Planning and Community Development, Melbourne. http://planningschemes.dpcd.vic.gov.au/schemes/greatergeelong (accessed on 1 July 2018).

CoGG (2017) *History of Council*. City of Greater Geelong, Geelong. http://www.geelongaustralia.com.au/council/article/item/8cbc3a413e4406f.aspx (accessed on 1 July 2018).

CoGG (2018) *Armstrong Creek: Whole of Growth Area*. City of Greater Geelong, Geelong. https://www.geelongaustralia.com.au/armstrongcreek/armstrong/article/item/8cfafd49ea31e3f.aspx (accessed on 1 July 2018).

Cust K, Cuttler R, Stirrat A (2013) *Food Access Needs Assessment*. Healthy Food Connect, Geelong. https://www.geelongaustralia.com.au/healthygeelong/documents/item/8d05a8cffc60d91.aspx (accessed on 1 July 2018).

Downton PF (2009) *Ecopolis: Architecture and Cities for a Changing Climate*. CSIRO Publishing, Melbourne.

DPI (2013) *Victoria GIS Data*. Department of Primary Industries, Melbourne.

Dubbeling M, Veenhuizen R, Zeeuw H (2010) *Cities, Poverty and Food: Multi-stakeholder Policy and Planning in Urban Agriculture*. Practical Action Publishing, Warwickshire.

Flannery T (2005) *The Weather Makers: How Man is Changing the Climate and What it Means for Life on Earth*. Atlantic Monthly Press, New York.

G21 GRA (2007) *G21 The Geelong Region Plan: A Sustainable Growth Strategy.* G21 Geelong Region Alliance, Geelong. http://www.g21.com.au/sites/default/files/resources/14.11_geelong_region_plan_fnl_webv2.pdf (accessed on 1 July 2018).

G21 GRA (2012) *G21 Regional Growth Plan Background Report.* G21 Geelong Region Alliance, Geelong. http://www.g21.com.au/sites/default/files/resources/g21_rgp_background_report_-_august_2012_-_low_res_0.pdf (accessed on 1 July 2018).

G21 GRA (2013) *G21 Regional Growth Plan.* G21 Geelong Region Alliance, Geelong. http://www.dtpli.vic.gov.au/planning/plans-and-policies/rural-and-regional-planning/regional-growth-plans/g21-regional-growth-plan (accessed on 1 July 2018).

GLA (2010) *Cultivating the Capital: Food Growing and the Planning System in London.* Greater London Authority, London.

Hackworth J (2007) *The Neoliberal City: Governance, Ideology, and Development in American Urbanism.* Cornell University Press, Ithaca.

Haney D (2010) *When Modern was Green: Life and Work of Landscape Architect Leberecht Migge.* Routledge, New York.

Hodgson K (2010) *Urban Agriculture: Growing Healthy, Sustainable Communities.* APA Planners Press, Chicago.

Houston P (2005) Re-valuing the fringe: some findings on the value of agricultural production in Australia's peri-urban regions. *Geographical Research* **43**(2), 209–223. doi:10.1111/j.1745-5871.2005.00314.x

idcommunity (2016) *City of Greater Geelong Population Forecast.* City of Greater Geelong, Geelong. http://forecast.id.com.au/geelong (accessed on 1 July 2018).

Ilieva R (2016) *Urban Food Planning: Seeds of Transition in the Global North.* Routledge, London.

Infrastructure Victoria (2017) *Victoria in Focus.* Infrastructure Victoria, Melbourne. http://www.infrastructurevictoria.com.au/sites/default/files/files/Fact%20sheet_Victoria%20in%20focus.pdf (accessed on 1 July 2018).

Jackson S (2016) *Geelong Food Hub Feasibility Study Report: Enabling Local Food Economies, Geelong Food Hub.* Deakin University, Geelong and Innate Ecology, Geelong.

Jones D, Beza BB (2018) Food-sensitive urban planning: Australian perspectives. In J Zeunert and T Waterman (Eds) *Routledge Handbook of Landscape and Food* (pp. 393–408). Routledge, London.

Kilbane S (2013) Green infrastructure: planning a national green network for Australia. *Journal of Landscape Architecture* **8**(1), 64–73. doi:10.1080/18626033.2013.798930

Kunstler J (2006) *The Long Emergency: Surviving the Converging Catastrophes of the Twenty-first Century.* Atlantic, London.

Leitner H, Peck J, Sheppard ES (2007) *Contesting Neoliberalism: Urban Frontiers.* Guilford Press, New York.

Morgan K (2013) Feeding the city: the challenge of urban food planning. In F Miazzo and M Minkjan (Eds) *Farming the City: Food as a Tool for Today's Urbanisation* (pp. 18–21). Valiz/Trancity, Haarlem.

Raja S, Born B, Russell J (2008) *A Planners Guide to Community and Regional Food Planning: Transforming Food Environments, Facilitating Healthy Eating.* American Planning Association, Chicago.

Roggema R (Ed.) (2016) *Sustainable Urban Agriculture and Food Planning.* Routledge, London.

Sheridan J, Larsen K, Carey R (2015) *Melbourne's Foodbowl: Now and at 7 Million.* Victorian Eco-Innovation Lab, Melbourne.

Sheridan J, Carey R, Candy S (2016) *Melbourne's Foodprint: What Does it Take to Feed a City?* Victorian Eco-Innovation Lab, Melbourne.

Soule D (2006) *Urban Sprawl: A Comprehensive Reference Guide.* Greenwood Press, Westport.

Spencer A, Gill J, Schmahmann L (2015) Urban or suburban? Examining the density of Australian cities in a global context. Paper presented at *The State of Australian Cities Conference,* 9–11 December 2015. Gold Coast, Queensland.

Trestrail C, Martin P, Mazur K (2013) *Agriculture and Fisheries in the Geelong Region of Victoria, 2013.* Department of Agriculture, Fisheries and Forestry, Canberra. http://data.daff.gov.au/brs/data/warehouse/9aa/regionalReports/201302/pb_regnrd9aa__0302013022Geelong1.0.0.pdf (accessed on 1 July 2018).

Urban Growth (2017) *Urban Growth.* City of Greater Geelong, Geelong. http://www.geelongaustralia.com.au/urbangrowth/default.aspx (accessed on 1 July 2018).

Victoria (2017) *Plan Melbourne 2017–2050: Metropolitan Planning Strategy.* Department of Environment, Land, Water and Planning, Melbourne.

Viljoen A, Wiskerke J (Eds) (2012) *Sustainable Food Planning: Evolving Theory and Practice.* Wageningen Academic Publishers, Wageningen.

Weller R (2009) *Boom Town.* University of Western Australia Press, WA.

Wilson E (2016) *Half-earth: Our Planet's Fight for Life.* Liveright, New York.

World Bank (2016) *Arable Land: % of Land Area.* World Bank, Washington, DC. http://data.worldbank.org/indicator/AG.LND.ARBL.ZS

Zeeuw H, Drechsel P (2015) *Cities and Agriculture: Developing Resilient Urban Food Systems.* Routledge, New York.

Zeunert J (2017) *Landscape Architecture and Environmental Sustainability; Creating Positive Change through Design.* Bloomsbury, London.

Zeunert J (2018a) Sustainability, resilience and regenerative agricultural landscape challenges. In J Zeunert and T Waterman (Eds) *Routledge Handbook of Landscape and Food* (pp. 231–252). Routledge, London.

Zeunert J (2018b) Dimensions of urban agriculture. In J Zeunert and T Waterman (Eds) *Routledge Handbook of Landscape and Food* (pp. 160–184). Routledge, London.

Zeunert J, Waterman T (Eds) (2018) *Routledge Handbook of Landscape and Food.* Routledge, London.

14

Emerging cultures

Kate Kerkin

> *Identity manifests itself on many levels, one of which is place ... When attachment to place grows, we start to identify ourselves with these places, both at a larger scale (nation, city, etc.) and at a smaller scale (neighborhood, workplaces, homes, rooms). This results in self-concepts that are based in part on place.*
>
> – Hauge (2007, p. 1).

The Geelong region comprises a diverse range of local cultures. Shackled with the stereotype of a manufacturing region, and located in the shadows of metropolitan Melbourne, the region has been represented through a range of stereotypes found elsewhere in Australia. The 'working poor' (Noakes 1992) and the 'isolated suburban dweller' (Richards 1990) are cultural references that have been applied to the Geelong experience. However, a range of local policies and networks currently support new and emerging cultures across the Geelong region. This chapter traces three emerging themes within the cultural landscape of the Geelong area: ethnicity, artistic practice and 'sea-change' identity. It highlights the ways in which local and regional organisations work to support new and emerging identities and the role that 'place' plays in the formation of these identities across Geelong (Giuliani 2003).

Increasing ethnic diversity

The City of Greater Geelong (CoGG) encompasses a diverse geographic area including coastal settlements, urban areas and rural townships. Like other areas across Australia, the social diversity of the Geelong region is rapidly changing. In the late 1990s the CoGG registered the trademark 'Geelong Smart Move', promoting the city as a destination for new residents. Playing on the anxiety of urban dwellers facing increasing mortgage costs and related lifestyle pressures, the campaign promoted the area as a location of choice that supports alternate work models and a balanced lifestyle. The number of people living in the Geelong region increased by 10.1% from 2011 to 2016 to 228 686 people, an additional 21 075. By 2036 it is estimated that the Geelong region will be home to 325 779 residents, a dramatic forecast anticipating a 42.4% population change between 2016 and 2036 – an additional 97 093 people living in the area (forecastid.com.au).

So, who are the residents of the Geelong region and how is the social profile of the city changing? In 2016, 37 916 people living in the Geelong region were born overseas, and

23 454 of them (10%) were born in a non-English speaking country. Birthplaces increasing most rapidly during recent census periods include India (+1244 people), China (+650 people), Afghanistan (+432 people), Philippines (+317 people) and Iran (+273 people). Changing trends in birthplace are reflected in the range of languages spoken: English and European languages continue to be the main language spoken at home by most residents but increasing numbers of people speak Arabic, Tagalog (Philippines), Cantonese and Vietnamese.

The legacy of post-World War II migration has had a huge impact on the cultural diversity of Geelong. Post-war migrants, mainly from European countries such as Italy, Greece, Yugoslavia, Poland, Hungary and Croatia, are now second- to fourth-generation Australians. Many of them worked in the factories of Shell, Ford, Pivot, International Harvester and Alcoa and the industrial areas of North Shore, Norlane and Corio. The children, grandchildren and great-grandchildren of these migrant families are now home owners, business owners and managers across the Geelong region. These people paved the way for the resettlement of refugees and asylum seekers from different parts of the world, who are now trying to establish their roots and young families in Geelong.

Global humanitarian trends are impacting on the social diversity of the Geelong region. Of those residents who were born overseas, many arrived in Australia during the 1980s and 1990s (18.1%). Geelong is also home to a significant number of newly arrived residents, with 16.9% of people born overseas arriving in Australia since 2011. Significant numbers have come from refugee backgrounds, including Karen, Karenni, Iraqi, Afghani, Congolese and Sudanese residents. How is the experience of being a refugee enacted in place in Geelong, and what does this say about the culture of the Geelong community?

While there are various discourses across Australia in relation to the role of refugees in our communities, significant support exists for these new residents in the Geelong region. This support reflects the complex processes that negotiate and support emerging local identities. In 2002 the CoGG formally declared the local government area a Refugee Welcome Zone. The *Multicultural Action Plan* (CoGG 2011) notes that 'Geelong continues to welcome refugees'. Similarly, the Colac Otway Shire, Borough of Queenscliffe and Surf Coast Shire are designated Refugee Welcome Zones, signing 'a commitment to the spirit of welcoming refugees, demonstrating compassion and embracing cultural diversity in local communities'.

A range of partnership arrangements further enhance refugee settlement needs across the G21 region. The Refugee and Asylum Seeker Policy and Advocacy Working Group, for example, aims to improve community connectedness and social inclusion for refugees and asylum seekers. The *Refugee and Asylum Seeker Service and Support Mapping Project* (G21 2014) lists over 100 services including men's sheds, financial counselling and youth services that support the settlement needs of newly arrived residents. Job clubs, food banks and clothing relief centres are co-ordinated with a range of local organisations, relying heavily on voluntary support often provided through church groups across the region. Diversitat (2018) is a key agency working to support the needs of asylum seekers and refugees across the Geelong region. It provides assistance including intensive support for the first six weeks, sourcing of long-term accommodation, and English language classes. Diversitat also offers a designated Art and Craft space, where people can use traditional skills and techniques to produce costume drawings and basket weaving, and showcase their culture through food, craft and dancing.

As well as these broader policy supports for refugee cultures in the Geelong region, the identity of refugee communities is closely linked to notions of 'place'. As happened with post-war migrants, the suburbs of Corio and Norlane have become home to many people who arrived in the region in recent years as refugees. Interviews with newly arrived refugee women reveal some of the ways in which this emerging community is negotiating these suburbs as 'home'. The Karen and Karenni playgroup space, provided at the Northern Bay Family Centre, offers a safe space to refugee women newly settled in Corio: 'it's your safe space, the language barrier is reduced, we all help each other, here we can overcome communication barriers, here the women relax and it's ok if they have problems, they know they can come here'. The space offers more than just playgroup time. Women bring in all types of requests – school enrolment information, Centrelink forms, updating electoral details, housing application forms – and are supported in completing these forms by the playgroup and centre workers. The group commenced in 2008 and 'has grown to become a place of family and cultural support that the women get from each other' (refer to 'Acknowledgements' at the end of this chapter for further details).

Various notions of place impact on the way in which newly arrived refugees contribute to their local area. One Karenni woman, who lived in a refugee camp in Thailand for 20 years, noted that:

> *I like Corio, it is similar to the camp that we had, the community, the trees, the flat area.*

A second woman noted that the growth of the Karen community provided an important support to her in a new place:

> *Some people told me Corio is a very bad place so I should be careful. I feel not so bad, Karen people, refugee people move here, big community not bad for me. I learn Australian now so I am not afraid.*

Other women use interesting ways to explore their new region:

> *Sometime I don't tell anyone, I travel to Marshall and Sunshine* [on the train] *and come back just to explore* [laughter] *when I can use my myki* [train ticket], *I can learn.*

These women also highlight some of the risks of being new to the area and their ways of subverting the challenges:

> *Two men in a car opened windows and shout at me but I think they know me so I smiled and say 'hi' and they close the door and drive away* [laughter].

Like elsewhere in Australia, these new residents bring new skills and experiences to the area:

> *I teach Karen language to Karen students at the secondary college Saturday mornings for three hours. They will lose their language, if we teach it they can use it, be interpreters.*

The activities of these new residents continue to change both the culture and identity of Corio as a place:

> *One day there will be more Karen people in Geelong, we will have our community and we look after our people ... Karen people who have been here ten years have their own business and people who get citizenship will have a vote one day. One day I can have my Karen people who live in the camp come here.*

Celebrating artistic practice

The role of art in changing urban and regional areas is a complex one. Various authors argue that artistic practices are a key aspect of cultural capital and are used to promote investment and tourism in areas undergoing economic change (Kearns and Philo 1993; Zukin 1995). Others argue that artistic practices play a strong role in the construction of local identities. Artists often question notions of personal and national identity, challenge stereotypes and present new ways of thinking about place (Hauge 2007). While it might be argued that Geelong is constructing art as a commodity through the development of new arts infrastructure, there is also strong evidence of artistic practices that are reconstructing cultural identity throughout the region.

Like the support of emerging ethnic groups above, a range of policies have been instrumental in promoting artistic practices and their connection to emerging cultures and identities. The G21 'Arts, Heritage and Culture Pillar' supports the promotion of arts, heritage and culture by 'developing artists, ideas and knowledge; building creative industries; acknowledging, celebrating and creating places and spaces; and, promoting cultural leadership'. These aims are enacted through the priority project of the Geelong Cultural Precinct which is creating a regional hub for arts and culture activities in Geelong, including: refurbishment of the Old Geelong Courthouse, a new Geelong Library and Heritage Centre, an expanded Geelong Performing Arts Centre, expansion of the Geelong Gallery, and public art and improved links to the train station in adjacent streets and public spaces (G21 2018).

Contributing to local cultural capital through such immovable heritage items reflects goals common to the place promotion of many post-industrial cities, with generic goals such as 'create an inclusive city, improving the liveability and lifestyle of the region'. It is claimed that the cultural precinct will:

> *create an estimated 390 local ongoing jobs after completion and support almost 700 Victorian jobs during construction ... provid*[e] *a cultural imperative for business and potential residents to relocate to the region; create a unique and distinctive artistic facility that will attract both domestic and international tourists ... all part of helping to ... regenerate Geelong's Central Business District* (CoGG 2013).

Tangible goals are evident in the utilisation trends of local cultural assets, with the Geelong Gallery experiencing 'strong growth in the past ten years, with visitation numbers doubling during this period'. The average annual attendances at the Geelong Performing Arts Centre have increased to 225 000, with claims that '1 in 2 households in Geelong and 1 in 3 of the households in the G21 region having attended GPAC in the past three years' (G21 2018).

A range of more ephemeral, or movable, artistic practices are also evident across the Geelong region. Unlike the arts practice above which focuses on place as building form, these movable artistic practices occur through social media and performances illustrating rich relations to place. They reflect the role of art as a cultural practice contributing to changing sociocultural dynamics and cultural mythologies, including 'notions of ancestry, self-esteem, belonging, prejudice, pride and hybridity' (van Eps 2010). This type of artistic practice involves engagement with communities of interest, democratisation of the relationship between creative practitioner and the public, and a sharing of expert and lay knowledge. It incorporates both collaborative and transgressive artistic actions (LSAD 2018).

The CoGG has been actively supporting the emergence of movable artistic practices through a range of programs and initiatives such as the Arts and Culture Assets Register, which maps a range of physical assets, including public art, memorials and spaces both built and natural, that are used as sites for performance and exhibition (CoGG 2018a). Other CoGG strategies include 'Geelong After Dark' with live music, pop-up art displays and light events occurring throughout the central city area, and the Geelong Arts and Culture Trails, involving apps that allow the user to 'unearth striking public artworks, creative spaces and exciting performing arts locations, and find new ways to connect with our Indigenous and post-settlement histories'. All of these activities promote new ways of knowing, exploring and being, in the place that is Geelong.

One recent artistic practice in the Geelong region that most clearly raises questions of identity and relationship to place is 'Connecting Identities', a three-year project designed to create links across a diverse municipality facing the challenges of change.

Connecting Identities is about:

- the land we walk, and the people that make up the CoGG
- the importance of local places and communities
- preserving memory, creating shared vision, and respecting individual expression
- art that is daring, listens, reflects, and challenges (CoGG 2018b).

Connecting Identities occurs through three elements. *Memory Bank – Connecting to Memory* involves the collection and recording, through film and written anecdotes, of stories that 'reveal the diverse personal meanings and associations of local places'. These recordings are made accessible through the Connecting Identities website and through locations where they can be listened to at their place of origin. 'Using a mobile phone, people will be able to listen to stories about the land on which they are standing … this aspect of the project is known as [murmur]'. *M2M Connecting … to Songline* similarly records representations of place and local identity, this time through the medium of song, 'a contemporary songline of music reflecting the changing landscape between the mouth of the Barwon River across country and city to the top of the You Yangs'. *Mountain to Mouth – Connecting to Land* is an extreme arts walk comprising 80 km, 12 km or 3 km art walks that draw on the memories and songs developed in the earlier project stages. This stage uses the symbol of water to connect places, meaning and memory across the municipality. Those who take part in the walk participate in 'forming a contemporary songline for others to follow' or 'simply take a moment to reflect on the gift of water, the memory it holds, the connection it makes between people and places, and the future it creates (CoGG 2018b).

Connecting Identities has been developed in collaboration with the *Wadawurrung* people, the Traditional Owners of much of the land within which the CoGG sits. The project reflects Geelong as a place of changing and emerging knowledge and culture, and creation of new relations to place. It presents a relationship to nature not as something to be mastered or transformed or 'turned to our advantage', but as an 'interlocutor and agent speaking to us in a language we are not always prepared to understand' (Kester 2010). The project draws on Indigenous knowledge, changing land use practices and emerging relations to environment to question the validity of national, regional and local identities within a modern hybrid society (Zhang and Barstow 2011).

Most recently, the CoGG community visioning project *Our Future* (CoGG 2017) produced a long-term community vision to guide the region's development over the next

30 years. The visioning process was a culmination of nine months of ongoing conversations with over 15 000 people from across Geelong. It describes a 30-year vision to ensure a thriving, inclusive and sustainable future:

> *By 2047, Greater Geelong will be internationally recognised as a* ***clever and creative*** *city-region that is forward looking, enterprising and adaptive, and cares for its people and environment.*

The Clever and Creative vision sets the scene for a new course in the city's history.

Other artistic practices are also spontaneously emerging across Geelong without the CoGG or other government endorsement, such as the Paper Mills Arts Precinct and the Federal Mills innovation and technology hub. These community-led initiatives seek to nurture and support a creative culture in Geelong through reclaiming abandoned industrial spaces and transforming them into cultural sites for creative practices.

A culture of returning

Coastal settlements are a strong feature of the Geelong region. Like other areas of Australia, these settlements have experienced recent population growth as people move away from large metropolitan areas seeking a 'sea-change'. This population trend has been occurring since the 1960s and has been linked to changing economies in major metropolitan areas. Burnley and Murphy (2004) conducted the most comprehensive study of this population trend and found that sea-changers comprise two distinct groups of people including 'free agents' and 'forced relocators'. They found that free agents included early retirees and working-age movers seeking to obtain the advantages of cheaper housing and high levels of amenity (Burnley and Murphy 2004). Lifestylers, on the other hand, were 'driven by anti-materialist attitudes and new age attitudes to life' (Burnley and Murphy 2004, p. 38). Forced relocators included people who relied on some form of income support and moved in greater numbers when metropolitan house prices were high (Burnley and Murphy 2004). Other research has noted the role of non-monetary factors in coastal migration. Walmsley *et al.* (1998), for example, identified the top three factors in migrant decision-making as 'more pleasant climate', 'more relaxed lifestyle' and 'attractive physical environment' (p. 114). Aesthetic and environmental considerations also emerge as driving factors in the search for coastal residential locations, including the 'love of the [Australian] bush' (Burnley and Murphy 2004, p. 149).

If these are some of the motivations for a move to a coastal location, what are the cultural implications of that move? In analysing this population trend, researchers have often concluded that metropolitan social geographies are being mapped onto non-metropolitan regions and that, as social conditions of the metropolis change, so too do the social conditions of non-metropolitan regions (Burnley and Murphy 2004; Newton *et al.* 1998). Others argue that new activities are emerging through recent population shifts to coastal locations. Salt, for example, has defined the sea-change phenomenon as part of a distinctive Australian culture facilitated by 'new values and concepts such as leisure, entertainment, lifestyle and retirement, aided by new social and financial arrangements such as superannuation' (Salt 2004, p. 2).

What cultural changes are evident in the coastal communities of the Geelong region? Answers to this question were sought through in-depth interviews with sea-changers living at Barwon Heads, Victoria, the home of the *Sea Change* television series and the icon of the sea-change phenomenon in Australia (see Notes at end of chapter).

A key theme in the Barwon Heads interviews related to differing forms of participation in community life. Early movers (EMs) revealed very active levels of participation in community life. Bronwyn, for example:

> *... joined the Residents Association to find out what was going on in the town ... Then ... formed the Diners Club ... to support the local traders ... because we felt that they would die before the next summer ... we also formed the Traders Association which later became the Traders and Tourism Association ... we started it so that we would have a voice ... we also set up the Barwon Heads film society* (EM Interview 4).

Recent movers (RMs) showed not only different levels of engagement in community activities but also different forms of participation in community life. There is a trend towards participating as consumers benefiting from the social contributions of the early movers:

> *Whatever's happening in the town we get involved in ... I support all the cafes ... having coffee there ... The hall has a film society so I try and go when I can ... I love our little market and that's always a real social occasion, you can go there on your own and you're bound to bump into lots of people* (RM Interview 2: 7e).

Early movers expressed frustration at the lack of community involvement by recent movers:

> *Regarding the school fundraising – in the final week before Easter another letter went home to all parents explaining the significance of the event. Subsequently, many new parents volunteered to help and the day was run very successfully ... So it seems that many of the newer parents were unaware of the need for their help, but when approached, they finally responded magnificently* (EM Interview 3).

One explanation for the lack of active community involvement by recent movers relates to the strong private subcultures that people create after a move to a coastal location. Kris, for example, talked about dinners, film nights and book clubs held in private homes:

> *We get together with other families and have dinners, drinks. There's a big group of us that get together and at the moment the boys have a film night, the dads get together every 5 weeks and watch a film and have discussion and have drinks so it ends up being quite a social night and us girls have a book club ... every 5 or 6 weeks and read a book and talk about it and drink champagne* (RM Interview 2).

These cultural activities played out in private spaces provide a network for meeting new friends in a new location:

> *It's lovely and I've met quite a few friends through that group because each person has invited someone else so you meet their friends* (RM Interview 2: 7b).

While such subcultures occur within private spheres, the efforts linked to their production replicate some of the broader community events. Film nights, for example, are accompanied by a data projector and a big screen:

We go around to each other's houses and you pick the book and you put on the wine and the food ... we'll get a data projector and a big screen. A lot of effort's put into the choice of film and usually there's something previewed before the film, either a doco or something like that that everyone walks away talking about (RM Interview 2).

This reproduction of some of the broader cultural activities occurring in the town reflects elements of community formation evident within new suburbs. Like Richards' famous study of social life in a new suburb, there is a:

... rotating around each other's houses and a sense that ... belonging was about all being in the same life stage ... they're all young couples ... all in the same boat ... It's people just very much like myself. Once you get to know them ... (Richards 1990, pp. 48–54).

A recurring theme in the formation of personal identities among both recent and early movers to Barwon Heads was their stated differentiation from tourists. As researchers have noted (Walmsley 2003), residents of coastal areas often describe tourists as 'disruptive' or an 'inconvenience'. In the Barwon Heads interviews, both recent and early movers spoke about the timing of their cultural activities in ways that separated them from 'tourists':

I found this summer ... when I took the children to the beach quite often it was 4 o'clock in the afternoon because by that time a lot of them had decided they'd had enough of the beach and gone home (RM Interview 4).

Late Sunday is the best time for surfing because all the cars start heading out to go home, frees it up. And in winter ... you mainly see the locals out there, that's good ... I'm not anti-tourists or anything but because you live here the less people around the better, pretty much (RM Interview 1).

Subcultures acted out in private domestic sphere and at non-tourist times, then, provide networks of support and markers of personal identity for recent movers to Barwon Heads. While some of these activities are shared by early movers, there is a notable separation of activity by recent movers who seek to develop new friendships and connect with others in similar positions to themselves, often creating their new subcultures within the private realm of their homes.

A final key theme emerging in the Barwon Heads interviews is the notion of 'returning'. Rather than leaving the city, the move from city to coast is seen as an ongoing part of a lifecycle that maintains connections to the coast:

I grew up ... in Ocean Grove ... when I went to uni I came back for the summers to keep surfing (RM Interview 2).

There's plenty of people who live in Melbourne who go surfing regularly ... coming down here a lot just ducking down after work (RM Interview 1).

The sense of returning to a preferred local culture is imbued with maintaining social connections and nostalgic associations with a coastal childhood:

... my parents are still down here and I'm still mates with the crew I went to primary school with (RM Interview 2).

... the kids are down at the river park and they're playing in the river or at the beach park instead of what we had in Preston which was pretty ordinary (RM Interview 2).

Conclusion

The influence of place on identity is evident through various elements of the culture of the Geelong region. Stories of newly arrived refugees creating connections to place, artistic practices making us reflect on our relations to land, nature and culture, and residents returning after periods of time in Melbourne maintaining their connections to a coastal lifestyle, are all stories that show a reciprocal interaction between people and their physical environment. As people affect places, and places influence how people see themselves (Hauge 2007), the cultures of Geelong continue to develop and emerge in interesting and complex ways.

Notes

This chapter refers to social data related to the G21 region including the following local government municipalities: City of Greater Geelong, Surf Coast Shire, Colac Otway Shire, Golden Plains Shire and the Borough of Queenscliffe. In 2016, 301 661 people lived in the G21 region, most of them (75.8% or 228 686 people) in the CoGG. The focus of this chapter then is on the geographical area of the CoGG.

The Barwon Heads interview data comprises several in-depth interviews conducted with 'early' movers (people who moved to Barwon Heads in the 1990s) and 'recent' movers (who moved to Barwon Heads in the early 2000s). Respondents were asked about their reasons for choosing Barwon Heads and their experiences of living in the town. The interviews sought to understand the reasons why people move to the area as well as their cultural experiences after the move. Six interviews were conducted in total, four with recent movers and two with early movers. All quotes are presented using pseudonyms rather than actual names. It should be noted that Barwon Heads is one of the most affluent areas in the CoGG, ranking as the third most advantaged local area according to the socio-economic index for areas (ABS 2011). This relative social advantage undoubtedly influences the sea-change experience in this location.

Acknowledgements

Thanks to Louise Jarvis, CoGG Supported Playgroup; Helen O'Connor, Northern Bay Family Centre; and the women of the Karen and Karenni Playgroup for their contributions to this chapter.

References

ABS (2011) *SEIFA 2011*. Australian Bureau of Statistics, Canberra. http://www.abs.gov.au/websitedbs/censushome.nsf/home/seifa2011 (accessed 1 March 2018).

Burnley I, Murphy P (2004) *Sea Change: Movement from Metropolitan to Arcadian Australia*. University of New South Wales Press, Sydney.

CoGG (2011) *Multicultural Action Plan*. City of Greater Geelong, Geelong. https://www.geelongaustralia.com.au/common/Public/Documents/8cefb3f6a4bf07e-Multicultural%20Action%20Plan.pdf (accessed 1 March 2018).

CoGG (2013) *Central Geelong Action Plan: Central Geelong Taskforce*. City of Greater Geelong, Geelong. https://www.geelongaustralia.com.au/common/public/documents/amendments/8d2d0901dd9c6bb-CentralGeelongActionPlan.pdf (accessed 1 March 2018).

CoGG (2017) *Our Future*. City of Greater Geelong, Geelong. https://www.geelongaustralia.com.au/our-future/default.aspx (accessed 1 March 2018).

CoGG (2018a) *Geelong Arts and Culture Trails App*. City of Greater Geelong, Geelong. http://www.geelongaustralia.com.au/arts/article/item/8d2182097cbefd4.aspx (accessed 1 March 2018).

CoGG (2018b) *Connecting Identities*. City of Greater Geelong, Geelong. https://www.geelongaustralia.com.au/connectingidentities/about.html (accessed 1 March 2018).

Diversitat (2018) *Diversitat*. https://diversitat.org.au/ (accessed 1 March 2018).

G21 (2014) *Refugee and Asylum Seeker Service and Support Mapping Project*. http://www.g21.com.au/sites/default/files/dmdocuments/Refugee%20%20Asylum%20Seeker%20Service%20and%20Support%20Mapping%20-%20May2014v6.pdf

G21 (2018) *G21 Arts, Heritage & Culture Pillar*. http://www.g21.com.au/arts-heritage-culture-pillar (accessed 1 March 2018).

Giuliani MV (2003) Theory of attachment and place attachment. In M Bonnes, T Lee and M Bonaiuto (Eds) *Psychological Theories for Environmental Issues* (pp. 137–170). Routledge, London.

Hauge AL (2007) Identity and place: a critical comparison of three identity theories. *Architectural Science Review* **50**(1), 44–51. doi:10.3763/asre.2007.5007

Kearns G, Philo C (Eds) (1993) *Selling Places: The City as Cultural Capital, Past and Present*. Pergamon Press, Oxford.

Kester G (2010) *Collaborative Practices in Environmental Art*. University of California, Los Angeles.

LSAD (2018) *Social Practice and the Creative Environment*. Limerick School of Art and Design, Limerick Institute of Technology. http://www.lit.ie/lsad/Courses/9/Pages/About%20The%20Course.aspx (accessed 1 March 2018).

Newton P, Flood J, Berry M, Bhatia K, Brown S, Cabelli A, Gomboso J, Higgins J, Richardson T, Ritchie V (1998) *Environmental Indicators for National State of the Environment Reporting: Human Settlements*. Australia State of the Environment (Environmental Indicator Reports). Department of Environment, Canberra.

Noakes F (1992) Working poor attacked. *Green Left Weekly*, Wednesday 25 November.

Richards L (1990) *Nobody's Home: Dreams and Realities in a New Suburb*. Oxford University Press, Melbourne.

Salt B (2004) *The Big Shift*. Hardie Grant Books, Melbourne.

van Eps M (2010) *Reconstructing Cultural Identity in Artistic Practice: Embracing Mixed Heritage through Framing Pride and Prejudice*. PhD thesis. Griffith University Open Access, Australia. https://www120.secure.griffith.edu.au/rch/items/b9172056-f119-4aa3-bf54-e75b478cb2e7/1/ (accessed 1March 2018).

Walmsley DJ (2003) Rural tourism: a case of lifestyle-led opportunities. *Australian Geographer* **34**, 61–72. doi:10.1080/00049180320000066155

Walmsley DJ, Epps WR, Duncan CJ (1998) Migration to the New South Wales North Coast 1986–1991: lifestyle motivated counterurbanisation. *Geoforum* **29**, 105–118. doi:10.1016/S0016-7185(97)00023-7

Zhang Y, Barstow C (2011) Encountering the third space: a study of identity and hybridity through trans-cultural artistic practice in Australia and China. Paper presented at the *4th Global Conference on Diasporas: Exploring Critical Issues*. Mansfield College, Oxford. http://ro.ecu.edu.au/ecuworks2011/425/ (accessed 1 March 2018).

Zukin S (1995) *The Culture of Cities*, Wiley Press, London.

15

The post-industrial landscape of Geelong

Matt Novacevski

We do not live in the best possible world. We are allowed, obliged, even, to think about the alternatives.

– Slavoj Zižek (2012, p. 77).

Geelong's post-industrial landscape remains at once a hovering question, existential challenge and vital opportunity. As this book goes to press, the city is at a turning point ...

Standing on one of the hills that surrounds Geelong and observing the city unfurling towards the oft-glistening Corio Bay, one may wonder what this place was like before the first plots of the Hoddle Grid were transposed onto the landscape.

The 'Hoddle grid', a common description of central Melbourne's urban form, was named for surveyor Robert Hoddle. What is less known is that three weeks after Hoddle completed his survey of Melbourne in 1838, he also led the survey of Geelong. Hoddle's grids became part of the emergence of Australian city planning, some 60 years after the first British landing in Australia.

The development of the first plots around Geelong's current city centre marked the beginning of a dramatic period of change and displacement that, through a nexus of timing, economics and geography, formed an industrial landscape. Geelong sat at a critical junction: a channel to the British Empire via Port Phillip Bay shipping, an emerging goldfield to the north and fertile agricultural hinterland to the west.

Coupled with the extractive ambition of Empire and the industrial revolution in Victorian England, the stage was set. In the first decade after its survey, Geelong became known as the 'pivot on which Port Phillip Bay's commerce turned' (Wynd 1971, p. 71). The city duelled with Melbourne for regional primacy, with both cities providing a departure point for miners and squatters eager to extract fortunes from the land, which was often seen as a vessel for exploitation and moulding.

Melbourne won this oft-bruising battle for primacy. During the mid 1800s, its population and those of Ballarat and Bendigo boomed, leaving legacies of Victorian architecture and early prosperity. Geelong's industry-propelled growth was a slower burn, based on processing primary products from the hinterland before heavy manufacturing developed in the 1900s.

Fig. 15.1. A city's past and present unfurls towards a gleaming bay. In the foreground are the spires of construction cranes next to an emerging office tower, while in the background is the now-defunct Port Henry aluminium plant. Photo: Donna Squire.

Industry has shaped Geelong's built landscape, urban form and identity. While the goldfields centres have grand Victorian buildings, Geelong's built fabric is more red brick and grit of mills, factories and woolsheds. The chimneys, silos and solidity of these buildings hint at the ambition of industrial modernism. But as the 20th century wore on, the strength of these structures thrusting from the earth proved a fig leaf for the fragility of the modernist paradigm that had defined Geelong's economy and identity. The gradual decline of manufacturing has recently been punctuated by the closure of the city's largest factories: the Alcoa aluminium plant in 2014, and the Ford Motor Co. factory in 2017 (Fig. 15.1).

These events have left the city pondering not only a steep economic transition, but also its sense of self. To channel German sociologist Ulrich Beck, the certainty of modernity that propelled Geelong's industrial rise has proven to be 'founded on quicksand' through economic change and a growing awareness of ecological limits (Beck 2009, p. 232). Yet just as Beck (2009) decries the lack of a new order to replace the destroyed certainty of a hollowed-out 'reflexive' modernity, a constructive new order to define a post-industrial landscape in Geelong has yet to emerge.

This chapter interrogates Geelong's journey towards a post-industrial landscape, drawing on Beck's theory of reflexive modernity and French philosopher Henri Lefebvre's theory of abstract space. These theories provide a prism through which the history of Geelong's industrial landscape, its present and the forces shaping its future may at once be understood.

An industrial landscape takes shape

Geelong's industrial landscape is a physical, economic and socio-cultural phenomenon. It may be seen as an assemblage of buildings, urban form, the natural environment and the stories and dynamics that have shaped the city today. It may also be seen as a social and economic creation defined variously by the demands of the British Empire and industrial production. Despite its solid built form, this landscape has proven fluid, marked primarily by a changing dialectic between production and consumption and three major turning points:

- the advent of primary processing in the 1800s
- the rise of Fordist modernism through the arrival of heavy manufacturing and the automotive industry in the early 1900s
- the end of a post-war boom, marked by exit of heavy industry and the rise of a landscape of consumption from the 1970s.

European settlement of the Geelong region began in earnest around 1836, with squatters running sheep on the land amid a mass displacement of Aboriginal people that is discussed in more detail in Chapters 4 and 5. This was a time when 19th-century capitalist and utilitarian doctrines were on the rise, casting their influence on new cities (Proudfoot 2000). Europe's industrial revolution reverberated across the oceans. Geelong became at once an economic engine and a transition point between 'mother country' and a colonial hinterland whose wealth the Empire was seeking to tap.

By 1841, three years after the city was surveyed, Geelong had sent its first load of wool to England and was seen as an ideal departure point for the diggings of Ballarat and Bendigo. By the end of the 1880s, an array of industrial processing operations had been established but the city lived most of all on the sheep's back, via the export and processing of wool. Large brick woolsheds continued to operate in the city centre until their closure at the end of the 1970s, when they became for a time 'relics of an eclipsed economic order' (Regan 2010, p. 511). Their closure was one of many signs across the city of the era of uncertainty and change to come.

As the 20th century approached, increasingly heavy industry brought increasingly heavy built forms, reflected in mills and factories. These remain scattered around the city as markers of Victorian industrial ambition – from the red brick woolsheds of the city centre, to the sheer scale of the bluestone paper mills at Fyansford which, upon their opening in 1878, were known as some of the largest and best in the southern hemisphere (Brownhill and Wynd 1990). Although the mills stopped producing paper in 1923, they retain an imposing presence with a physicality and materiality that was replicated in other industrial structures:

> *The main buildings, several of them two storeys, were all built in coursed bluestone, locally quarried, with brick internal partitions and iron roofs. The walls were generally from 12 to 18 inches* [30–46 cm] *thick and the general appearance of the mill, atop the precipitous northern bank of the river, was that of a fortress* (Milner 1985, pp. 19–20).

The Fyansford Mills are particularly distinctive for the way they sit in the landscape. As Geelong's industrial base began to diversify from the 1900s, more modern buildings became more distant from their surroundings. The city was chosen after a nation-wide search to host a woollen mill to supply fabric for World War I military uniforms, which led to the Federal Woollen Mills being built north of the city centre in dramatic red brick. But it was the arrival of the Ford Motor Co. in 1925 that fundamentally changed Geelong, starting a love affair with the car that continues to define the city today. Ford's Geelong base was the first factory set up by an overseas car-maker in Australia, and its factory's deployment of then-innovative techniques and the building's classical façade are recognised for their heritage significance (Heritage Victoria 2017). In 1932, Ford's Geelong-based automotive illustrator Lewis Bandt invented the ute, which quickly became an Australian icon (Townsend 2008). Although the Ford factory closed in 2017, the company, its cars and the building fronting the main highway running through Geelong remain essential threads in the city's story.

For much of the 20th century, industrial growth and population growth went hand in hand. As prosperous goldfields centres like Ballarat and Bendigo limped into the 1900s, Geelong's population overtook them, doubling from 25 943 in 1901 to 52 408 in 1942, before doubling again by 1970 (Wynd 1971). This post-war boom brought a new scale of construction, with Shell deciding to develop a bayside oil refinery in the city's north at Corio in 1953 and Alcoa starting work on an aluminium smelter and rolling mill on a waterfront site at Point Henry in 1961 (Brownhill and Wynd 1990). These industrial behemoths practically form bookends at the north-west and south-east of Geelong's Corio Bay coastline and, by 1985, they were the city's second and third largest employers behind Ford (Brownhill and Wynd 1990).

Up until the 1970s, the city hummed, pumping for all appearances 'as a perfect organ' (Townsend 2008, p. 194). This vision of Geelong as poster city of Fordist modernism can be seen in a video produced in 1966 by the Australian government's then Department of Immigration as part of the *Life in Australia* series (NFSA 2018) to attract continental European migrants with the promise of factory jobs, education, housing and modernist prosperity.

The film tells an idealised story of the institutions and institutionalising nature of Fordism – an ideal that had become entrenched in the city's image of itself through industrial production, gendered social roles, the nuclear family and the ability of the worker to also be a consumer via a regular wage. These elements were interpolated into the urban landscape via wide streets designed for cars that rolled off the production line, and detached houses on almost uniform rectangular allotments (Colacino 1966). This promise of the modernist dream was not only a selling pitch: it grew new suburbs around and inside Geelong's industrial belt, lending a sprawling quality to the city that became supercharged as the 21st century approached.

Certainty collapses

As the whir of industry accelerated from the 1950s to the 1970s, significant portions of Geelong's early built heritage were lost (Brownhill and Wynd 1990). Fast capital flowed into the city while globalisation, changes to production methods and microeconomic reform eroded the city's manufacturing base.

As the spectre of industrial decline rose, the then Geelong Regional Commission developed an ambitious *City by the Bay* plan, which sought to revitalise the city centre by introducing an urban form marked by a mass of largely internalised multi-storey centres and tourism-based structures supported by large blocks of car parking, parkland on the water's edge and glass-covered pedestrian links (Regan 2010). Never lacking in ambition, *City by the Bay* is notable for its pastiche of imported ideas that suggested an intent to 'gather as many elements as possible from waterfront developments around the world and (roll) them all together' (Regan 2010, pp. 512–513).

Fig. 15.2. The Yarra Street pedestrian skyway, which has privatised a prized north-facing waterfront view and replaced it with symbols of abstract space. Photo: Donna Squire.

Economic recession and the collapse of political support meant *City by the Bay* was largely shelved. One key thread of the plan, however, came to redefine the city. The opening of the Market Square Shopping Centre in 1986 with public funds brought a decisive turn in the city's form, entrenching the pattern of 'mall and sprawl' development that continues today. Two years later, a second internalised centre, Bay City Plaza (now Westfield Geelong) opened next to Market Square Shopping Centre. The twin malls are notable for the introduction of an urbanism of hyper-consumption, a semi-privatised form of highly administrated space and the demolition of a once-vibrant and walkable section of the city's grid, leading to losses of permeability, public street-fronts and character. According to Gray (2015, p. 218), this fragmentation of 'the traditional intimate urban structure ... led to a deterioration of the city's social cohesion, sense of place and economic prosperity'.

Where factories and mills once dominated Geelong's urban landscape, the car-dependent internalised shopping centre has become one of the cityscape's enduring images. Some 20 years after Market Square opened, it and the adjacent Westfield Geelong occupied nearly half of the 101 900 m^2 of retail floor space in Geelong's central business district (Essential Economics 2006). After 2006, Westfield Geelong added another 17 000 m^2 (or 90 shops) in an expansion that included a pedestrian skyway that covered a prized north-facing waterfront view along Yarra Street (Fig. 15.2). This solid block now forms a barrier between much of the city centre, and the steadily rejuvenated public space along the city's waterfront.

In the suburbs, internalised shopping centres comprised some 77% of sub-regional retail space outside Geelong's CBD (Essential Economics 2006). Satellite towns and outer suburbs such as Leopold and Waurn Ponds are based on highway-fronting internalised centres, with little or no street-fronting retail activity or opportunities for alternatives to car transport. In 2013, Waurn Ponds Shopping Centre added 15 000 m^2 of retail space fronted by 700 new car parking spaces (Australian Unity 2013) while a five-fold, $72 million expansion of Leopold's shopping mall cut into farmland and now separates the town from Geelong (Vicinity Centres 2015). In these areas, suburban streets run to highways, which run to shopping malls and beyond.

This continued mall development is supported by ever-increasing suburban sprawl, which includes the new suburb of Armstrong Creek, reported to be Victoria's largest growth area with a planned population exceeding 50 000 (G21 GRA 2013). More greenfield growth areas are planned to the north and west of the city, their ambitions expressed foremost in numbers via population targets.

Meanwhile, the end of manufacturing at Alcoa's Point Henry plant in 2014 and the Ford Motor Co.'s factory in 2017 leaves two large, high-profile and vacant sites that sit like allegories for the questions facing the city. Of the industrial giants, only the flickering flame of the Corio oil refinery remains.

The grand abstraction

Geelong's landscape shift from the 1970s through to the present parallels trends in the western world, whereby consumption has escalated in spite of industrial giants closing, breaking up or moving to cheaper labour markets (Gleeson 2014).

This is not a post-industrial or even post-modern age, but 'a reflexive stage of modernity' (Beck 2009). While the western world grapples with crisis phenomena such as the fading borders of national societies, mass unemployment, economic globalisation, ageing societies and an ecological crisis that 'calls for a revision of industrial society's exploitative conception of nature', the very elements of modernisation that have caused these crises have proven resilient and even resurgent, continuing to influence society in new ways (Beck 2009).

Modernity's seemingly 'unstoppable victory march' has become a crisis point in itself (Beck 2009), paralleled by the drive of capital to reform space in its own image and for its own ends through processes Soja (2010) describes as 'creative destruction'.

The spatial implications of this capitalist drive, coupled with an increasingly reflective modernity, are reflected in French philosopher Henri Lefebvre's (1991) idea of 'abstract space', first identified in his 1974 book, *The Production of Space*. Abstract space is described by Goonewardena *et al.* (2008) as space in which modernism's alienation of the rhythms of everyday life is inscribed. Abstract space is at once a tool of capital and a spatial manifestation of what Žižek (2015, p. 146) describes as the inner antagonism of capitalism, 'the incessant revolutionizing of its own material conditions', a 'mad dance of its unconditional spirit of productivity'. Yet abstract space is not just a product or a force of capitalism's reproduction – the act of abstraction implies a stripping away of meaning, a process that alters lived experience by hollowing out relations between people and place (or, in Lefebvre's terms, the abstraction of lived practices of people inhabiting a space). This alienation may be seen, for example, in the isolation and narrowed experience of place promoted by individual car transport (Lefebvre 1991).

The most significant characteristics of abstract space in Geelong's context may be summarised as:

- its dissolution, incorporation or replacement of distinctive objects ranging from trees or nature to whole villages and sections of cities, as seen in the creation of the Market Square Shopping Centre (Lefebvre 1991)
- its drive to achieve conformity via homogenisation – erasing existing objects, narratives or points of difference in its path to form a *tabula rasa* (Lefebvre 1991). Abstract space works towards homogenisation by dominating and asphyxiating 'whatever is conceived within it and then strives to emerge', 'destroying historical conditions ... and any such differences that show signs of developing' (Lefebvre 1991, p. 370)
- its artful repression, manifested by imposing hierarchies, segregation or erasure (Lefebvre 1991)
- its general focus on prohibitive messages (via gates, barriers or signs), with the exception of consumption, which it seeks to promote (Lefebvre 1991)
- its prioritising of the quantitative (measurement and administration) over qualitative elements like lived experience (Lefebvre 1991).

Abstract space is both an institution and institutionalising force of capitalist modernism: Prigge (2008) describes a process where Fordist urban planning, dominant in post-World War II Geelong, institutionalises not only the landscape, but the 'dominant ideology of urban everyday life' via the promotion of a 'functionalised architecture of the city' that shatters a utopian vision of progress.

As the 21st century approached, Geelong's landscape story increasingly became one of an accelerated churn of abstract space in spite of industrial decline. The *City by the Bay* plan and the creative destruction that formed Market Square Shopping Centre betrayed a form of revitalisation underpinned by a conception of existing place as *tabula rasa*. These approaches were underpinned by a process of urbanism that saw erasure as a prerequisite to 'revitalisation', a key characteristic of abstract space. Likewise, the development of cookie-cutter internalised shopping centres on Geelong's fringes reflects a view of existing place as a clean slate, to be manipulated in the name of capital via ubiquitous forms of development that draw not from their surroundings and context, but from the ersatz consumption-based 'anywhere' beloved of global capital. Friedmann (2010, p. 157) speaks as much to this process of abstraction-via-erasure as to its results when he describes a form of violence that 'destroys established patterns of human relationships'.

The acts of erasure and erosion that have accompanied Geelong's internalised shopping centres reveal much about abstract space and the reflexive industrial landscape of Geelong. Just as the process of abstraction is fluid, the city centre's twin malls have not remained static. They have grown their footprint and stripped bare their surroundings. The latter is demonstrated in the floundering fortunes of the despised Market Square Mall, a strip of public space retained on the southern side of the centre as a mostly pedestrianised thoroughfare. The centre's edge on to the pedestrian mall consists not of lively street-fronting shops, but expanses of blank walls and loading docks that undermine the pedestrian realm (Gray 2015). This is powerful in physical and symbolic terms: the shopping centre's turning of its back on the city outside suggests a curious dialectic formed by the radiating impacts of abstract space on its surroundings, and its inward pull on capital and activity. It also sends a strong message of abstract space's valorisation of consumption and conformity above all else.

The expansion of Westfield Geelong in 2008 by way of a skyway provides a further understanding of abstract space's multi-faceted attacks on the public commons (Westfield Group 2008). A legacy of Geelong's city centre grid layout is the waterfront views on many north–south running streets as the city slopes towards the Bay. The Yarra Street skyway has blocked one of these prized viewlines, effectively privatising a defining cityscape feature to allow the shopping centre to grow into adjacent blocks. In subsequent years, advertising signage sprouted on and around the skyway, replacing the viewlines of place with further solicitations of consumption as if to further underline the point.

This thread of expanding centres of hyper-consumption in Geelong's city centre is replicated on the city's fringes, in the ongoing expansion of internalised shopping centres at Leopold and Waurn Ponds. Rather than addressing their surrounds, these centres front major highways and have limited pedestrian access. This dynamic places the car-dependent beacon of consumption at the centre of suburbs and city centre alike. The continued rise of internalised shopping centres precipitates landscapes of alienation, uniform in their banality and bombardment of solicitations to consume (Bailey 2013; Langman 1992). Thus, the dynamic of homogenisation, consumption and regulation that drives the internalised shopping centre arguably places the internalised shopping centre as Geelong's ultimate expression of abstract space.

It is instructive at this point to refer to the history of the shopping mall itself – a utopian construct of Austrian refugee Victor Gruen intended to insulate the city from itself, by providing respite from natural conditions and the grittiness of the street (Cohen 1996). The echoes of Gruen may well be observed in the *City by the Bay* plan. One may pose the question of whether Gruen's malls became an immediate agent of abstract space, or whether abstract space, as is its wont, appropriated and abstracted the mall itself for its own ends.

Beyond abstraction

To paraphrase Beck and Lefebvre, Geelong's reflexive industrial landscape is one of continued and intensified abstraction. If the city is to seek out a truly post-industrial landscape, it must accept what Zižek (2012) describes as an obligation to think about alternatives. This involves overcoming the asphyxiating tendencies of abstract space and its underpinning networks of power and capital. This of course raises more questions: how? And to what end?

For Lefebvre, the response to abstract space involves a 'right to difference' that 'implies no entitlements that do not have to be bitterly fought for' (Lefebvre 1991, p. 396). In other words, pushing back against the homogenising tendencies of abstract space involves both conscious decisions and action. To loosely quote the Beastie Boys, a 1980s hip-hop band, avoiding abstraction involves a 'fight for the right to be different'. In a capitalist system, the idea of distinctiveness has an economic end explained by English academic David Harvey's idea of 'monopoly rent'. Monopoly rent picks up on a paradox within abstract space: while capital seeks to reproduce itself via abstract space with its homogenising tendencies, it also seeks out monopoly rents via the:

> *distinctive local initiatives, uniqueness, authenticity ... originality and all manner of other dimensions to social life that are inconsistent with the homogeneity presupposed by commodity production* (Harvey 2012, p. 109).

In other words, capital seeks out specialisations and prized elements that cannot be found elsewhere in order to appropriate them for its own ends. This dialectic between homogenisation and distinctiveness exposes a paradox of dysfunction at the heart of abstract space. Equally, distinctive spaces that provide opportunities for monopoly rent imply tension with capital (Christopherson 2000): there must be opportunities for 'divergent and to some degree uncontrollable developments that can be antagonistic to [capital's] own smooth functioning' to emerge and grow (Harvey 2012, p. 109).

Monopoly rent explains an economic imperative for cities to specialise in being themselves. But if we channel Lefebvre truly, distinctiveness becomes more than an economic imperative, it becomes an essential part of 'the right to the city' – with all the baggage that such a right entails. Harvey (2012, p. xv) warns that 'the right to the city is an empty signifier ... everything depends on who gets to fill it with meaning'. The notion of distinctiveness therefore cannot be considered a destination in itself, but a guidepost. The type of distinctiveness demanded to fulfil a broad right to the city must emerge from a nexus between the many voices and stories of the city, a sense of time where history permeates the present, along with the landscape and built environment. Distinctiveness cannot be imposed, but it can be fostered to emerge from what is already there, channelling and conflating the past and present into the future.

Towards the post-industrial ...

The journey towards a post-industrial landscape remains unclear, but there are signs in the city's faded industrial precincts, spaces all but abandoned by rationalist administration. In the Fyansford Paper Mills, a creative enterprise has flourished with galleries, a café, artists' studios and other small-scale creative businesses inhabiting the bluestone buildings. The paper mills' renovation has focused on restoring the built elements and stories of the past, through meticulous refurbishments of building details and new murals recounting the labour-intensive paper production that once happened inside. This has created a space where industrial production has given way to smaller scale, amid a field of narratives and echoes of the past (Gray *et al.* 2017). In the once-abandoned waterfront woolstores, Deakin University has developed a city-centre campus that has juxtaposed industrial patina with modern academia and learning, while the National Wool Museum has operated since 1987 in another former woolstore. These acts of habitation and creation have invested these unique spaces with new stories and meaning, turning industrial relics into richer palimpsests and weaving them once again into the fabric of a city that had all but turned its back.

These stories of creative rejuvenation make it clear that rust-bucket heritage is, to draw from Harvey (2012), a source of monopoly rent for Geelong that might be nurtured. They also hint at a direction for a new paradigm of renewal that draws on a triad of history, landscape and grassroots creativity, and does not neglect the fine-grained and small-scale details that create opportunities for meaning.

As Geelong confronts its future, its city centre is a cacophony of construction through government-funded capital projects, in which jackhammers compete with cars to dominate the soundscape. New office towers rise and compete with public architectural megaprojects for the skyline. Some of these, such as the recently completed Geelong Library and Heritage Centre and, to a lesser extent, the upgraded Kardinia Park Stadium, provide potential repositories of memory, story and accumulated experience that re-tell and create anew the public experience of life in the city. Other mooted big projects, such as office towers and a convention centre on the waterfront, risk merely serving to entrench the patterns of abstraction by potentially privatising public space in the pursuit of fast outside capital, repeating some of the story of Market Square. The critical test of these megaprojects is how, or indeed if, they contribute to the new narratives and interpretations of place necessary to define a post-industrial landscape and if indeed they contribute to the notion of public life in Geelong.

Clearly, a cargo cult-esque fascination with big construction in itself cannot redefine Geelong's reflexive industrial landscape. The way to defining a post-industrial landscape may not be clear, but what is clear is that it demands a fundamentally new way of thinking about the process of city-making in Geelong. This process must not see landscape as *tabula rasa* but as a construct replete with stories and cues for the future. This process of city-making must work at the fine-grained scale as well as the big picture, looking at the spaces between buildings and the stories that lurk around them as much as it does the built fabric. Lefebvre's vision of distinctiveness involves practical action, via everyday acts at a variety of scales. It makes us all responsible for Geelong's post-industrial landscape, and it demands that we re-form city-making to draw from the richness of the world around us. As Lefebvre asks, 'why should spaces created by virtue of human understanding be any less varied, as works or products, than those produced by nature?' (Lefebvre 1991, p. 397). A worthy thought next time one gazes from a hill at the city unfolding by the bay.

A hovering question

Geelong's post-industrial landscape remains at once a hovering question, existential challenge and vital opportunity. As this book goes to press, the city is at a turning point, facing a choice between the emergence of a reflexive industrial landscape of consumption through continued abstraction, or rethinking its process of urbanisation. The theoretical framework formed in this chapter, based on Lefebvre and Beck, demonstrates both the urgency of this task, and what may happen if the status quo is allowed to persist.

Along with a myriad of socio-economic, environmental and indeed existential challenges posed by Geelong's deindustrialisation, the closure of the city's largest industrial sites presents an imperative to explore a new mode of revitalisation that focuses on place as not a blank canvas, but an already rich work ready for new layers of meaning. This is a fundamental and complex task of transforming the idea of transformation, for a city ensconced in ever-accelerating cycles of consumption-based mall-and-sprawl urbanism, planning as though cars continue to roll off the production line.

Yet as this book demonstrates, Geelong is a richer place than the forces of abstract space suggest. The echoes of thousands of years of history and stories of landscape await new interpretation and amplification. It is time for a process of city-making that rises to the situation. We may yet see the existential challenge of a post-industrial landscape as an opportunity.

References

Australian Unity (2013) *Waurn Ponds Expansion Commences*. http://www.australianunity.com.au/health-insurance/sitecore/content/about-us/home/news-and-views/news-folder/waurn-ponds-expansion-commences (accessed 1 March 2018).

Bailey M (2013) The central focus of my life: Roselands and youth culture in the 1960s. *Sydney Journal* **4**(10), 112–116. https://researchers.mq.edu.au/en/publications/the-central-focus-of-my-life-roselands-and-youth-culture-in-the-1 (accessed 1 March 2018).

Beck U (2009) *World at Risk*. Polity, Cambridge.
Brownhill W, Wynd I (1990) *The History of Geelong and Corio Bay, with Postscript 1955–1990*. Wilks & Co., Melbourne.
Christopherson S (2000) The fortress city: privatized spaces, consumer citizenship. In A Amin (Ed.) *Post Fordism: A Reader*, 5th edn (pp. 409–427). Blackwell Publishers, Oxford.
Cohen L (1996) From town center to shopping center: the reconfiguration of community marketplaces in postwar America. *American Historical Review* **101**(4), 1050–1081. doi:10.2307/2169634
Colacino A (1966) *Geelong (Life in Australia Series)*. Australian Commonwealth Film Unit, Department of Immigration, Canberra. https://www.youtube.com/watch?v=BfFYc4C_JCA (accessed 1 March 2018).
Essential Economics (2006) *City of Greater Geelong Retail Strategy*. City of Greater Geelong, Geelong. https://www.geelongaustralia.com.au/common/public/documents/amendments/8ce70eeea3d8008-Essential%20Economics%20-%20Retail%20Strategy%20Final%207%20(June%202006).pdf (accessed 1 March 2018).
Friedmann J (2010) Place and place-making in cities: a global perspective. *Planning Theory and Practice* **11**(2), 149–165. doi:10.1080/14649351003759573
G21 GRA (2013) *G21 Regional Growth Plan 2013*. Geelong Regional Alliance, Geelong. http://www.g21.com.au/g21-regional-growth-plan-2013 (accessed 1 September 2018).
Gleeson B (2014) *The Urban Condition*. Routledge, London.
Goonewardena K, Kipler S, Milgrom R, Schmid C (2008) Abstract. In K Goonewardena, S Kipfer, R Milgrom and C Schmid (Eds) *Space, Difference, Everyday Life: Reading Henri Lefebvre*. Routledge, London.
Gray F (2015) The misanthropes, larrikins and mallrats of Market Square: an enduring public space dilemma in central Geelong. In *SAHANZ 2015: Architecture Institutes and Change. Proceedings of the 32nd Annual Conference of Society of Architectural Historians Australia and New Zealand*. 7-10 July 2015, Sydney (pp. 218–230).
Gray F, Garduño-Freeman C, Novacevski M (2017) Milling it over: Geelong's new life in forgotten places. *Historic Environment* **29**(2), 58–70.
Harvey D (2012) *Rebel Cities: From the Right to the City to the Urban Revolution*. Verso, London.
Heritage Victoria (2017) *Ford Motor Company Complex*. Victorian Heritage Database. http://vhd.heritage.vic.gov.au/places/result_detail/12548?print=true&style=heritagevic (accessed 1 March 2018).
Langman L (1992) Neon cages: shopping for subjectivity. In R Shields (Ed.) *Lifestyle Shopping: The Subject of Consumption* (pp. 43–71). Routledge, London.
Lefebvre H (1991) *The Production of Space*. Blackwell Publishing, Malden, UK.
Milner P (1985) The Barwon Paper Mill, Fyansford. *Heritage Australia*, Autumn, 18–21.
NFSA (2018) *Life in Australia: Geelong [1966]*. National Sound and Film Archive. https://www.youtube.com/watch?v=BfFYc4C_JCA (accessed 1 March 2018).
Prigge W (2008) Reading the urban revolution: space and representation. In K Goonewardena, S Kipfer, R Milgrom and C Schmid (Eds) *Space, Difference and Everyday Life: Reading Henri Lefebvre* (pp. 46–61). Taylor & Francis, London.
Proudfoot H (2000) Founding cities in nineteenth-century Australia. In S Hamnett and R Freestone (Eds) *The Australian Metropolis: A Planning History* (pp. 11–26). Allen & Unwin, Sydney.
Regan A (2010) Not building the 'City by the Bay': redeveloping the Geelong CBD, 1980 to 1993. In D Nichols, A Hurlimann, C Mouat and S Pascoe (Eds) *Green Fields, Brown Fields, New Fields: Proceedings of the 10th Australasian Urban History, Planning History Conference*. University of Melbourne, Melbourne (pp. 507–518).
Soja E (2010) *Seeking Spatial Justice*. University of Minnesota Press, Minneapolis.

Townsend H (2008) Born to run. *Griffith Review: Cities on the Edge* **20**, 193–200. https://griffithreview.com/editions/cities-on-the-edge/
Vicinity Centres (2015) *Gateway Plaza Development*. https://www.gatewayplazaleopold.com.au/centre-info/development (accessed 1 March 2018).
Westfield Group (2008) History plays its part: Geelong's heritage has not been forgotten as Westfield has evolved. *Geelong Advertiser*, 17 July 2008, p. 4.
Wynd I (1971) *Geelong: The Pivot*. Geelong Historical Society, Geelong.
Žižek S (2012) *The Year of Dreaming Dangerously*. Verso, London.
Žižek S (2015) *Trouble in Paradise: From the End of History to the End of Capitalism*. Penguin Books, London.

16

Event, fall, return and the transformation of sites: a diagnosis for Point Henry

Gavin Keeney and Owen O'Carroll

All true art is in fact nothing but an attempt to transmit the sensation of ecstasy.

– Pyotr D. Ouspensky (1931, p. 133).

The condition ...

Landscape as microcosm has a strange way of smacking us over the head from time to time – of knocking us for a loop, as the old expression goes. It is that sentiment that needs to enter or re-enter landscape design and urban planning. A certain amount of humility is, then, in order. Oddly or not, it is students who can perceive this most readily, insofar as they are not shackled in advance with representational systems or post-rationalisations that are drawn from architectural and design biases. Equally, the introduction of spectacle and anti-spectacle can alter perception. Performance-based media, while not the goal of such experiments (as it often is in the art world), offers a clearly accessible pathway to a total eclipse of design rationality, opening the insights required to alter the path of architectural interventions and hence the rational reality of landscape.

It can be hard sometimes to see such post-rationalisation as the 'art work', assuming that 'art' is effectively an enduring transmission. Yes, as part of an 'event', post-rationalisation can be undoubted and visible, and can offer relays and interpretations in the hands of a guide. But post-rationalisation can never be a duplicate, a simulacrum for the real deal. If spectacle is what appears, subjective, interpretative and conjectural in time, if the image needs culturing, then surely the doing of the culturing of imaging in the event is far greater and more meaningful than its post-rationalisation. 'Intentionality of end' is an art form in its own right or rite – one of the most beautiful, probably, it may be said, far beyond pretence. This is where 'design' or, better, '*dasein*' lies.[1] In landscape the performance of design (the 'spectacle'), against the sheer vastness of landscape's '*dasein*', opens a pathway to an insight of its reality.

Sites have an enormous backstory, a backstory that often includes a pre-history that is only valid in reference to the anthropocentric view that is the bias of human culture. Urban post-industrial sites can offer classic palimpsests redefined as heritage or turned to other, perhaps even 'immemorial' purposes.[2] Immemoriality is hardly a fictive gesture – the land in and for itself holds such promises in escrow.

The conflicts of landscape urbanism, where ecological precepts are combined with the pretences of urbanisation and conventional land use, whether passive or active, are one example of how event, and the genealogy of palimpsests, may be read over time – intentionally, haphazardly or by default. What generally intervenes in such cases is the past or future presence of infrastructure, with the ecological or dynamic principles of 'native' systems reduced to a network of proto-events that privilege 'nature' although, in fact, it is the bias of urban systems that holds sway. In the case of Geelong's Point Henry, a site tied to Aboriginal dreamlines through a record that is ostensibly only kept in the minds of the Aborigines, this pre-history is part of a 30 000-year 'memory' of the agency of the north-facing peninsula.

The decommissioned Alcoa aluminium smelting plant, a relic of Australian industrial heritage, sits awaiting its uncertain future as memorial, as ruin, as memory. Projects undertaken in the spring of 2014 by students at Deakin University sought, from within this complex configuration of use, disuse and possible re-use, access to the immemorial reserve the site preternaturally embodies. The natural and semi-natural histories of the peninsula are irrelevant, not extant as such, and many are speculative or imaginary. They are a weightless version of 'history' – or the backstory.

The abandoned Alcoa facility sits at the south end of Port Phillip Bay on a north-facing spit of low-lying land that serves as bridge in time and space between an armature of the south Australian coastline and the inland sea of the Bay reaching towards Melbourne, first site of adequate fresh water found by early colonists. As part of a mythology, the peninsula and the inland sea, the low-lying point, plus shallow watery passage (passable at low tide), is perceived as a causeway, the symbolic crossing point to a distant mountain range (the You Yangs) suggesting telluric forces at play from point to peak.

Projects such as Geelong's post-industrial revitalisation thus emerge from a mixture of memory and history, often imagined, retold and re-created, and do not necessarily invalidate the claims of design's intentions. 'Writing' from memory (true or imagined) is an art – 'design' in such cases is actually a subtle form of benign and informed neglect, operating somewhere between memory and history. Paul Ricoeur's (2000a, b) sense that neither memory nor history is a valid force in its own right to claim ownership of design or place, is an admission that we have interpretive freedoms in regard to the agency of memory or history. And if memory and history in Geelong's case are not the same thing – the former being primarily subjective, while the latter is at least intended to be objective – such projects enter into dialogue with mythic reserves of the site and place, an internal and external form of conversation.

This might be one of the few truly valid positions to offer in late-Anthropocene land planning and landscape-architectural design. The traditional biases noted in landscape-urbanist causes are sufficient to formally and purposively negate the figurative 'crossing' of the sites in question, yielding yet again to the very immemorial resources that have come to reside in Alexander Pope's well known and well abused trope (borrowed from Pliny the Younger) of genius loci (Hunt 1988).[3]

An antidote to anthropocenic landscapes: commons, ownerships and reality

The proliferation of theories and the circular discourses they engender, in both academic and professional circumstances, appear endless. Russian philosopher Pyotr Demianovich Ouspensky, surveying the sociologically biased landscape of what he calls pseudo-science from 1911 to 1929, while looking for older sources of knowledge that might short-circuit those that characterise modernism, or even the humanist agendas of modernity, captures this circularity in a particularly damning diagnosis. Its antidote is a return to a level of cognition that embraces systems that modernity would otherwise render obsolete as forms of mystification:

> *Where then is the way out? And is there a way out? It must be recognised that nobody knows. One thing is certain, that not one of the ways out offered to humanity by its friends and benefactors is a way out in any sense. Life becomes only more entangled and more complicated, but even in this entanglement and in these complications it does not take any new forms but endlessly repeats the same infinitely old forms* (Ouspensky 1931, p. 511).

This diagnosis gives ways to an inferred antidote that would shatter many of the premises of the pseudo-scientific approaches to problem-solving today, foremost because Ouspensky is invoking what would be called irrational sources by claiming that man has turned his back on a vital part of our heritage. This pattern can be seen in design disciplines as they cannibalise the past, fall prey to eviscerating history, authorise and enforce enslavement in the prevailing idioms of the day, and generally disown that which paradoxically underwrites the very existence of the fashionable discourses that are 'of the moment':

> *Having ceased to believe in divine help and in the superman, people for a long time believed in man, in creative personality, in a leader, in genius, in a hero. But gradually the belief in personality in history began to fade away, and in the place of personality people began to believe in theories. Now people believe that with the aid of theories or by following theories it is possible to alter and reconstruct life completely. At no other time in history have theories had such power and such wide recognition as in our time* (Ouspensky 1931, pp. 511–512).

This suggests that the circularity of the pseudo-scientific method and theory of design is approaching exhaustion, and that the 'Capitalocene' – as a broader category of exploitation – is forced to search elsewhere for new resources. The latest trend of appropriating Indigenous knowledge (not Indigenous peoples) into the circuit of Capital by forcing such knowledge into the public domain, is but one symptom of a search for an antidote (Moore 2015; Soper 2016). How often have we seen the simplistic deployment of misunderstanding representing new design and fashion, the alleged contact of sensorial experience in landscape and space?

Any number of versions of 'Landscape Urbanism' illustrates this questionable and contestable appropriation, usually under the name of 'emergence'. These pseudo-ecological systems have little to do with ecological systems *per se* and far more to do with attempts to capture ambient or dynamic systems in the charmed circles of urban design for the benefit of Capital's extraction through urban systems.

Any sincere attempt to alter the processes of urbanisation that are increasingly and artificially complex and entropic would require a complete rethinking of the idea of the public domain and/or the commons.[4] This latter term, while currently the most fashionable element of urban design speak, actually hides the much older idea of 'no ownership'. If no ownership has become public ownership, and if public ownership has devolved to public–private stewardship of the urban environment – with the aim of shunting cost and risk onto the private domain, a shift which monetises public space and levies overt and hidden fees on its users – then the process does, in fact, need to be reversed. How that is to be done is a question both for urban design and for what is left of the guardians of so-called public resources and public freedoms.

The return

If the city is to be considered a 'garden' – a complex set of visual and infrastructural elements and occasional ruins – it is instructive to consider how gardens function. The categorisation of gardens is a quite clumsy affair, historically or otherwise: formal or informal, wild or tamed, baroque or modern, classical or neoclassical and so on. These terms are always applied retrospectively. If the post-modern condition is the privileging of the proverbial 'now', when everything is present all at once and historical time or the march of progress has effectively collapsed, gardens are the perfect embodiment of such a sentiment. They are notoriously unstable in terms of their owners' intent and constantly changing montages of effects and affect – of materiality and phenomenon.

It is possible, therefore, that the best description of gardens is 'cinematic', as might cities be (there is a rather inspired literature on the cinematic aspects of urbanism, both from the design and the observational points of view). As gardens transform with the seasons, or as they evolve in a type of time-lapse or stop-motion photography, most often recorded in the eye of the beholder but also captured in amateur and professional photography, they reveal dynamic principles of exchange that resemble metaphysical-ecological precepts. Perception slides towards apperception. There are moments, for example, when the garden 'enters' the house, as light enters a camera obscura, and the image of the outside appears on the inside. This image is sometimes moving; for example, miniature videos playing on a kitchen wall, or the filtered light of late day casting a spectral, animated 'net' over a dusky interior. The apparatus of camera or house actually mimics perception itself, plus the subjective states induced in any real contact with the world at large.

This classic perspectivity (purposive or otherwise) is the equivalent of an art project that records the city at large through the perceptive or perspectival lens of the fenestration of buildings. There is no end to the examples of how architects 'play' with this trans-effectivity. Architectural and landscape-design discourses privilege these phenomenal effects through the polarities of inside/outside, light/darkness, moment/season, etc. – all such intuitions often appropriated towards misreadings of canonical works of modernism (symptomatic of historical post-rationalisations). For example, 'when and where' doors are quite literally 'blown off' the building mass can suggest accord with an external environment. In this regard, the Barcelona Pavilion must forever be considered Exhibit A (Quetglas 2001).

The value of landscape is, therefore, always in the eye of the beholder. The appreciation of gardens is an intensely personal affair, in an existentially profound way, even when they are public gardens. As incredibly complex assemblages, gardens require constant maintenance or, in the case of a wild or naturalistic garden such as Geelong's Point Henry, the apparent opposite – intentional benign neglect. Intentionality is the key. The art of the garden requires, regardless of provenance and regardless of location or purpose, an editioning strategy that resembles the choices made to produce a book or a work of art, with the result 'archived' in the living milieu of the setting. Urban landscape is not dissimilar.

These moving images of gardens are also moving hieroglyphs, personalised codes that disintegrate in a higher register or zone that becomes intensely felt – an emotional or ethical electrical charge kicking in where a relationship between 'things' and concepts and 'dasein' emerges. In effect, and certainly as affect, these hieroglyphs signal across time and transformations the editioning strategy of the landscape, the garden, as book or work of art. In the urban setting they would signal the real presence of real public space, freeing the image from the artefact, setting it adrift in a sea of inner speculations – a micro-world held in tension with a larger 'outside' or 'other' that is the mark of the 'Il y a' (as conceptualised by Maurice Blanchot and Emmanuel Levinas).[5] This 'Il y a' is, in Lacanian exegesis, the spectral form the Real takes, an impossible Other that never arrives but nonetheless seems to exist. This is the 'crossing', the shallow hollow, the knee-deep abyss, and – for Geelong's Point Henry – the moment of transformation.

Geelong's landscape might then be emotionally or existentially kicked further 'upstairs' or 'upstream' (quotidian perception coupled with renewed apperception), to a zone where far-fetching or highly moral speculations appear, occur or wage war. Often such speculations are only tangential to the landscape or the landscape designer, intimating a sense that something miraculous is at play, even if in the Lacanian worldview that miraculous thing at play is the proverbial 'ding' (time as thing).

All of this is to say that the aesthetic hides the moral, and vice versa, and landscapes observed are landscapes felt. The historical, if it is to have any value, is found embedded in the time senses of the garden or city, with the verb 'as found' to be the first point of entrance or purchase (point of departure or leverage) for gardens or cities as editioned works. 'To be' in the life of the garden is 'to change'.

'To be' in the life of Geelong is 'to change'. This possibility of change affects the designer and the subjective field that produces landscape as artefact – the 'outside', which situates and re-naturalises representational time, turns towards the interior (the garden or the image of the garden) and the 'inside' turns towards the exterior (the world beyond the garden and the existential fold in which it sits and/or operates). Extend this to the wider landscape and Point Henry comes into view. Its spectral significance is based on its being a low-lying, north-facing landmass at water's edge, plus tidal flats, plus distant promontory. It is, needless to say, singularly cinematic in the extreme.

Notes

Ouspensky defines 'ecstasy' as 'the highest degree of emotional experience'. What Ouspensky is instilling through this remark is the necessity of transcending mere intellectual engagement in the production of works of art. He adds, 'Emotional life must also evolve, in certain comprehensible forms. And the chief change in man must come precisely from the evolution of emotional life' (Ouspensky 1931, pp. 133–134). Ouspensky's bias in this context is the individual person, not the collective. Yet it is implied that an individual will impact the collective in the manner of increasing emotional intelligence through works of one order or another.

In terms of Blanchot and Levinas' approach to 'Il y a', the 'Il y a' is impersonal and generally represents a threat to subjects:

> *There remains after this imaginary destruction of things, not something, but the fact that there is [il y a]. The absence of all things returns like a presence: like a site wherein everything has sunken, like an atmospheric density, like a plenitude of emptiness or like the murmur of silence. There is, after this destruction of things and beings, the 'field of forces' of existing, impersonal. Something that is neither a subject, nor a substantive. The fact of existing, which imposes itself when there is nothing more. And it is anonymous: there is no one and nothing that might take this existence upon itself. It is impersonal like 'it is raining' or 'it is hot.' An existing that returns, whatever be the negation by which one sets it aside. This is something like a pure existing that cannot be remitted.*[6]

Endnotes

1 The word 'dasein' came from the writings of German philosopher Martin Heidegger (1889–1976) who developed his own theory of being (which separated him from other philosophers), that he called Dasein. Dasein is derived from the verb 'dasein', which simply means 'to exist'. Heidegger perceived that dasein consisted of two parts: (1) it is the human being, and (2) it asks the question of itself 'what is being?' Under this definition, dasein is everything: 'the primary phenomenon is not the world, space, time, or history, but dasein's being in the world, dasein's spatiality, dasein's temporality, or dasein's historicity'. Thus, dasein is neither just an insect, nor a God, but in between (Inwood 1997, p. 89).

2 'Immemorial' means extending or existing since beyond the reach of memory, record or tradition.

3 '*Genius loci*' was a Latin term pertaining to the classical Roman religion wherein a *genius loci* was the protective spirit of a place. Today, *genius loci* is often used to describe a place's distinctive atmosphere or 'spirit of place', rather than a guardian spirit. Eighteenth-century English poet Alexander Pope (1688–1744) elevated *genius loci* to a key principle in garden and landscape design by the term's use in the *Epistle IV, to Richard Boyle, Earl of Burlington* that argues that landscape designs should *always* be adapted to the context in which they are located.

4 'Entropic' means having a tendency to change from a state of order to a state of disorder.

5 *Il y a* is a French expression that means 'there is, there are' or literally 'it has there' that, as an example, can be expressed in a sentence as 'There are some problems with this arrangement'.

6 See Levinas (1948) *Le temps et l'autre*, pp. 25–26, cited in Levinas (2003, p. 25).

References

Ho S (2014) *Point Henry*. Video still. https://vimeo.com/147295123 (accessed 1 March 2018).

Hunt JD (1988) *The Genius of the Place: The English Landscape Garden 1620–1820*. MIT Press, Cambridge, MA.

Inwood M (1997) *Heidegger: A Very Short Introduction*. Oxford University Press, Oxford.

Levinas E (1948) Le temps et l'autre. In *Le choix, le monde, l'existence* (pp. 25–26). Arthaud, Paris.

Levinas E (2003) *On Escape: de l'évasion* (trans. B Bergo). Stanford University Press, Stanford, CA.

Moore JW (2015) *Capitalism in the Web of Life: Ecology and the Accumulation of Capital*. Verso, London.

Ouspensky PD (1931) *A New Model of the Universe: Principles of the Psychological Method in its Application to Problems of Science, Religion, and Art* (trans. RR Merton). Kegan Paul Trench Trubner & Co., London.

Quetglas J (2001) *Fear of Glass: Mies van der Rohe's Pavilion in Barcelona*. Birkhauser, Basel.

Ricoeur P (2000a) L'écriture de l'histoire et la répresentation du passé. *L'école des hautes études en sciences sociales*. http://cmb.ehess.fr/95 (accessed 1 November 2017) [abstract for a paper of the same title presented by Ricoeur at XXIIe Conférence Marc-Bloch, June 13 2000].

Ricoeur P (2000b) L'écriture de l'histoire et la répresentation du passé. *Annales* **55**(4), 731–747. doi:10.3406/ahess.2000.279877

Soper K (2016) Capitalocene. *Radical Philosophy* 197. https://www.radicalphilosophy.com/reviews/individual-reviews/capitalocene (accessed 1 November 2017).

17

The promise of vision-making a city: a perpetual journey

John Rollo and Yolanda Esteban

A city is not gauged by its length and width, but by the broadness of its vision and the height of its dreams.

– Herb Caen (cited in OECD 2015, p. 119).

Introduction

This chapter is about the vision-making of Geelong with a particular focus on its inner-city area as defined 180 years ago in the initial town allotment plan by surveyors Robert Hoddle and W.H. Smyth in 1838. It concentrates on the four decades between 1975 and 2017. To assist readers, Table 17.1 sets out a chronological inventory of the studies and reports relevant to the planning of Geelong's central business district (CBD).

The four decades between 1975 and 2017 are defined by two key documents. The first is *Geelong: City by the Bay* (GRPA 1975), a 25-year CBD strategy developed in 1975 by the Sydney-based consultancy, The Planning Group, for the Geelong Regional Planning Authority, which, at the time, was anticipating a population projection within the region of some 400 000 people by 2001. The second is the 30-year community-led vision *Greater Geelong: A Clever and Creative Future* (CoGG 2017), which looks forward to 2047 when it is anticipated that the growth projections of the early 1970s will eventually be realised. *Greater Geelong: A Clever and Creative Future*, in all but name, is embedded with the hallmark of a 21st-century 'smart' cities armature readily adaptable to a metrics of ever-evolving information and communications technologies. In addition to these two documents, the chapter presents no fewer than 22 other vision studies. In recounting changes in the governance structure of the city, and hence the balance between top-down approaches and the more inclusive horizontal-based community-led strategies, the analysis of these documents also appears to reflect changes in the political, economic and social atmosphere of the times. Mapping different sets of aspirations, as Geelong transitions from a strong large-scale manufacturing focus, through a protracted period of economic transition and self-reflection, to emerge as a dynamic, novel and highly progressive 21st-century city, the various visions also seem to reflect changing attitudes with respect to identity, esteem and resilience. The chapter opens with the current vision, *Greater Geelong: A Clever and Creative Future* (CoGG 2017).

Table 17.1. Chronology of Geelong visions

Report	Date
Geelong: City by the Bay – Geelong Central Business District Study, Geelong Regional Planning Authority	June 1975
Geelong Central Area Study: Preferred Strategy Public Discussion Paper, Geelong Regional Commission	February 1979
Geelong Central Area Strategy, Geelong Regional Commission	January 1981
Geelong: City by the Bay, Geelong Regional Commission	December 1981
Geelong Regional Development Strategic Plan, Geelong Regional Commission	December 1985
Geelong Urban and Township Development Strategy, Geelong Regional Commission	December 1985
Geelong Region: Framework for the Future, Development Strategy, Geelong Regional Commission	February 1986
City by the Bay: Foreshore Precinct Project, Geelong Regional Commission	January 1988
Directions: The Geelong Regional Development Strategy, Geelong Regional Commission	June 1988
Geelong: City by the Bay – The Bay Link, Geelong Regional Commission	July 1988
Geelong and Foreshore Draft Structure Plan: Discussion Document, City of Greater Geelong	March 1990
Geelong Regional Local Government Review: Economic Assessment of Local Government in the Geelong Region, Geelong Regional Commission	September 1990
Directions: The Geelong Regional Development Strategy – Five-year Implementation Plan 1991/92, Geelong Regional Commission	June 1992
Central Geelong Foreshore and Environs Framework, City of Greater Geelong	July 1993
Geelong 2010: A Preferred Vision, Vol. 1 – Report, City of Greater Geelong	October 1994
Geelong 2010: A Preferred Vision, Vol. 2 – Community Consultation Information Bank, City of Greater Geelong	October 1994
Waterfront Geelong Design and Development Code, City of Greater Geelong	1996
Geelong Western Wedge Framework, City of Greater Geelong	April 2005; updated September 2005
Geelong Transit City Station Precinct Urban Design Framework, City of Greater Geelong	July 2008
Central Geelong Urban Design Guidelines, City of Greater Geelong	October 2008
Central Geelong Vision: Design Studio Report Draft, City of Greater Geelong, Deakin University	March 2013
Central Geelong Action Plan, Central Geelong Task Force, City of Greater Geelong	December 2013
Geelong City Centre: City in the Park, City of Greater Geelong	July 2015
Revitalising Central Geelong Action Plan Department of Environment, Land, Water and Planning	2016
Greater Geelong: A Clever and Creative Future, City of Greater Geelong	August 2017

A clever and creative future

On 30 August 2017 a time capsule was buried at the base of the steps outside the east façade of Geelong's City Hall, marking the centenary of its construction. The time capsule contained three things – a copy of that day's edition of the local newspaper, the *Geelong Advertiser*, a token selection of present-day Australian currency, and both a printed and digital copy of *Greater Geelong: A Clever and Creative Future* (CoGG 2017). This visionary document, which lit the spark to implement a 30-year community-led series of aspirations on what the people of Greater Geelong want, had been developed over six months of consultation and workshopping contributed to by a staggering 16 000 people from the Greater Geelong region. Surprisingly, the *Greater Geelong: A Clever and Creative Future* (CoGG 2017) does not contain a graphically enhanced strategic master plan, or a series of hyper-realistic computer-generated renderings featuring futuristic visons of what the city might be in three decades. Instead, it presents a narrative of nine aspirations assessed against five elements that were collectively acknowledged by workshop participants as making a successful community. These are underpinned by no fewer than 37 indicators, or measures, each with a corresponding baseline. These measures provide the hallmark of a smart cities framework, readily adaptable to the constantly evolving information and communications technologies that will not only increase operational efficiency and improve the quality of services and the community's welfare but, through information-sharing and data-sensing, will allow the vision to remain current and responsive into the future. The indicators include aspects such as the percentage of tree canopy coverage across the city and access to public open recreational space, the proportion of journeys to work made by walking, cycling and public transport, and other sorts of data that will be progressively compiled relating to community participation rates, the diversity of business structures, and the drive to ensure that all Greater Geelong postcodes are above the bottom 20% of socio-economic indices (CoGG 2017).

The nine aspirations include:

1. a prosperous economy that supports jobs and education opportunities
2. a leader in developing and adopting technology
3. that creativity drives culture
4. a fast, reliable and connected transport network
5. a destination that attracts local and international visitors
6. people feel safe wherever they are
7. an inclusive, diverse, healthy and socially connected community
8. sustainable development that supports population growth and protects the natural environment
9. development and implementation of sustainable solutions (CoGG 2017, p. 11).

The five elements of a successful community are that it is:

1. connected
2. sustainable and resilient
3. prosperous
4. creative
5. designed for people (CoGG 2017, p. 11).

The challenge is to position the community so that:

By 2047 Greater Geelong will be internationally recognised as a clever and creative city-region that is forward looking, enterprising and adaptive, and cares for its people and environment. People have told us what they value about the region, how they rate it against the recognised elements of a successful community, and shared ideas about how to address challenges now and in the future (CoGG 2017, p. 1).

Given the city's recent winning bid to be named a UNESCO Creative Cities Network, City of Design, in October 2017, it would seem that the shift to a community-inspired way of thinking is off to an excellent start. Indeed, *Greater Geelong: A Clever and Creative Future* is not to be read like a council's three-year strategic plan, where various local government departments are required to meet certain key performance criteria. Neither is it a politically charged 'build it' or 'fix it' concept, which are often rolled out for short-term political gains.

While it has been facilitated by the city, council is not the responsible authority to make change happen. Rather, as Andrew Kerr, the CEO of the City of Edinburgh stated in an interview with City AM radio regarding the *2050 Edinburgh City Vision*, it is the community who shares the responsibility 'developed, built and shared with residents, businesses, and public agencies' (City AM 2016) 'the greater part of the legacy of Edinburgh is not its buildings, but its people' (City of Edinburgh 2017, p. 1). Hence it encompasses a step change in the way that the community both collectively and as individuals think about the future. It therefore is not a structured top-down approach, but something that strives to be inclusive of the whole community and allows ideas to be nurtured from a grassroots level, facilitating a more broad-based approach to change. Acknowledging that ideas and positive change can occur at a local scale mean that perception can be shifted from obligation and at times imposed, to that of being locally owned, as Margaret Ledwith and Jane Springett write in *Participatory Practice: Community-based Action for Transformative Change*:

Participation is transformative. It is central to the purpose of community development in striving for a just, healthy and sustainable world ... A participatory worldview sees life from a horizontal perspective, juxtaposing everything and everybody as a world in common, in all its difference and diversity.... [and] *is predicated on human worth, not as some fanciful utopian dream, but as a component of an ecosystem in fragile balance* (Ledwith and Springett 2010, p. 189).

This process of transformative change is not new to the Geelong community – a similar process had been entered into almost a generation previous in October 1994 with the development and release of *Geelong 2010: A Preferred Future* (CoGG 1994). This vision engaged a very similar method to *Greater Geelong: A Clever and Creative Future* (CoGG 2017). Launched on the back of a seminar presented by Dr Peter Ellyard, Executive Director of Preferred Futures, on the theme of 'Choosing our Future', the vision was directed by a steering committee chaired by the then Vice Chancellor of Deakin University, Professor John Hay, and involved over 2000 people and organisations in the extensive public consultation process. This was arranged around a wide program of forums, 'Grouputer' sessions and community response forms. It should be noted that this was one of the first significant community participation programs to have taken place on a city-wide scale.

It was inspired by the work of Christopher Alexander and the Oregon Experiment in 1975 (Alexander *et al.* 1975) and probably had a positive influence in encouraging other local government authorities to engage in similar public consultative processes after the state-wide amalgamation of Local Government Areas during the early 1990s. More recently, Dr Phillip Roös further explored and applied the fundamentals of *A Pattern*

Language (Alexander *et al.* 1977) and *The Nature of Order* (Alexander 2001–05) in his research with communities along the Great Ocean Road coast, where local knowledge is used to inform generative patterns and codes, similar to Alexander's work in the Oregon Experiment (Alexander 2001–05; Neis *et al.* 2012; Roös 2017, 2014).

The extraordinary success of the engagement process followed by the *Geelong 2010: A Preferred Future* (CoGG 1994) team compares with that of only three years earlier when the *Geelong and Foreshore Draft Structure Plan* (CoGG 1990), developed by Ratio Consultants for the City of Geelong in March 1990, received fewer than 50 responses from the community survey.

The narrative for the 2010 vision presents 11 aspirations built on the extensive database of responses from the consultation process. The aspirations include:

1. a new community – a caring and cohesive Geelong
2. an ecologically sustainable Geelong
3. generating jobs
4. an Arts and Cultural Centre
5. an Education and Research Centre
6. a Centre for Tourism
7. a major transport hub
8. an international wool capital
9. a Centre for Smart Component Manufacturing
10. creating a healthier Geelong
11. re-empowering our Elders (CoGG 1994, p. 9).

The unique aspect of the report was that it crafted the narrative for each aspiration in a retrospective process. Using imagination and projecting forward, each chapter is written in the past tense as if the time of authorship was 2010 and the description was an account of how the city and its community had achieved that aspiration:

> *In 1994 we were just emerging from a long period of recession which had affected our region with particular severity. Geelong had acquired a somewhat depressing self-image from media preoccupations with its problems: closure of industry, unemployment, the failure of the local Pyramid financial institution. We were continually told we had a 'victim mentality', that the old 19th century title of 'Sleepy Hollow' was again applicable to our city, that our best and brightest, including many of our young people were leaving for better opportunities elsewhere, and that our long history of divisiveness amongst civic leaders had created intractable problems which were barriers to development. We had failed to change, modernise and adapt ourselves, and as a result our destiny was being shaped by others ... there were sceptics who regarded the concept of inviting a community to imagine its preferred future as indulgence and mere fantasy: something far removed from the pragmatic, 'economic rationalist' attitudes of those times. We intended to prove that they were wrong, and to demonstrate that the active involvement and participation of the community was essential to the ultimate success of any strategy for the future .... We invited the community to tell us the basis for developing strategies for change: what they most liked and valued about the city, and what they most disliked* (CoGG 1994, pp. 11–12).

The opportunity to engage in *Geelong 2010: A Preferred Future* (CoGG 1994) was provided in 1993 when the state government of Victoria reshaped the Local Government

Areas through an amalgamation process which combined six former councils in the region into a single entity.

The vision was driven by three goals to prepare the city for the 21st century:

1. to build an innovative culture in Geelong – a culture which would be creative, learning, collaborative, intelligent and enterprising and would be constantly seeking to begin doing new things relevant to the times
2. to establish an infrastructure and enterprise to produce products and services, which we anticipated would be needed nationally and globally in 2010
3. to create and sustain a high quality of living (CoGG 1994, p. 14).

Having undergone a significant setback after the collapse of the Pyramid Building Society in 1990, *Geelong 2010: A Preferred Future* helped lift the morale and spirit of the Geelong community by enabling them to dream of a more vibrant and prosperous future. With a 2000-strong community-supported vision, it undoubtedly had a significant influence in attracting both federal and state government support to re-invigorate the waterfront through the purchase of the Dalgety and Denny Lascelles Woolstore buildings by Deakin University in 1994 (Deakin University 2018).

Predicating a vision

> *Vision: (Noun) 2nd definition - 'The ability to think about or plan the future with imagination or wisdom'* (OUP 2018).

Planning in the 18th and 19th centuries, in the western world, was a product of the Enlightenment and the predominance of scientific and rational thinking. The concept of nation-building and imperial expansion, of vision-making towns, cities and colonies on a continent that had only been circumnavigated 35 years earlier by Matthew Flinders, was to design and imagine a future that was well beyond the lifetime of those working on the vision. Structured on the mapping undertaken by Assistant Surveyor W.H. Smythe, and the tartan grid subdivision made up of smaller streets inserted between the main principal streets on an east–west axis developed by Robert Hoddle (Surveyor in Charge at Port Phillip), the resulting 1838 plan provided the framework on which the morphology of the city still presents itself today (see Fig. 17.1). Linking the waterfronts of Corio Bay and the Barwon River, the plan presented a rectangle of 36 blocks on a north–south axis, bounded by Barwon Terrace in the south and Corio Terrace in the north, with Swanston Street defining the east boundary and Gheringhap Street the west. Its ideas were grand and the future full of opportunity.

After the first land sales on 14 February 1839, the population grew steadily during the 1840s from 545 in 1838 to 8291 by 1851. Melbourne at the time had a population of 23 000, while the whole of the Port Phillip region, then still a district of the colony of New South Wales, had 77 000. Between 1851 and 1853 Geelong's population nearly tripled and when gold was discovered near Mount Buninyong it peaked at 23 314. From then on growth in Geelong was relatively slow and steady, taking another 80 years to double its population to just over 40 000 by 1933. Hence, it was not until 1910 that Geelong received 'city' status. By the end of World War II, population in the region had increased to around 66 000 (1947). With post-war migration during the 1950s and 1960s population boom, the city's population grew rapidly at a rate of over 3.3% pa over a seven-year period. It had grown to 89 000 by1954 and 110 000 by 1961. Coopers Lybrand,

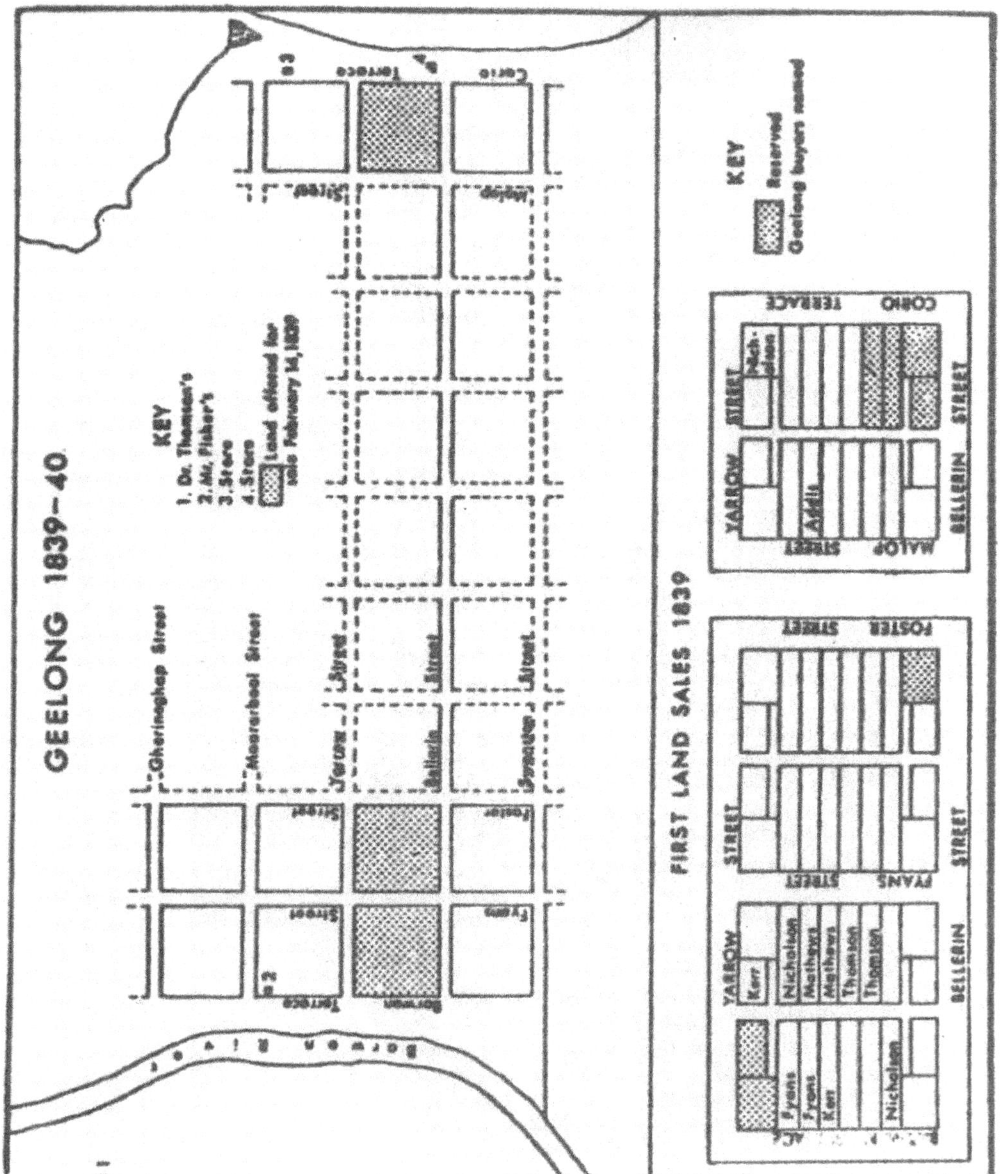

Fig. 17.1. 'Geelong 1839–40', illustrating the areas sold at the first land sales. Source: GAG (1985, p. 3).

in a report prepared for the Geelong Regional Commission (GRC 1993), categorised Geelong's growth around four phases:

- 'The hinterland served by the city was initially developed, e.g. pastoral expansion of the Western District in the 1840s' (GRC 1993, p. 8).
- 'The scale of the hinterland served by the city was extended through the development or application of new technology, e.g. closer settlement of the Wimmera

Mallee and Southern Riverina and the establishment of railway linkages between these areas and Geelong' (GRC 1993, p. 8). The industrial waterfront expanded considerably with the development of railway sidings in the 1890s curving from the main line out onto Cunningham Pier. Cutting underneath Mercer Street, the line swung past Corio Street, crossed Gheringhap Street and inscribed a crescent through the Hoddle grid out onto Cunningham Pier. Further sidings were located up Clare Street to service woolstores on Brougham Street at the Denny Lascelles Bow Truss building in 1911.

- 'The Geelong Region itself fell within the "economic hinterland" of an expanding industry, e.g. the globalisation of US and European manufacturing companies between 1925 and 1963' (GRC 1993, p. 8). The Ford Motor Co. established an assembly plant in Norlane in 1925 and the Phosphate Co-operative formed in the same year that packaged its brand as 'Pivot'; Pilkington Glass followed in 1937, and International Harvester commenced operations in 1939. In 1940 the grain terminal was completed, and in 1946 the CSIRO opened its textile research laboratories in Belmont, and Rylands Pty Ltd began manufacturing wire. New employment anchors were introduced to the region – Shell in 1954, Alcoa in 1963 and Huyck in 1966.
- 'The Geelong Region achieved sufficient critical mass to support net additional services, e.g. Deakin University and, to some extent, the Australian Animal Health Laboratory (AAHL)' (GRC 1993, p. 8).

The promise of vision-making

Of the 20+ reports which are reviewed in this chapter, certain patterns seem to emerge and appear to present glimpses of the way people of the time were thinking about the city of Geelong and its future. While some of the reports present a clear concept of what the city might 'look' like, they are largely based on an ideal, a concept of what could be. Others are much more aspirational and become more reflective of the feelings and needs of the community and of ways to give a sense of direction, especially during times of economic downturn. On the other hand, some take a retrospective viewpoint and draw on population forecasting and economic data as a base for guiding strategy. The Australian Academy of Science, in a think-tank on a vision for Australia's future population, suggested that:

> *... when drawing on net population growth, births deaths and migration patterns, going out 10 or 20 years is a more accurate yardstick than projecting 50 years, and that future projections beyond a generation are much more difficult to predict with accuracy* (AAS 2012, p. 4).

Earlier vision reports developed under the Geelong Regional Planning Authority (1961–75) appeared to draw on the Commonwealth Bureau of Census and Statistics (CBCS 1971) population forecasts, which introduced population projections in some early studies. However, there appears to be a shift in forecasting between 1970 and 1976 when the CBCS was replaced by the Australian Bureau of Statistics (ABS). An example can be seen in the projected national population estimates for 2001. Probably anticipating similar rates of growth to those experienced during the post-war period marked by higher immigration and higher fertility rates, the 1971 CBCS 30-year population forecast estimated Australia's population to range between just under 21 million (60 000 net migration per annum) and 24 million people by 2001 (140 000 net migration per annum). However, by 1976 it soon emerged that the high growth rates of the 1950s and 1960s were not going to continue. The

ABS 25-year projections presented a more conservative forecast ranging between 19 million (75 000 net migration & NRR = 0.1) and 21 million (125 000 net migration & NRR = 0.1), indicating a shortfall of 2–6 million people. It is interesting to note that the actual national population census in 2001 was 19 603 500, which confirmed the lower-range forecast of the 1976 ABS projections (ABS 1976). Similarly, it appears that there was a significant regrouping of projections by different state and local government agencies between the two periods. Hence the extensive urban analysis and strategic thinking undertaken by The Planning Workshop, in producing the *Geelong: City by the Bay – Central Business District Study* (GRPA 1975), were likely based on CBCS data and therefore developed strategies to accommodate an anticipated population forecast of 400 000 people in the Geelong region by 2001.

By the time 2001 arrived, actual ABS data for the census period revealed that the 1975 projection was almost 250 000 persons short of the design target. Indeed, the G21 Planning for Growth model in 2010 indicated that Geelong would not be approaching the magnitude of 400 000 people until 2050. Note that the current estimated growth forecast by 2040 for the City of Greater Geelong is 388 599 people (G21 GRA 2012). Subsequent vision documents developed under the Geelong Regional Commission (GRC 1975–93), working with ABS data generated in the mid 1980s, had significantly revised projections. Perhaps based on the substantial drop in net migration into the Geelong region between 1976 and 1981 (GRC 1993), the Geelong Regional Commission's informed forecasting regarding population modelled over a 20-year timeframe in 1986 was quite accurate.

The Geelong region, under the oversight of the Geelong Regional Commission, comprised the pre-amalgamation Local Government Areas of the City of Geelong, City of Geelong West, City of Newtown, Shire of Corio, Shire of Bannockburn, Shire of Barrabool, City of South Barwon, Shire of Bellarine, and the Borough of Queenscliffe.

The authors of this chapter have been able to cross-correlate present-day City of Greater Geelong population catchment areas with the 1986 Geelong region data by only including the 1986 Local Government Areas that correlated with the current City of Greater Geelong boundary (GRC 1986).

Several of the visions developed by the various authorities in Geelong since 1975 have had an economic development focus. Perhaps the most seminal of these is the 1993 vision *Determinants of Growth in the Geelong Region* developed by Coopers Lybrand for the Geelong Regional Commission (GRC 1993), where a significant amount of highly developed demographic modelling was undertaken to inform an understanding of how the population would change, including indicators of employment and family structure. The report also benchmarked other urban centres of a similar population size, notably Newcastle (pop. 432 640; ABS 1991), Wollongong (pop. 239 940; ABS 1991), Gold Coast–Tweed (pop. 274 210; ABS 1991), Hobart (pop. 179 020; ABS 1991) and Canberra–Queanbeyan (pop. 315 420; ABS 1991). At the time, the main manufacturing enterprises in the Geelong region included Ford (3700 employees), Alcoa (1540), Godfrey Hirst (780), Shell (750), Asta Aircraft (550) and Pilkington (450), which together demonstrated greater industrial diversity and a larger spread of anchor employers than either Newcastle or Wollongong.

Developed on an extensive set of economic and demographic data, the GRC (1993) report became the yardstick for what was regarded as a considered piece of analytical research. The report does not include any illustrative material, such as aerial montages or clever artistic renderings, but rather presents a scientific model called the 'assessed value and the contribution of the determinants of growth in the Geelong region'. Based on two matrices, the model enabled a means of examining the 'capacity to affect growth' against the 'probability of affecting growth'. These matrices reflect the rigour with which the

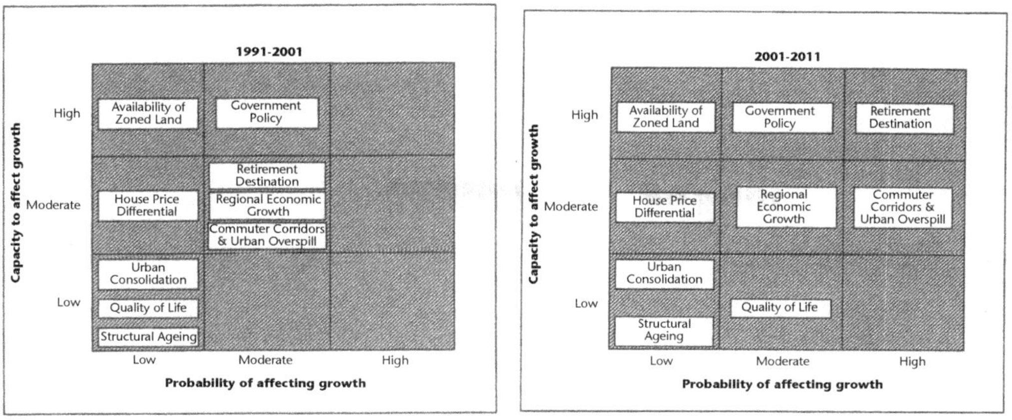

Fig. 17.2. 'Assessed value and the contribution of the determinants of growth in the Geelong region' model, 1991–2001 and 2001–11. Source: GRC (1993, pp. 62–63).

report investigated the potential for growth in the Geelong region between 1991 and 2011. The model was evaluated according to low, medium and high implied measures and included the following nine factors earmarked as the 'determinants' of growth:

1. availability of zoned land
2. government policy
3. retirement destination
4. house price differential
5. regional economic growth
6. commuter corridors and urban overspill
7. urban consolidation
8. structuring ageing
9. quality of life (GRC 1993).

The matrices (Fig. 17.2) assigned a value to each of the determinants with respect to two 10-year timeframes, 1991–2001 and 2001–11. It appears that the report was used extensively throughout the 1990s and underpinned both *Geelong 2010: A Preferred Future* (CoGG 1994) and the period during the amalgamation of the smaller Local Government Areas into the City of Greater Geelong. None of the other studied visions for the Geelong region appear to have engaged in this level of rigour in their analysis. Interestingly, the report includes some rather sage predictions. The most notable being the:

> *... development of communications technology allowing workers more flexibility in working from home via personal computers, modems, faxes and video conferencing facilities* (GRC 1993, p. 55).

22 visions in 42 years

There appear to be three types of vision documents:

- design-based
- community-focused
- economic-driven.

While some of the vision reports fall clearly within one of these three categories, others are often a combination of two and a few will encompass all three. The flavour or direction of each type is often directed by the discipline focus of the project team charged with its delivery. Hence, if a project team was based within a specific discipline then the strengths of the vision report will probably be driven from a particular agenda. Alternatively, if the project team is cross-disciplinary, involving a collaboration from a range of disciplines, then the report may be more comprehensive and more inclusive of a broader range of issues. However, increasing complexity may result in confusing and possibly conflicting outcomes or recommendations.

- Design-based vision documents present a clear concept of what the city might 'look' like; they are largely design-led and present an idea of what the future could be. They are often driven by a team of professional designers, and sometimes underpinned by ambitious economic and population growth targets.
- Community-focused vision documents are a much more aspirational type, and appear more reflective of the 'pulse' and needs of the community, especially during times of social and economic change. These are often directed by a project team focused on public participation, and draw on the fields of psychology and social planning.
- Economic-driven vision documents appear quite removed from the participatory approach. They often adopt a retrospective viewpoint and draw on population forecasting and important economic data as a base for guiding strategy. Because of this, they are often more abstract and hence less accessible by the wider community.

Each type of report has its place and contributes to advancing the city. While vision-making is important to keep alive, reports such as business strategies and economic development plans are crucial, for without them dreaming will struggle to become reality. Of the 22 documents reviewed, 12 were design-based, two were community-focused and five were economic-driven. There were also two that successfully addressed both a design and economic focus. Only one of the visions, the *Geelong and Foreshore Draft Structure Plan* (CoGG 1990), attempted to address all three categories, but it would appear that time constraints on the delivery of the document may have limited the effective integration of the different agendas.

The chart in Plate 17.1 maps the 22 visions released by the relevant authoritative body charged with developing the City of Geelong since 1975. The transition between these authoritative bodies is of particular interest when observing the overall pattern of vision-making in the region. When we analyse this pattern, we can see that there is an overlap during the early 1990s that pertains to the period in which Local Government Areas were being amalgamated and the City of Greater Geelong was being established.

There are four key periods during which more than one vision was released in a calendar year – 1981, 1988, 1993 and 2013. The Geelong Regional Commission was seemingly prolific in 1988, with three vision documents released in that year alone. This is no surprise as 1988 was a significant year in the development of central Geelong with the establishment of the National Wool Museum as part of the Australian bicentennial celebrations, the opening of Bay City Plaza (now Westfield) shopping plaza, and the planned development of a third retail and hospitality zone along the waterfront, outlined extensively in the *City by the Bay – The Bay Link* vision (GRC 1988a) (Figs 17.3 and 17.4). The plan involved several public use spaces, food and entertainment hubs and tourism destinations that provided an uninterrupted connection from the centre of the city to the water's edge.

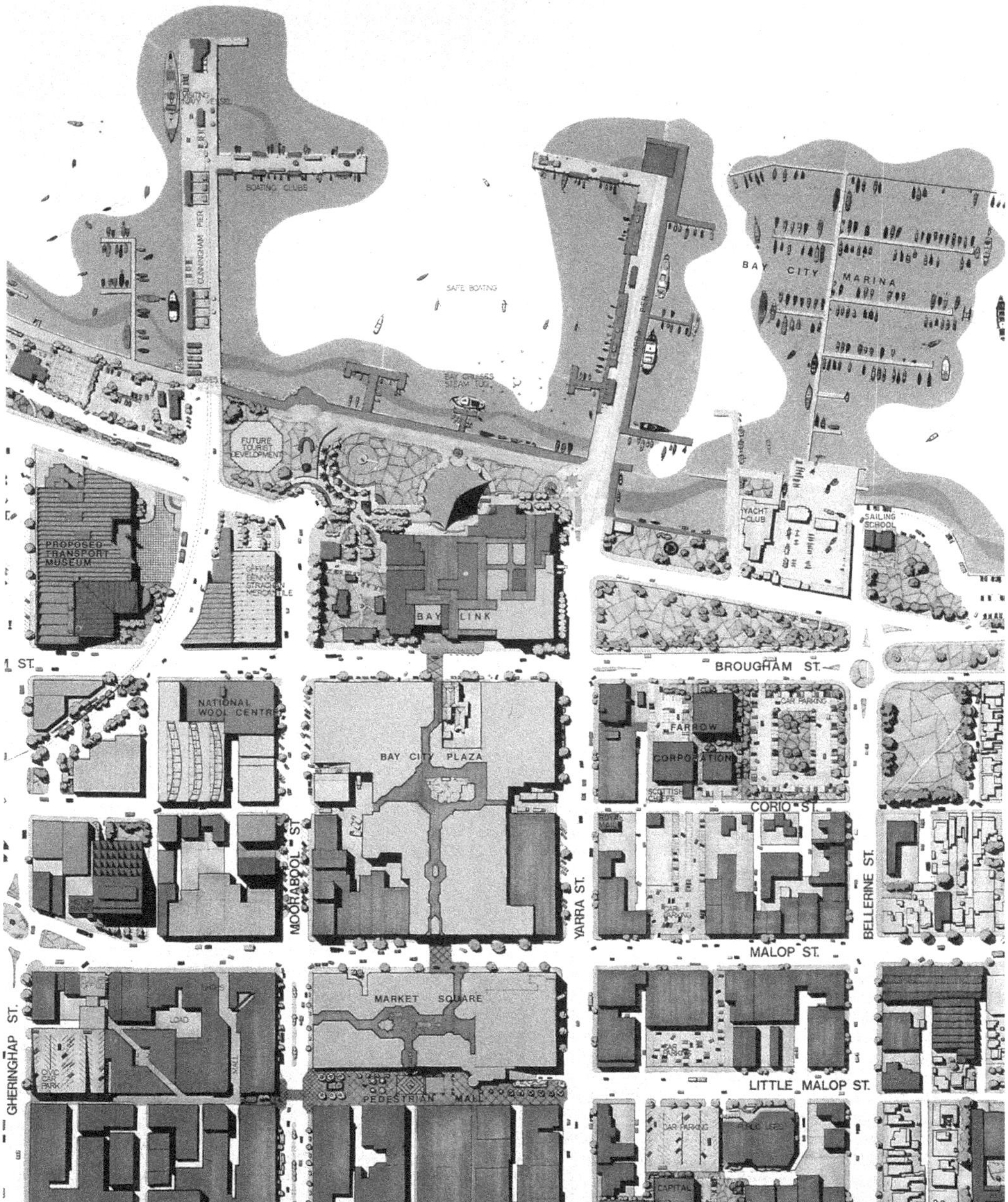

Fig. 17.3. Plan: *Geelong: City by the Bay – The Bay Link.* Source: GRC (1988a).

Another observation to note is the drop in frequency of vision-making during the mid to late 1990s and into the first decade of the 2000s. The City of Greater Geelong established a Major Projects department during this time as resources were deployed from vision-making to implementation. Two decades (1975–96) of promoting and refining ideas to

Fig. 17.4. Perspective looking south: *Geelong: City by the Bay – The Bay Link.* Source: GRC (1988a).

enable the city to embrace its north-facing aspect and redevelop the waterfront were to become a reality. With a series of swift adaptive re-use designs for expanding Deakin University into the Dalgety and later Lascelles Woolstores, and reconfiguring Steampacket Place to an international public recreation space, both the landscape architecture practice Taylor Cullity Lethlean (Lethlean 2013) and the Geelong-based architectural practice McGlashan Everist won commissions to implement key aspects of these aspirations. With public support and civic pride growing through the transformation of the waterfront, new strategic initiatives were beginning to emerge. The Department of Sustainability and the Environment, now the Department of Environment, Land, Water and Planning, supported an investigation with the City of Greater Geelong into the revitalisation of the 'Western Wedge', a mix of old industrial and residential areas wedged between the rail corridor of Geelong West, Corio Bay and central Geelong. Energy was being channelled into consolidating vision-making with the development of a new suite of urban design guidelines and urban design frameworks (CoGG 2005, 2008a, b). The focus was on encouraging development opportunities across the city within a new regime of land use opportunities, connections and building heights, guiding the city from three, four and six storeys in 1979 (GRC 1979) to six, eight and 10 storeys by 2008 (CoGG 2008a, b).

The chart also includes the various Urbanheart Surgery studios that have focused on key development challenges within the City of Greater Geelong. The Urbanheart Surgery program, pioneered in the School of Architecture and Built Environment at Deakin University (Rollo 2004), has been active in engaging with the city on numerous occasions over the past 20 years. The program was designed with the purpose of allowing local government authorities to bring key projects into the university and to allow students to assist with projecting ideas, exploring both their implications and opportunities, five and

10 years into the future. Over 1500 students have passed through the studio since 1999. Urbanheart was designed to immerse students within a design research culture that allowed them to engage in critical discourse by working on strategic design projects in three areas critical to Victoria's future development: metropolitan urbanism, urbanism on the periphery and regional urbanism. During the 2009 and 2010 workshops, the focus of the Urbanheart Surgery program was centred on housing provision in the centre of Geelong and on the physical connections between the waterfront, the city centre and the Barwon River. The images in Figs. 17.8 and 17.9 presents a series of permutations developed by students exploring the relationships between the Geelong Railway Station precinct, the Arts and Cultural precinct and the Brougham Street spine including the Deakin University precinct.

Several CBD-targeted programs saw the development of student visions and ideas for the Western Wedge (CoGG 2005) (see Plate 17.2): inner-city student accommodation (2009), the revitalisation of the city's urban morphology through the reinvigoration of streets and laneways (2010) (see Plate 17.3), and the development of a city 'heart' (2015). Other Urbanheart Surgery programs have focused on areas of Geelong outside the immediate CBD area, such as High Street in Belmont (2012, 2017) and Waurn Ponds Shopping Centre (2017). It is important to note that, for most of these contributions, the program did not align itself with the official publication of a vision by council. In 2015, the program collaborated with the City of Greater Geelong on the City Heart strategy while they were working on *City in a Park*, a vision developed by Steven Thorne and Robert Adams for the City of Greater Geelong (CoGG 2015).

Out of the 22 reports reviewed only four seemed to present key aspirations which looked beyond planning and design outcomes to address considerations relating to economic, social/cultural and environmental objectives. These were:

1. *Directions: The Geelong Regional Development Strategy* (GRC 1988b)
2. *Geelong 2010: A Preferred Future* (CoGG 1994)
3. *Central Geelong Action Plan* (CoGG 2013b)
4. *Greater Geelong: A Clever and Creative Future* (CoGG 2017).

When comparing the aspirations of the four vision documents, it is interesting to note the significant differences relating to the emphasis placed on the type of objectives presented in each, which may be a reflection of the issues and concerns that were considered pertinent at the time. Plate 17.4 presents a series of pie-charts that reflect the weighting of the aspirations with respect to economic (blue), social/cultural (red) and environmental (green) concerns. These weightings have been derived from Table 17.2, which lists each of the aspirations in the four vision reports.

The *Directions: The Geelong Regional Development Strategy* (GRC 1988b) report presents no fewer than 15 aspirations – 40% earmark economic objectives, 20% address social/cultural and a somewhat surprising 40% address environmental concerns. Contrast these weightings with the aspirations developed six years later in *Geelong 2010: A Preferred Future* (CoGG 1994) which published 11 aspirations, and there appears a distinct shift in emphasis. While economic is still high at 45.5% of the objectives, there is a significant increase in social/cultural with an increase from three to five objectives, also 45.5% of the weighting. However, there is an equally significant decline in environment concerns with a reduction from six objectives to one, a fall of over 30% down to 9% of the total weighting. It would appear that the impact of the economic downturn experienced during the intervening years weighed heavily on the community. The community-led vision of *Geelong*

Table 17.2. Comparative aspirations study of four key Geelong visions

Directions: The Geelong Regional Development Strategy **(GRC 1988)**	***Geelong 2010: A Preferred Future*** **(CoGG 1994)**	***Central Geelong Action Plan*** **(CoGG 2013)**	***Greater Geelong: A Clever and Creative Future*** **(CoGG 2017)**
E Concentrate on developing specific strengths of the regional economy, particularly in relation to manufacturing, port related investment, research and tourism	**SC** A new community – a caring and cohesive Geelong	**E** Stimulate greater levels of investment in central Geelong and unlock the development potential of under-utilised land	**E** A prosperous economy that supports jobs and education opportunities
E Develop orientation to rest of the world, in particular markets in western Pacific Basin	**V** An ecologically sustainable Geelong	**SC** Maximise utilisation of existing public transport infrastructure to provide greater transport choice	**E** A leader in developing and adopting technology
V Develop orientation to Port Phillip Bay and the Western District of Victoria	**E** Generating jobs	**V** Create a vibrant, lively shopping, *al fresco* dining and event space environment with bay views, northerly aspect, active edges and connections to surrounding areas	**SC** That creativity drives culture
V Promote natural assets to outsiders.	**SC** An Arts and Cultural Centre	**SC** Build a strong learning and student culture and experience. Integrate opportunities for expansion of the Gordon and Deakin campuses with the revitalisation of central Geelong	**E** A fast, reliable and connected transport network
E Adopt a least-cost approach to expansion of urban Geelong	**SC** An Education and Research Centre	Increase the number of people living in the city, making it a safer, more vibrant and mixed use area	**SC** A destination that attracts local and international visitors
SC Retain long-term options for alternative quality accommodation close to Geelong	**E** A centre for tourism	**E** Make best use of government and private land assets as a catalyst to leverage investment and activity in central Geelong	**SC** People feel safe wherever they are
V Encourage intensive coastal settlement only in designated centres	**E** A major transport hub	**E** Capitalise on Geelong's premier health facilities and services for employment and investment	**SC** An inclusive, diverse, healthy and socially connected communit
V Maintain rural environment on Bellarine Peninsula and Geelong hinterland	**E** An international wool capital	**SC** Create an integrated, attractive and well-designed pedestrian and cycling network which supports healthy living and public safety	**V** Sustainable development that supports population growth and protects the natural environment

Table 17.2. (*Continued*)

E Attract quality tourist developments in appropriate locations	**E** A Centre for Smart Component Manufacturing	**V** Develop a distinctive and enduring network of green spines and spaces that support living, working and walking, commencing with the greening of Malop Street	**V** Development and implementation of sustainable solutions
V Protect significant natural features of the region	**SC** Creating a healthier Geelong	**SC** Build on the annual calendar of events and activities which attract more people and increased investment to central Geelong	
V Encourage growth to take account of conservation and heritage considerations to achieve sustainable development	**SC** Re-empowering our Elders		
SC Encourage a co-ordinated approach to human services planning and vocational training			
SC Promote self-determination and independence in regional community life			
E Promote benefits of Geelong as a region and discourage provincial town image			
E Promote the manufacturing sector to maintain a stable workforce			

E = economic; SC = social/cultural; V = environmental aspirations from four different reports.

2010: A Preferred Future may well have served as much more than a strategic report, in that it gave the community a voice. The process itself, which was carried out over months of workshopping in schools, open community forums and interest groups, facilitated a sense of self-repair through which the community was not only able to reaffirm its resilience but was able to regain a sense of pride about the prospect for the future.

Moving forward 20 years, the *Central Geelong Action Plan* (CoGG 2013b), developed by the City of Greater Geelong, presented 10 aspirations. Social/cultural increased its weighting to 50% with five objectives, while economic levelled out at three or 30%, and environment at two or 20%. There are many factors that drive strategy or vision development – economic transition, shift in the role and function of Central Activity Areas, changes in retailing practices, growing community and demographic diversity, maturing housing market, changes in media interest relating to shop vacancies (a strong driver of the *Central Geelong Action Plan*) – the list is extensive. One of the strengths of the *Central Geelong Action Plan* is that rather than establish a new plan it consolidated ideas and strategy elements from previous work and built on them with community engagement to test actions and opportunities. What distinguishes this document from previous visions is that it presented a staged implementation strategy underpinned by a comprehensive action list of priority projects linked to series of planning and design outcomes on a 15-year time-frame to 2028. Its successor policy, the *Revitalising Central Geelong Action Plan* (DELWP 2016), echoed this approach.

Looking forward 25 years to 2047, the community-led vision, *Greater Geelong: A Clever and Creative Future* (CoGG 2017), presents nine aspirations that reflect a similar weighting of objective type as the *Central Geelong Action Plan*. That is, four social/cultural, three economic and two environment-focused. However, what distinguishes this document from all previous reports is its 'Smart' underpinning, with an extensive range of benchmarks and indicators that will hopefully help maintain its currency over time.

Rediscovering past ideas

With the more traditional 20th-century design-led models, where the architect, urban designer, economic and strategic planner often left the community behind in the vision-planning process, there are several key reports which produced quite extraordinary spatially focused ideas. Often underpinned by clear economic and demographic data, a vision is nonetheless only relevant provided the modelled projections are eventually realised. The urban analysis of *Geelong: City by the Bay* (GRPA 1975) was very thorough and the identification of the weaknesses and threats of a city in significant industrial transition enabled the Planning Consultancy to produce a strategy that is still highly relevant today. It earmarks opportunities and strengths, yielding key projects, some of which have only recently started to come to fruition (see Figs. 17.5, 17.6 and 17.7). Projects include:

- introducing 10 000 residents into the central activities area – this target was introduced again by the *Central Geelong Action Plan* (CoGG 2013b)
- turning the focus of the city to the waterfront and developing adaptive re-use of the woolstores
- intensifying the commercial and retail core of the city and reinvigorating Market Square as the focus of town centre life
- introducing a more comprehensive public space and pedestrian-oriented strategy within core business district (Figs. 17.8 and 17.9)
- expanding the tree canopy, and linking Johnstone Park with the Botanic Gardens with a tree-lined boulevard along Malop Street, and a similar intensive suite of green connections between the Barwon River and Corio Bay along Gheringhap, Moorabool, Yarra and Swanston Streets.

The concept of linking Johnstone Park with the Eastern Botanic Gardens along Malop Street was a key recommendation in *City by the Bay* (GRPA 1975). Indeed, the diagram in Fig. 17.7 and the elaborate dyeline plan in Fig. 17.6 clearly illustrate the green spine in all but name, and more importantly consider this ecological initiative in the context of the wider city plan by proposing three additional north–south green spines linking the waterfront with the environs of the Barwon River. This would appear to strengthen the *City in the Park* concept (CoGG 2015).

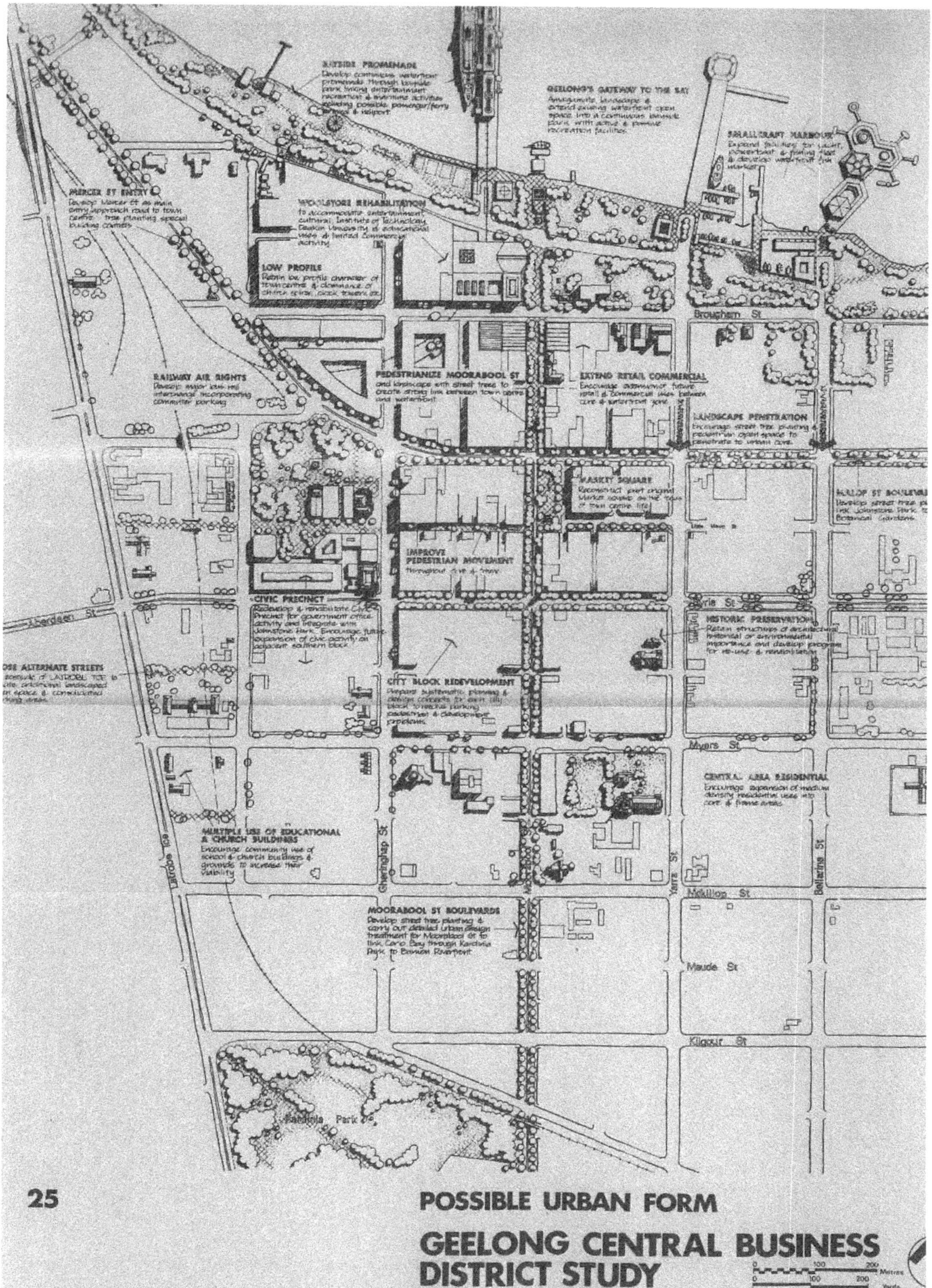

Fig. 17.5. Geelong CBD study – possible urban form. Source: GRPA (1975).

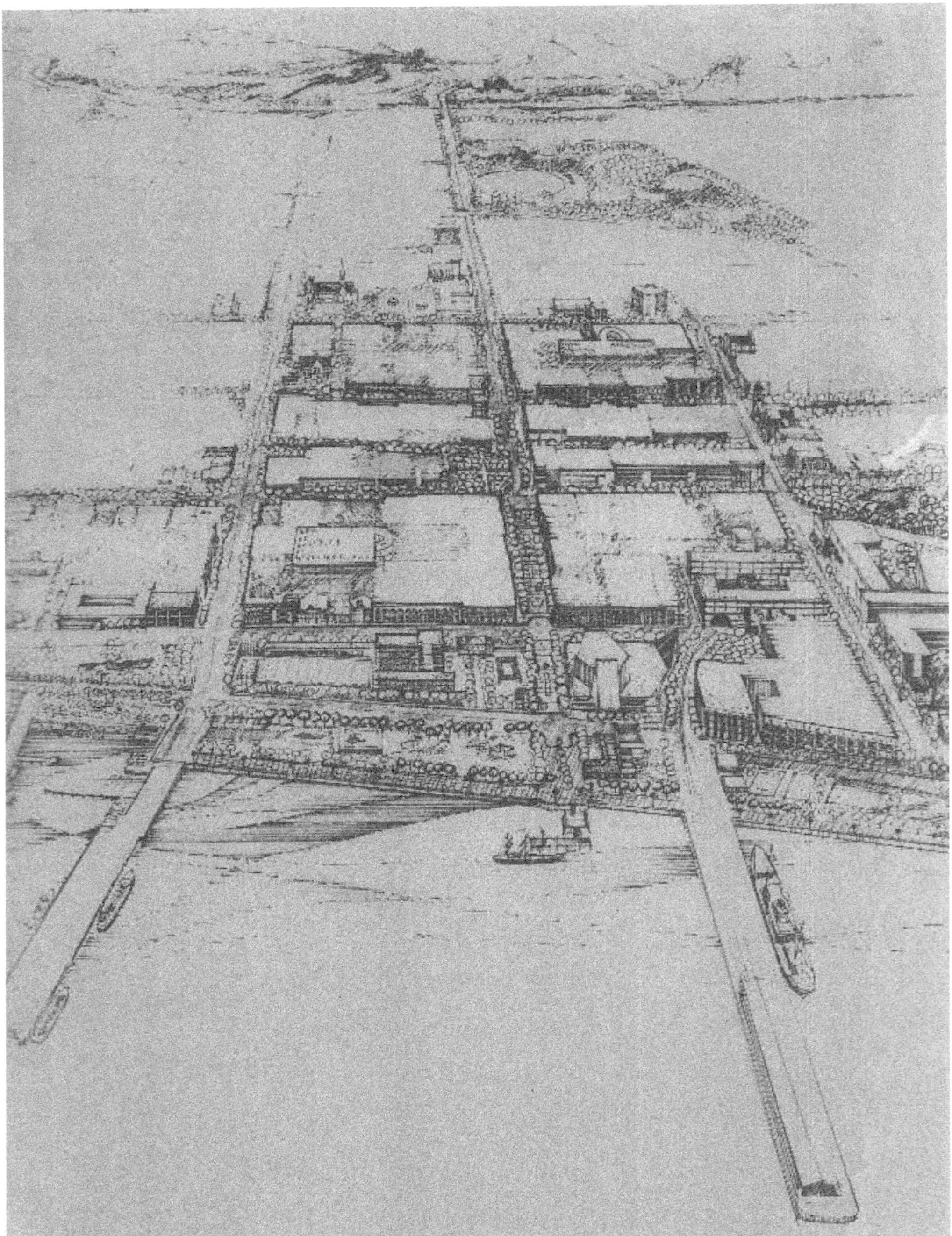

Fig. 17.6. Waterfront precinct looking south showing east–west spine linking Johnstone Park with the city centre and north–south spines from waterfront to Barwon River. Source: GRPA (1975).

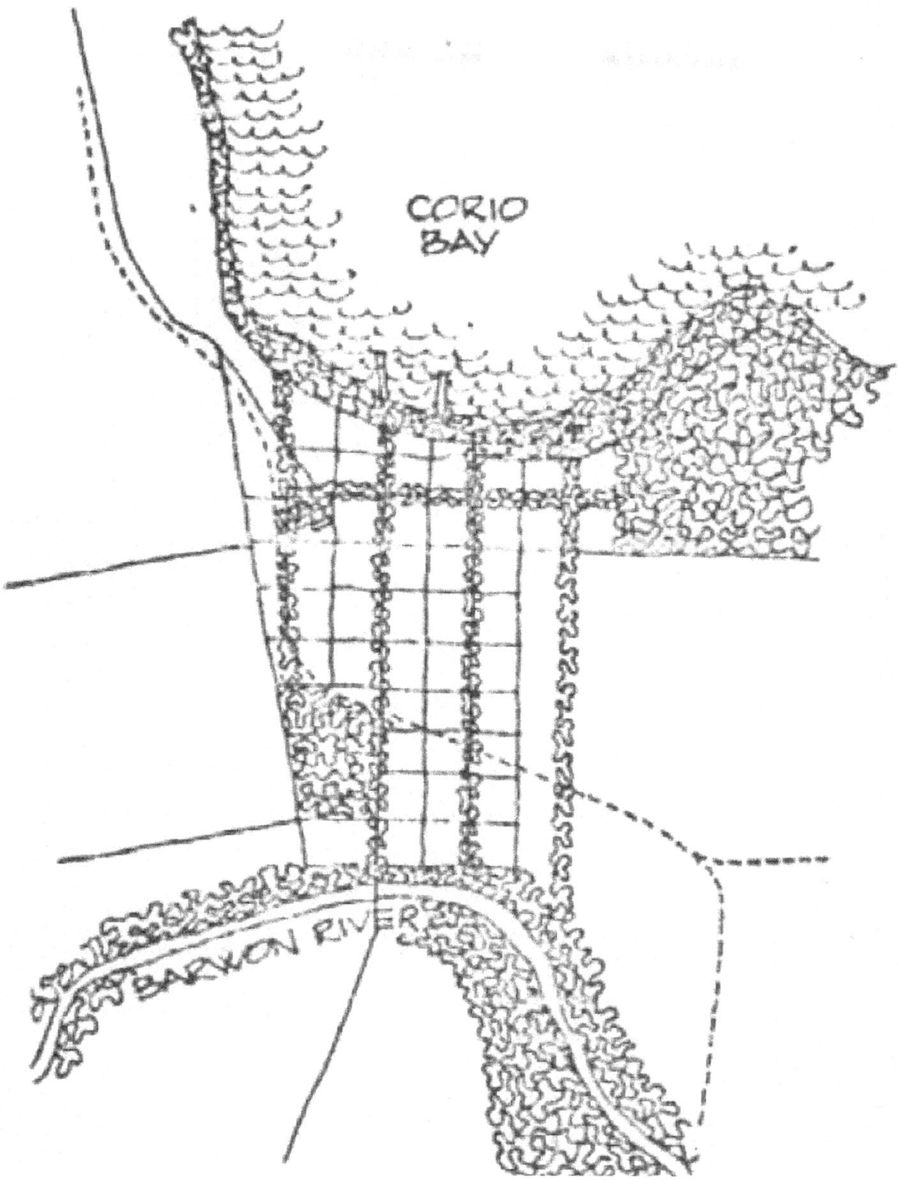

Fig. 17.7. 'Open space structure', design analysis diagram showing the east–west Malop Street link between Johnstone Park and the Botanical Gardens, and three additional north–south green spines linking the waterfront with the environs of the Barwon River. Source: GRPA (1975).

City by the Bay was quickly followed up by the *Geelong Central Area Study* (GRC 1979). Similar to the 1975 strategy, the *Geelong Central Area Study* presented a significant amount of urban analysis data with a focus on viewlines and building height studies. With Myers Street elevated 30 m above sea level, running just below the dominant ridgeline that shelters Corio Bay, a gentle slope falls away – north to the Bay and south to the Barwon River. This

Fig. 17.8. View sheds study. Source: GRC (1979).

provides prominent view corridors to the water and the You Yangs, and an orientation that allows winter sun to penetrate deep into the city blocks with the change in topography. Working with this unique set of traits, the report presented an extensive sequence of measured drawings of the six central city blocks, defined by Gheringhap, Moorabool and Yarra Streets on the north–south axis, and Brougham, Malop, Ryrie and Myers Streets running east–west. With each set of drawings being carefully developed, plot ratios, height lines and set-backs reducing overshadowing were explored and recommendations for maximum heights of three-, four- and six-storey buildings were carefully plotted (Fig. 17.8). Similarly, the study presented a range of options for the rail siding which ran beneath Mercer Street, curved past Corio Street, crossed Gheringhap Street and inscribed a crescent through the Hoddle grid out onto Cunningham Pier. This was identified in the *Geelong Central Area Study* as providing a unique green crescent link that opened up the Hoddle grid to establish a distinctive pedestrian desire-line between Johnstone Park, the arts, civic

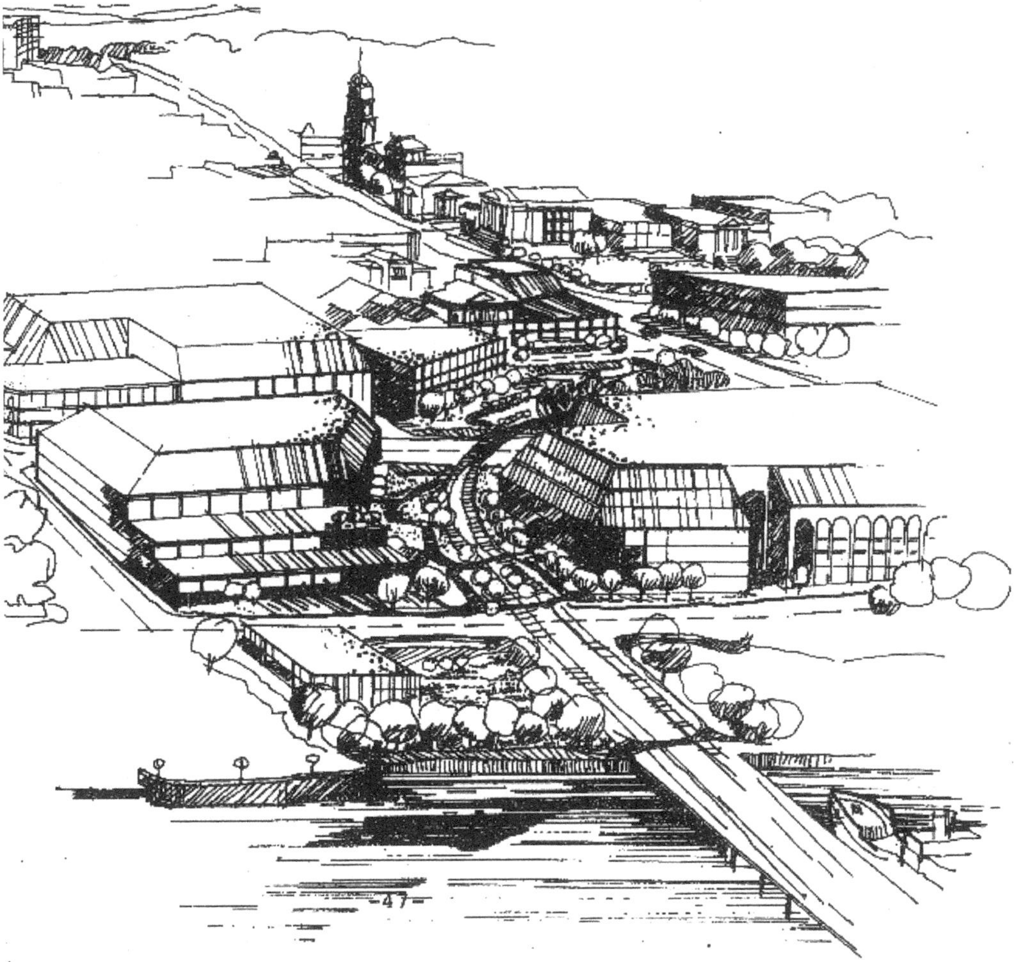

Fig. 17.9. Green link, Johnstone Park to Cunningham Pier. Source: GRC (1979).

and station precincts, and the waterfront (Fig. 17.9). It was maintained in vision documents through to the mid 1990s but was rationalised out of the City Plan with the insertion of a three-storey car park that realigned the grid between Corio and Gheringhap Streets.

Conclusion

> *We know Geelong is at a crossroads, we're the largest regional economy in our State, but we're playing on a world stage in a fast changing economic, social and technology landscape. We have to work very hard just to stand still … and if we are to drive the economic, social and cultural growth that will secure our region's future then we must go faster, work harder and most importantly we must work together* (den Hollander 2018; Deakin University 2018).

After having reviewed the extensive body of rich material presented in the 22 reports that inspired this chapter, the authors are both amazed and humbled at the thousands of hours and numbers of people who have been involved in their development. The foundations of all these visions are fairly consistent, suggesting that good ideas are always good ideas just waiting for the right circumstances coming together to achieve them. Spanning over a generation of vision thinking and dreaming, of significant milestones in economic and social highs and lows, there are several reflections that come to mind:

- the importance of having aspirations, not simply planning and design outcomes
- valuing focus studies and background documents that approach an understanding of the city from its social, economic, cultural and environmental perspectives
- the importance of being able to monitor the progress of achieving those aspirations
- the opportunities gained by having people working on vision documents who are completely outside a Melbourne-centric view of the world and can come to a workshop with a completely different set of eyes and experiences
- valuing the work, imagination and commitment of the professionals working in local government, and the valuable ideas that emerge from within
- the importance of having a balanced multi-disciplinary team working on visions, and in particular ensuring that they have an agreed set of aspirations and values which effectively integrates the various agendas
- having a realistic and achievable implementation strategy that is responsive to economic and social change
- understanding that change does not happen immediately, that city-making goes well beyond political cycles and hence requires sound bipartisan support
- the importance of recognising the value of community engagement, in particular the community-led vision, not only from the notion of developing realisable objectives, but in particular the extraordinary sense of spirit and ownership that it offers all members of the community.

All too often a new regime comes in and all that was developed previously is forgotten. It is important to acknowledge the work and thinking that has gone before. Hence the authors of this chapter hope that these vision documents will someday be appropriately archived and made accessible for all. Not only are there some amazing ideas still hidden in their pages, but they provide a fascinating glimpse for future generations of the evolving dreams and concerns of a city and its community during the period in which they were written. The continuum they form is an ever-evolving record of the perpetual journey of vision-making Geelong.

Acknowledgements

The authors gratefully acknowledge the support and assistance of the following organisations: Department of Environment, Land, Water and Planning; Geelong Art Gallery; Departments of Planning Strategy and Urban Growth, Economic Development, Governance and Legal Services, and Strategy and Program Delivery; and the City of Greater Geelong.

References

AAS (2012) *Australia's Population: Shaping a Vision for our Future*. Australian Academy of Science, Canberra. www.science.org.au/events/thinktank/thinktank2012/documents/recommendations.pdf (accessed 1 November 2018).

ABS (1976) *Projected Population, Australia, 1981–2021*. Australian Bureau of Statistics, Canberra.

ABS (1991) *Census of Population and Housing, 1991*. Australian Bureau of Statistics, Canberra. http://www.abs.gov.au/AUSSTATS/abs@.nsf/mf/2101.0/ (accessed 1 November 2018).

ABS (2012) *Australia's Population: Shaping a Vision for Our Future. Theo Murphy High Flyers Think Tank*, Adelaide, 26–27 July, 2012. Theo Murphy Australia Fund and Australian Bureau of Statistics, Canberra.

Alexander C (2001–05) *The Nature of Order: An Essay on the Art of Building and the Nature of the Universe. Book Two: The Process of Creating Life*. Centre for Environmental Structure, Berkeley, CA.

Alexander C, Silverstein M, Angel S, Ishikawa S, Abrams D (1975) *The Oregon Experiment*. Oxford University Press, New York.

Alexander C, Ishikawa S, Silverstein M, Jacobson M, Fiksdahl-King I, Angel S (1977) *A Pattern Language: Towns, Buildings, Construction*. Oxford University Press, New York.

CBCS (1971) *Projected Population of Australia, 1972–2001*. Commonwealth Bureau of Census and Statistics, Canberra.

City AM (2016) *Q&A with Andrew Kerr: How Can You Help Shape Edinburgh's Future?* Radio broadcast, 11 October 2016, 7am. Edinburgh, UK.

City of Edinburgh (2017) *2050 Edinburgh City Vision, One Year On*. City of Edinburgh, UK.

CoGG (1990) *Geelong and Foreshore Draft Structure Plan*. Report prepared by Ratio Consultants for City of Greater Geelong, Geelong.

CoGG (1993) *Central Geelong Foreshore and Environs Framework*. City of Greater Geelong, Geelong.

CoGG (1994) *Geelong 2010: A Preferred Future*. City of Greater Geelong, Geelong.

CoGG (1996) *Waterfront Geelong Design and Development Code*. Report prepared by Keys & Young Consultants for City of Greater Geelong, Geelong.

CoGG (2005) *Geelong Western Wedge Framework*. Report prepared by Planisphere and Jones & Whitehead for City of Greater Geelong, Geelong. https://www.geelongaustralia.com.au/common/public/documents/8cb84ccd1469095-12%20page%20sep%2005%20(lowres).pdf

CoGG (2008a) *Geelong Transit City Station Precinct Urban Design Framework*. Report prepared by Hassell, Parsons Brinckerhoff and Charter Keck Cramer for City of Greater Geelong, Geelong. https://www.geelongaustralia.com.au/common/Public/Documents/8cc3d63ba922bdb-Geelong%20Station%20Precinct%20Urban%20Design%20Framework.pdf

CoGG (2008b) *Central Geelong Urban Design Guidelines*. City of Greater Geelong, Geelong. https://s3-ap-southeast-2.amazonaws.com/ehq-production-australia/5a9c41eae74a460fa312aeb1f8a1814f24c609bf/documents/attachments/000/011/280/original/Central_Geelong_UDG.pdf?1397547460

CoGG (2013a) *Central Geelong Vision: Design Studio Report Draft*. Deakin University and City of Greater Geelong, Geelong.

CoGG (2013b) *Central Geelong Action Plan*. Central Geelong Task Force, City of Greater Geelong, Geelong. https://www.geelongaustralia.com.au/common/public/documents/amendments/8d2d0901dd9c6bb-CentralGeelongActionPlan.pdf

CoGG (2015) *Geelong City Centre: City in the Park*. City of Greater Geelong, Geelong.

CoGG (2017) *Greater Geelong: A Clever and Creative Future*. City of Greater Geelong, Geelong. https://www.geelongaustralia.com.au/common/Public/Documents/8d4d2ad3b2b24c1-Greater%20Geelong%20-%20A%20Clever%20and%20Creative%20Future%20Report.pdf

Deakin University (2018) *Our History at a Glance*. http://www.deakin.edu.au/about-deakin/reputation/history (accessed 1 March 2018).

DELWP (2016) *Revitalising Central Geelong Action Plan*. Department of Environment, Land, Water and Planning, Melbourne. https://www.planning.vic.gov.au/__data/assets/pdf_file/0021/95250/Revitalising-Central-Geelong-Action-Plan.pdf

den Hollander J (2018) Clever and Creative Deakin Lecture Series: Additive Manufacturing (3D Printing). 8 March 2018.

G21 GRA (2012) *G21 Regional Growth Plan Background Report*. G21 Geelong Regional Alliance, Geelong.

GAG (1985) *The Changing Face of Geelong*. Geelong Art Gallery, Geelong.

GRC (1979) *Geelong Central Area Study: Preferred Strategy Public Discussion Paper*. Geelong Regional Commission, Geelong.

GRC (1981a) *Geelong Central Area Strategy*. Geelong Regional Commission, Geelong.

GRC (1981b) *Geelong: City by the Bay*. Geelong Regional Commission, Geelong.

GRC (1985a) *Geelong Regional Development Strategic Plan*. Technical Paper no. 9(g). Geelong Regional Commission, Geelong.

GRC (1985b) *Geelong Urban and Township Development Strategy*. Report prepared by Neilsen Associates for Geelong Regional Commission, Geelong.

GRC (1986) *Geelong Region: Framework for the Future – Development Strategy*. Geelong Regional Commission, Geelong.

GRC (1988a) *Geelong: City by the Bay – Foreshore Precinct Project*. Geelong Regional Commission, Geelong.

GRC (1988b) *Directions: The Geelong Regional Development Strategy*. Geelong Regional Commission, Geelong.

GRC (1988c) *Geelong: City by the Bay – The Bay Link*. Geelong Regional Commission, Geelong.

GRC (1990) *Geelong Regional Local Government Review: Economic Assessment of Local Government in the Geelong Region*. Report prepared by EIR Consulting for Geelong Regional Commission, Geelong.

GRC (1992) *Directions: The Geelong Regional Development Strategy – Five-year Implementation Plan 1991/92*. Geelong Regional Commission, Geelong.

GRC (1993) *Determinants of Growth in the Geelong Region*. Report prepared by Coopers Lybrand for Geelong Regional Commission, Geelong.

GRPA (1975) *Geelong: City by the Bay – Geelong Central Business District Study*. Report prepared by Planning Workshop for Geelong Regional Planning Authority, Geelong.

Ledwith M, Springett J (2010) *Participatory Practice: Community-based Action for Transformative Change*. Policy Press, Bristol.

Lethlean P (2013) *It's Hard Getting Messy when you're Compositional*. PhD thesis. RMIT University, Melbourne. https://researchbank.rmit.edu.au/eserv/rmit:160723/Lethlean.pdf (accessed 1 March 2018).

Neis HB, Brown GA, Gurr JM, Schmidt A (2012) *Generative Process, Patterns, and the Urban Challenge*. PUARL Press, Portland, OR.

OECD (2015) *The Metropolitan Century: Understanding Urbanisation and its Consequences*. OECD Publishing, Paris.

OUP (2018) *Oxford Living Dictionaries: Vision*. https://en.oxforddictionaries.com/definition/vision (accessed 1 March 2018).

Rollo J (2004) Urbanheart: a design research forum. *Architectural Research Quarterly* **8**(3/4), 325–331. doi:10.1017/S1359135504000338

Roös PB (2014) Regenerative-adaptive design: patterns to coastal resilience. In *Proceedings of the 3rd Practical Responses to Climate Change Convention*. 25–27 November 2014, Melbourne. Engineers Australia, Canberra.

Roös PB (2017) *Regenerative-adaptive Design for Coastal Settlements: A Pattern Language Approach to Future Resilience*. PhD thesis. School of Architecture and Built Environment, Deakin University, Geelong. www.dro.deakin.edu.au/eserv/DU:30103436/roos-regenerativeadaptive-2017.pdf

18

Land use planning challenges facing the Geelong region in the next 10–20 years

Kirsten Kilpatrick

> *Cities are fantastically dynamic places, and this is strikingly true of their successful parts, which offer a fertile ground for the plans of thousands of people.*
>
> – Jane Jacobs (1961, p. 14), *The Death and Life of Great American Cities.*

Geelong is wonderfully diverse – from its coastline, to the rural hinterland, suburbs and city centre – and it is poised to be a contemporary and dynamic city in the future. The inward tide of people will bring new ethnic and cultural depth, demand for new and more frequent services and of course housing. Growth in Geelong is inevitable and, as the city stretches along the coast and into the surrounding pastoral plains, its economic prosperity, through jobs and investment, will need to be carefully balanced against the potential social and environmental impacts.

The city is evolving, maturing and certainly becoming more sophisticated. Taller buildings are increasingly defining the skyline, evidenced by the construction of the Worksafe and NDIS office towers, and more houses emerge in the expanding city fringe. As the city rides its third population boom, a 'Big Geelong' will bring new challenges for governments, planners and the wider community.

Geelong has struggled with its identity for many years but as Victoria's second city it is increasingly positioning itself as a national economic powerhouse. In the past, Geelong was recognised as the gateway to the Great Ocean Road and an industrial town. Over recent years, Geelong has proven its resilience, particularly when the large-scale employers Alcoa and Ford first wound back then ceased operations altogether. A long-term unified 'Clever and Creative' vision will assist in addressing the future land use challenges in Geelong.

Government intervention has heavily influenced Geelong's recent reinvention but private developers and visionaries are the lifeblood of the new Geelong. Relocation of state and federal government services has increased jobs in professional services and demand for all housing types across the city. Investment in the open spaces, streetscape and most notably the Geelong waterfront through *Vision 1* (Vision 2 Committee 2013) and *Vision 2* (CfG 2018), and the Green Spine project, has also altered the aesthetic and created a welcoming and attractive environment. Clearly, the combination of lifestyle, world-class education, health services and proximity to many attractions has changed the perception of Geelong (CoGG 2017).

Geelong was a city of 273 000 people in 2015 (ABS 2018), and it continues to grow. The current growth rate is nearly 3%, which is beyond the 1.6% growth recorded in 2011 (DELWP 2016).

Rapid population growth in Melbourne and its western suburbs has led to many people moving outside the metropolitan areas. Victoria's population is projected to grow from 5.5 million to 10.1 million by 2051, of whom 8 million people will live in Melbourne and only 2.1 million in Victoria's regions including Geelong (DELWP 2016).

Plan Melbourne 2017–2050 identifies a '20-minute city' as a key strategy to preserve liveability and community in the metropolis of Melbourne (DELWP 2015). The ideal 20-minute city is a reality in Geelong: driving distances are generally no more than 20 minutes from the city centre, to the coast, rural areas and the other surrounding areas between. Geelong is a logical destination for many, as it has been throughout history. Located ~70 km from Melbourne's CBD, Geelong is close enough to commute to Melbourne (if necessary) and is an independent and thriving urban centre. It's accessible by road, rail, sea and air and as people commute and work in different ways Geelong is poised to attract more people.

Between 2011 and 2016, neighbouring Wyndham City Council's population grew by 37% and Geelong by only 11% (ABS 2018). The rural break between Geelong and Werribee is under increased pressure as Melbourne continues to grow, but as Geelong seeks to retain its identity there is limited desire to support contiguous urban development. This equally applies to Geelong's surrounding townships. The Western Plains grassland between Melbourne and Geelong has preserved the settlement break between the growing cities, but how long will this last? Avalon Airport, wedged between the two centres, will increasingly become a key economic node which has been reinforced through its status as an international airport. Direct flights to Asia commenced in late 2018, and distribution and logistics businesses, such as the $40 million Cotton On development, are located within the precinct (Cotton On 2018).

As Geelong's population reaches 300 000, it is important to reflect on the history of the city and the surrounding area. Lessons from the past will define pathways for the future and also connect today's community with the city's history.

Geelong was gazetted as a town in 1838, but its first population boom was during the Victorian gold rush in the 1850s when it was the 'seaport of the gold district' (Wynd 1971). By 1851, the population of Geelong had grown enormously as migrants came looking for gold around the Ballarat goldfields (Wynd 1971). But by 1861, the infamous 'Sleepy Hollow' label had been attached to Geelong (Wynd 1971).

The second wave of population was the post-war period from the late 1940s to the 1960s. Geelong was ensconced as a manufacturing centre, and the traditional textile companies were joined by global companies Shell, Alcoa and Ford during this period. Its growth exceeded that of the regional centres of Bendigo and Ballarat, as people followed the jobs. The Australian dream of the quarter-acre block with a detached single-storey house was achievable in the growing suburbs of Geelong during this period. Holiday homes at Torquay, Portarlington and Ocean Grove, while still a luxury, were affordable options for many (Wynd 1971).

The current and third population boom for Geelong is different. In the past, people came to Geelong for jobs. Today, with technological improvements and faster connections between Melbourne and the world, Geelong residents can work anywhere. The rise of 'fly-in fly-out' workers, the ability to work from home and the growing professional services sector are driving housing demand. People are drawn to Geelong, the hinterland

and the coast, away from the intensity of the inner city or blandness of the suburbs. Geelong can offer a lifestyle that is more in touch with nature, but with important amenities and access to the city (ABS 2018).

Geelong is Victoria's second-largest city and rightfully owns the title 'Second City'. It should not be denied its role as one of Australia's, and potentially the region's, most important cities (Almeida Correia and Denham 2016). As Melbourne grows to become a truly global city, equal in population and scale to London, Geelong will have an increasingly important role in defining its own identity and a fundamental supporting role in maintaining liveability. Its growth should be harnessed and channelled into innovative and sustainable development outcomes. The great cities of the world embrace growing populations; however, it is important that, as cities become more successful, disadvantaged citizens are supported and the entire city – not just the wealthy – prospers.

Manufacturing continues to be a key employer in Geelong, with over 10% of the city's workforce still employed in the sector. It will continue to have an important role in the future of Geelong but its workforce will become more highly skilled and rely more heavily on technology and design. The growth of health, education and professional services sectors will continue, reinforced through the influx of companies which will leverage the existing and new government enterprises coming to Geelong.

Geelong is a great city and is poised to become a greater city, but a deep understanding of its history, environment and culture will provide a solid foundation for its future. The remnants of Geelong's Victorian past are scattered throughout the city. The architectural relics, such as the Geelong Gaol, need to be adapted to a contemporary use while preserving the integrity of the buildings.

The increased demand for housing on the urban fringe in Armstrong Creek, Lara, Lovely Banks and Batesford will house thousands of new residents. The older, established suburbs of Geelong will also change. Multi-storey apartment living, now only in concentrated pockets in central Geelong, will surround shopping centres of Highton, Belmont, Bell Post Hill and Waurn Ponds. The density in the townships of Ocean Grove and Torquay will become much greater. The inner city will be the central hub of employment, high-rise housing and transport connections which will service the entire city, but co-created nodes will also involve economic, retail and community centres that will connect to the city through corridor development.

Geelong will be a large, global city in the future. It will be spatially more expansive, but there will be more intensive development concentrated around the inner suburbs. This will be challenging, as residents will become increasingly vocal and resist change. In the past, the community has passively witnessed changes in the urban landscape, vocalising concern only in extreme circumstances.

The environment will be paramount as Geelong continues to grow. Sustainability and achieving the right balance will be vital with the increased demand for water, energy and waste management. Geelong will develop innovative solutions such as integrative technology and new developments for energy and waste management. Wind turbines and solar energy production will become mandatory, and a cost-effective part of the urban fabric of the city.

Planning has a crucial role in shaping the new Geelong. Policy-makers and politicians will be under increasing pressure to balance the interests of current and future residents (DELWP 2018). This will be more difficult with greater growth pressures (G21 GRA 2013, 2018). There will be more pressure on financial resources. Funding of schools, transport and basic infrastructure will demand more from the public purse. Greater congestion will require rail systems and advanced transport such as a rail network connecting the suburbs to the city centre, to Melbourne, coastal communities and other regional centres. This of course brings other challenges in managing the city. As has been experienced in the past, the introduction of rail into a community changes the appeal and demand for housing. Opening up new residential areas and coastal communities will attract many new residents, particularly those who continue to work in Melbourne. A high-speed rail connection would completely transform these places. The cost will be either unprecedented growth in housing prices, which will erode affordability, or sprawling suburbs and towns.

The growing Geelong will need to breathe. Open spaces are the lungs of the city and planning legislation has provided an important statutory framework for the provision of parks and open space. New developments and intensification of land uses will require more open spaces in areas which may have not been well provided in the past. The open spaces should fulfil the dual roles of drainage and protection of important environmental assets. The cities of the future should maximise space, not just for housing but for infrastructure. Schoolyards are also community parks, buildings are community halls and libraries are for all members of the community.

Cultural expression will define the big Geelong, through design, performance and public art. Geelong is now a UNESCO City of Design (CoGG 2018) and supports the creative industries including artists, makers, performance-based arts, world-class architecture and process design. The Geelong Library and Heritage Centre is now a landmark in Geelong, for a multitude of reasons. It defines the Geelong landscape at several approaches to the city and provides a community space in which to learn and reflect on the city's romantic past. Its most precious attribute is the public access to the high balcony. The spectacular views from the balcony allow visitors and residents to enjoy, absorb and appreciate the wonderful urban landscape of Geelong. The building is a physical symbol of the 'clever and creative' Geelong.

Geelong is already embracing new technology – again being clever and creative – and this is affecting the settlement of the city. More people are working from home, supported by improvements in internet, NBN and changing attitudes to the mobile workplace. Inevitability, technology will shape the future of the city. The smart city of Geelong will have technology embedded in all facets of city life – parking, transport, work, engagements, sport and shopping, to name a few. People will still commute to work and central services such as retail and larger-scale community facilities such as libraries and arts centres, but local areas will also be vibrant and active spaces. The local café will replace the corner store as the hub in the local area. Land use planning will need greater flexibility to adapt to the changing desires of the community. Pop-up cafes and wine bars should not be seen as catalysts for social dysfunction but as small commercialised community hubs that unite and foster the local sense of community that could otherwise be lost as more people engage through technology.

'Cities are fantastically dynamic places' – this is true for Geelong (Jacobs 1961). Geelong's growth must be supported to allow it to flourish. From the inner city to the growing suburbs, Geelong will need to have better transport, embrace technology and provide a sustainable, active environment. Policy-makers and city leaders will be challenged by pressure from the community, developers and governments in their attempt to shape a big Geelong. The land use challenges in the future will be how to create a city that is liveable, accessible and affordable and that allows all citizens to share and prosper from its growth.

References

Almeida Correia J, Denham T (2016) *Winning from Second: What Geelong can Learn from International Second Cities.* Committee for Geelong, Geelong. http://citiesprogramme.org/wp-content/uploads/2016/12/winning-from-second-web.pdf (accessed 1 June 2018).

ABS (2018) Australian Bureau of Statistics, Canberra. http://www.abs.gov.au/ (accessed 1 June 2018).

CfG (2018) *Central Geelong Vision 2.* Committee for Geelong, Geelong. http://www.committeeforgeelong.com.au/key-initiatives/vision-2/ (accessed 1 March 2018).

CoGG (2017) *Greater Geelong: A Clever and Creative Future.* City of Greater Geelong, Geelong, https://www.geelongaustralia.com.au/clevercreative/default.aspx, accessed 1 June 2018.

CoGG (2018) *UNESCO City of Design.* City of Greater Geelong, Geelong. https://www.geelongaustralia.com.au/clevercreative/article/item/8d531b5add016db.aspx (accessed 1 June 2018).

Cotton On (2018) *Cotton On.* Cotton On, Geelong. https://cottonon.com/au/ (accessed 1 June 2018).

DELWP (2015) *Plan Melbourne 2017–2050.* Department of Environment, Land, Water and Planning, Melbourne. https://www.planmelbourne.vic.gov.au/

DELWP (2016) *Victoria in Future 2016.* Department of Environment, Land, Water and Planning, Melbourne. https://www.planning.vic.gov.au/land-use-and-population-research/victoria-in-future-2016 (accessed 1 June 2018).

DELWP (2018) *Greater Geelong Planning Scheme.* Department of Environment, Land, Water and Planning, Melbourne. http://planning-schemes.delwp.vic.gov.au/schemes/greatergeelong (accessed 1 June 2018).

G21 GRA (2013) *Regional Growth Plan.* G21 Geelong Region Alliance, Geelong. http://www.g21.com.au/sites/default/files/resources/g21_regional_growth_plan_-_april_2013_-_low_res_0.pdf (accessed 1 June 2018).

G21 GRA (2018) *G21 Geelong Regional Alliance.* G21 Geelong Region Alliance, Geelong. http://www.g21.com.au/ (accessed 1 June 2018).

Jacobs J (1961) *The Death and Life of Great American Cities.* Random House, New York.

Vision 2 Committee (2013) *Central Geelong Vision: Design Studio Report 25 March 2013, draft.* School of Architecture and Built Environment, Deakin University and the Vision 2 Committee, Geelong.

Wynd I (1971) *Geelong: The Pivot – A Short History of Geelong and District.* Cypress Books, Melbourne.

19

Ensuring a quality future for the tapestry of Geelong

David S. Jones and Phillip B. Roös

> *Geelong is Victoria's second city, and proudly so. However, second does not mean second rung or second place; it means Geelong enjoys all of the benefits of being within arm's reach of Melbourne but free from the constraints of big city living. It means being grounded by Geelong's rich legacies, yet experiencing the freedom that comes from being surrounded by open space and open water. More than that, it is about having the unbridled ability to think clearly and create a life with wellness at its core.*
>
> – Almeida Correia and Denham (2016, p. iv).

In surveying Geelong and the Geelong region, it is evident that this landscape possesses many natural and cultural qualities and values that have enabled and influenced its present context.

Its locational legacy on Corio Bay and its indirect location on the intersection of the Barwon and Moorabool River valleys have structured its historical, physical and urban development, and continue to do so today. While considerable emphasis has been placed upon Geelong City's unique north-facing Bay location, recognised by the former Geelong Regional Planning Authority and the Geelong Regional Commission (McLean 2006) in their strategies and economic initiatives, the valleyscapes and coastal characteristics of the city and its surrounding region have received little insightful and co-ordinated vision. The former is the blue 'front-door mat' to Geelong whereas the latter is its green 'backyard'.

The wider locational legacy is Geelong's proximity between the undulating and scenic Bellarine Peninsula, the volcanic pastoral plains of the Western District that brought its 'golden fleece' roots, and the coastal landscapes of the Otway–Great Ocean Road corridor. All three landscapes inform and have historically scaffolded the psyche of Geelong and its economic and cultural pursuits.

Within this fabric, Geelong has witnessed major economic and development challenges over the last 25 years that have caused major stress for many families. Included in the stresses are the collapse of major industrial and financial entities, and a lethargic political attitude toward providing quality and stable planning and infrastructure in the region by often Melbourne-based politicians and governance administrators. The buoyancy of a new Geelong, driven by technology innovations, new health services and roles, new transit node roles, the growth of Deakin University as an education and research node, and the escalating urbanisation in the region as people seek out new affordable home ownership or sea-change residences, have refashioned Geelong as the key regional growth driver for Victoria.

Glimpses of insightfulness and innovation abound around Geelong today, signalling this transformation. Ventures in Norlane with housing and community bonding, strategies by the National Disability Insurance Scheme (NDIS) to roll out new service infrastructures, Deakin University's ventures into carbon fibre technology (Deakin 2017), the execution of parts of *Vision 2* with Malop Street and Johnstone Park landscape designs (Sharman *et al.* 2013), additional arterial road and road projects such as Baanip Boulevard and Drysdale Bypass (CoGG 2018b), Waurn Ponds rail duplication and the Geelong Ring Road, the drive of Warralily (Warralily 2018) and Armstrong Creek greenfields, the positioning of Avalon International Airport, and Geelong's designation as Australia's first UNESCO City of Design (CoGG 2018c), all evidence components of this innovation-informed transformation.

Such components and patterns are typical of changes unfolding in various second-tier cities around the world as they reinvent themselves away from their historical blue-collar industrial roots. Places like Eindhoven, Newcastle (Australia) (Westbury 2015a, b), Detroit (CDN 2018), Dundee, Cleveland, Pittsburgh, Richmond (US), Bristol, Liverpool and Sheffield are all new 'second cities' (Almeida Correia and Denham 2016).

The pace of self-reinvention and transformation is escalating. It is causing deterioration of Geelong's qualities and values and putting major stress upon existing infrastructure to manage this growth. There is concern about the compromise of the region's environmental qualities and its capacity in responding to climate change challenges, and community and professional practice uncertainty as to a co-ordinated 'plan' and vison despite the advances of G21 (G21 GRA 2018) and the recently established Geelong Authority (DELWP 2018).

In the threads of this publication is a recognition of these transformations, this change, but also a keen sense of concern about the consequences of this reinvention. They include the need for a sensitive visionary strategy for the region, unfettered by economic imperatives, which mitigates this reinvention against deterioration of its cultural and environmental qualities including Indigenous, visual, biodiversity, water, coast and vegetation attributes. These are the very qualities that are attracting new people to the region, and what existing families treasure and enjoy.

A clear conclusion from this book is the lack of clarity of a vision, and its sense of co-ordinated execution unfettered by political interference.

A second conclusion is that while considerable attention has been paid to the Corio Bay edge, recharting it as a major venue and identity-maker for Geelong, there is a dearth of equivalent attention being paid to the Barwon and Moorabool River corridors. This is significant – if we were in Melbourne, Boston, Singapore (HDB 2018; Tan *et al.* 2018; Wong 2018) or Adelaide (RTC 1979; Mugavin 2004), such a linear corridor would have been already appropriated as a green tree-planted recreation venue, a water quality-treating/flood-mitigating venue, a visually delightful refuge and a 'green necklace' encircling the urban expanse. The clues are there in the legacy of the Barwon edge around Belmont, Queens Park, Barwon Falls and the little enclaves around Lake Connewarre but the patches are not united. Geelong does not possess a 'green necklace' to match its 'blue necklace'.

Third, Geelong's existing fixed transit network is struggling to service this transformation. It is struggling to keep pace in its tangible commuter network service role, and has disregarded its internal network responsibilities. Geelong's intangible network of former rail corridors and new freeway and arterial road corridors are little strategically incorporated into Geelong's environment enhancement strategies like what was achieved in Adelaide (Mugavin 2004; RTC 1979). A clear radial network needs to be built to create an internal Geelong rail network, rather than it being simply a subservient Melbourne-commuter service.

A fourth conclusion is that the centre of Geelong is starting to successfully transform the central business district from a tired urban landscape to a green urban design-invigorated place; placelessness is becoming place-making (CoGG 2018a; DELWP 2016). There is a danger that multiple commercial centres will dilute a logical service and functional city structure, and all 'second cities' have recognised that their CBD is the central driver of urban design initiatives and successes (CfG 2018). Policy change and the increasing influx of Deakin's students and new office and hospitality complexes is peopling and validating the execution of Geelong's urban design initiatives.

Finally, this growth, and the strategy-making to inform and guide the growth, need to better grasp and be accountable for the region's unique environmental qualities. Ecologically responsive (McHarg 1969) or regenerative design (Roös 2017) or quality sustainable environmental planning for communities, as embodied in UNESCO's Sustainable Development Goals (UNESCO 2018), is not being witnessed nor executed. It is evident throughout this book that the importance of the rich natural environment of the Geelong region needs to be acknowledged. Further, in the context of a changing climate and warming global environment, it is clear that we need to find narratives to mitigate and adapt to the impacts of change.

In this regard, we offer these thoughts and the narratives in this publication, as both warnings and answers to current and future generations of leaders and participants in the Geelong region. We, as a community of thinkers and residents in the region, are concerned about what is transpiring and the mixed directions this place is heading towards. We collectively want to ensure that the spirit of this place is conserved, that its deterioration of qualities is mitigated, and that we offer a legacy to our children that demonstrates that we did listen, we did ask, we did act, and that we sought to execute the best innovations and actions we could muster.

References

Almeida Correia J, Denham T (2016) *Winning from Second: What Geelong can Learn from International Second Cities*. Committee for Geelong, Geelong. http://www.committeeforgeelong.com.au/current-projects/winning-from-second-international-research-paper/ (accessed 1 March 2018).

CDN (2018) *Detroit: UNESCO City of Design*. Cities of Design Network. http://www.designcities.net/city/detroit/ (accessed 1 March 2018).

CfG (2018) *Central Geelong Vision 2*. Committee for Geelong, Geelong. http://www.committeeforgeelong.com.au/key-initiatives/vision-2/ (accessed 1 March 2018).

CoGG (2018a) *Revitalising Central Geelong*. City of Greater Geelong, Geelong. https://revitalisingcentralgeelong.com.au/ (accessed 1 July 2018).

CoGG (2018b) *Drysdale Bypass: Amendment C369 to the Greater Geelong Planning Scheme*. City of Greater Geelong, Geelong. https://www.geelongaustralia.com.au/amendments/item/8d4c84408aeb358.aspx (accessed 1 March 2018).

CoGG (2018c) *UNESCO City of Design*. City of Greater Geelong, Geelong. https://www.geelongcityofdesign.com.au/ (accessed 1 March 2018).

Deakin (2017) *Future of Carbon Fibre is Here*. Deakin University, Geelong. http://www.deakin.edu.au/research/research-news/articles/future-of-carbon-fibre-is-here (accessed 1 March 2018).

DELWP (2016) *Revitalising Central Geelong Action Plan*. Department of Environment, Land, Water and Planning, Melbourne. https://www.planning.vic.gov.au/__data/assets/pdf_file/0021/95250/Revitalising-Central-Geelong-Action-Plan.pdf (accessed 1 March 2018).

DELWP (2018) *Geelong Authority*. Department of Environment, Land, Water and Planning, Melbourne. https://www.planning.vic.gov.au/project-panels/panels-and-committees/Geelong-Authority (accessed 1 March 2018).

G21 GRA (2018) *G21 Geelong Regional Alliance*. G21 Geelong Region Alliance, Geelong. http://www.g21.com.au/ (accessed 1 March 2018).

HDB (2018) *Homes at One with Nature: Biophilic Design for Greater Sense of Place, Better Health and Well-being, and Enhanced Quality of Life*. Housing and Development Board, Singapore. https://www.hdb.gov.sg/cs/infoweb/press-release/homes-at-one-with-nature (accessed 20 July 2018).

McHarg I (1969) *Design with Nature*. John Wiley & Sons, Washington, DC.

McLean GA (2006) The history of the Geelong Regional Commission. Unpublished PhD thesis. Deakin University, Geelong.

Mugavin D (2004) Adelaide's Greenway: River Torrens Linear Park. *Landscape and Urban Planning* **68**(2–3), 223–240. doi:10.1016/j.landurbplan.2003.07.002

Roös PB (2017) *Regenerative-adaptive Design for Coastal Settlements: A Pattern Language Approach to Future Resilience*. PhD thesis. School of Architecture and Built Environment, Deakin University, Geelong. www.dro.deakin.edu.au/eserv/DU:30103436/roos-regenerativeadaptive-2017.pdf

RTC (1979) *River Torrens Study: A Coordinated Development Scheme*. Report prepared by Hassell Planning Consultants and Land Systems for River Torrens Committee, Adelaide.

Tan PY, Liao K-H, Hwang YH, Chua V (2018) *Nature, Place & People: Forging Connections through Neighbourhood Landscape Design*. World Scientific Press, London. https://www.worldscientific.com/worldscibooks/10.1142/10879 (accessed 20 July 2018).

UNESCO (2018) *UNESCO and Sustainable Development Goals*. https://en.unesco.org/sdgs (accessed 1 March 2018).

Warralily (2018) *Warralily News*. http://warralily.com.au/ (accessed 1 March 2018).

Westbury M (2015a) *Creating Cities*. Niche Press, Melbourne.

Westbury M (2015b) *Bespoke: Rise of the Makers*. Episode 1 of 3, ABC, Melbourne. http://www.abc.net.au/tv/programs/bespoke/ (accessed 1 March 2018).

Wong D (2018) Nature to play bigger role in HDB estates. *Straits Times*. https://www.straitstimes.com/singapore/housing/nature-to-play-bigger-role-in-hdb-estates (accessed 20 July 2018).

Index

www.ingramcontent.com/pod-product-compliance
Ingram Content Group UK Ltd.
Pitfield, Milton Keynes, MK11 3LW, UK
UKHW061957041026
15591UKWH00016B/1048

9 780643 103603